T0336871

Modern Quantum Field Theory
A Concise Introduction

Quantum field theory is a key subject in physics, with applications in particle and condensed matter physics. Treating a variety of topics that are only briefly touched on in other texts, this book provides a thorough introduction to the techniques of field theory.

The book covers Feynman diagrams and path integrals, and emphasizes the path integral approach, the Wilsonian approach to renormalization, and the physics of non-abelian gauge theory. It provides a thorough treatment of quark confinement and chiral symmetry breaking, topics not usually covered in other texts at this level. The Standard Model of particle physics is discussed in detail. Connections with condensed matter physics are explored, and there is a brief, but detailed, treatment of non-perturbative semi-classical methods (instantons and solitons).

Ideal for graduate students in high energy physics and condensed matter physics, the book contains many problems, providing students with hands-on experience with the methods of quantum field theory.

Thomas Banks is Professor of Physics at the Santa Cruz Institute for Particle Physics (SCIPP), University of California, and NHETC, Rutgers University. He is considered one of the leaders in high energy particle theory and string theory, and was co-discoverer of the Matrix Theory approach to non-perturbative string theory.

Modern Quantum Field Theory

A Concise Introduction

Tom Banks

University of California, Santa Cruz
and Rutgers University

CAMBRIDGE
UNIVERSITY PRESS

CAMBRIDGE
UNIVERSITY PRESS

University Printing House, Cambridge CB2 8BS, United Kingdom

One Liberty Plaza, 20th Floor, New York, NY 10006, USA

477 Williamstown Road, Port Melbourne, VIC 3207, Australia

314-321, 3rd Floor, Plot 3, Splendor Forum, Jasola District Centre, New Delhi - 110025, India

79 Anson Road, #06-04/06, Singapore 079906

Cambridge University Press is part of the University of Cambridge.

It furthers the University's mission by disseminating knowledge in the pursuit of education, learning and research at the highest international levels of excellence.

www.cambridge.org
Information on this title: www.cambridge.org/9780521850827

First published 2008

A catalogue record for this publication is available from the British Library

Library of Congress Cataloging in Publication data
Banks, Tom.
Modern quantum field theory : an introduction / Tom Banks.
p.   cm.
Includes bibliographical references and index.
ISBN 978-0-521-85082-7
1. Quantum field theory.   I. Title.
QC174.45.B296   2008
530.14′3–dc22
2008027150

ISBN 978-0-521-85082-7 Hardback

Contents

1 Introduction

1.1 Preface and conventions

This book is meant as a quick and dirty introduction to the techniques of quantum field theory. It was inspired by a little book (long out of print) by F. Mandl, which my advisor gave me to read in my first year of graduate school in 1969. Mandl's book enabled the smart student to master the elements of field theory, as it was known in the early 1960s, in about two intense weeks of self-study. The body of field-theory knowledge has grown way beyond what was known then, and a book with similar intent has to be larger and will take longer to absorb. I hope that what I have written here will fill that Mandl niche: enough coverage to at least touch on most important topics, but short enough to be mastered in a semester or less. The most important omissions will be supersymmetry (which deserves a book of its own) and finite-temperature field theory. Pedagogically, this book can be used in three ways. Chapters 1–6 can be used as a text for a one-semester introductory course, the whole book for a one-year course. In either case, the instructor will want to turn some of the starred exercises into lecture material. Finally, the book was designed for self-study, and can be assigned as a supplementary text. My own opinion is that a complete course in modern quantum field theory needs 3–4 semesters, and should cover supersymmetric and finite-temperature field theory.

This statement of intent has governed the style of the book. I have tried to be terse rather than discursive (my natural default) and, *most importantly, I have left many important points of the development for the exercises. The student should not imagine that he/she can master the material in this book without doing at least those exercises marked with a* *. In addition, at various points in the text I will invite the reader to prove something, or state results without proof. The diligent reader will take these as extra exercises. This book may appear to the student to require more work than do texts that try to spoon-feed the reader. I believe strongly that a lot of the material in quantum field theory can be learned well only by working with your hands. Reading or listening to someone's explanation, no matter how simple, will not make you an adept. My hope is that the hints in the text will be enough to let the student master the exercises and come out of this experience with a thorough mastery of the basics.

The book also has an emphasis on theoretical ideas rather than application to experiment. This is partly due to the fact that there already exist excellent texts that concentrate on experimental applications, partly due to the desire for brevity, and partly to increase the shelf life of the volume. The experiments of today are unlikely to be of intense interest even to experimentalists of a decade hence. The structure of quantum field theory will exist forever.

Throughout the book I use natural units, where $\hbar = c = 1$. Everything has units of some power of mass/energy. High-energy experiments and theory usually concentrate on the energy range between 10^{-3} and 10^3 GeV and I will often use these units. Another convenient unit of energy is the natural one defined by gravitation: the Planck mass, $M_P \approx 10^{19}$ GeV, or reduced Planck mass, $m_P \approx 2 \times 10^{18}$ GeV. The GeV is the natural unit for hadron masses. Around 0.15 GeV is the scale at which strong interactions become strong. Around 250 GeV is the natural scale of electro-weak interactions, and $\sim 2 \times 10^{16}$ GeV appears to be the scale at which electro-weak and strong interactions are unified.

I will use non-relativistic normalization, $\langle p|q \rangle = \delta^3(p - q)$, for single-particle states. Four-vectors will have names which are single Latin letters, while 3-vectors will be written in bold face. I will use Greek mid-alphabet letters for tensor indices, and Latin early-alphabet letters for spinors. Mid-alphabet Latin letters will be 3-vector components. I will stick to the van der Waerden dot convention (Chapter 5) for distinguishing left- and right-handed Weyl spinors. As for the metric on Minkowski space, I will use the West Coast, *mostly minus*, convention of most working particle theorists (and of my toilet training), rather than the East Coast (mostly plus) convention of relativists and string theorists.

Finally, a note about prerequisites. The reader must begin this book with a thorough knowledge of calculus, particularly complex analysis, and a thorough grounding in non-relativistic quantum mechanics, which of course includes expert-level linear algebra. Thorough knowledge of special relativity is also assumed. Detailed knowledge of the mathematical niceties of operator theory is unnecessary. The reader should be familiar with the Einstein summation convention and the totally anti-symmetric Levi-Civita symbol $\epsilon^{a_1 \ldots a_n}$. We use the convention $\epsilon^{0123} = 1$ in Minkowski space. It would be useful to have a prior knowledge of the theory of Lie groups and algebras, at a physics level of rigor, although we will treat some of this material in the text and Appendix G. I have supplied some excellent references [1–4] because this math is crucial to much that we will do. As usual in physics, what is required of your mathematical background is a knowledge of terminology and how to manipulate and calculate, rather than intimate familiarity with rigor and formal proofs.

1.1.1 Acknowledgements

I mostly learned field theory by myself, but I want to thank Nick Wheeler of Reed College for teaching me about path integrals and the beauties of mathematical physics in general. Roman Jackiw deserves credit for handing me Mandl's book, and Carl Bender helped me figure out what an instanton was before the word was invented. Perhaps the most important influence in my grad school years was Steven Weinberg, who taught me his approach to fields and particles, and everything there was to know about broken symmetry. Most of the credit for teaching me things about field theory goes to Lenny Susskind, from whom I learned Wilson's approach to renormalization, lattice gauge theory, and a host of other things throughout my career. Shimon Yankielowicz and Eliezer Rabinovici were my most important collaborators during my years in Israel. We learned a lot of great physics together. During the 1970s, along with everyone else in the field, I learned from the seminal work of D. Gross, S. Coleman, G. 't Hooft, G. Parisi, and E. Witten. Edward was a friend and a major influence throughout my career. As one grows older, it's harder for people to do things that surprise you, but my great friends and sometimes collaborators Michael Dine, Willy Fischler, and Nati Seiberg have constantly done that. Most of the field theory they've taught me goes beyond what is covered in this book. You can find some of it in Michael Dine's recent book from Cambridge University Press.

Field theory can be an abstract subject, but it is physics and it has to be grounded in reality. For me, the most fascinating application of field theory has been to elementary particle physics. My friends Lisa Randall, Yossi Nir, Howie Haber, and, more recently, Scott Thomas have kept me abreast of what's important in the experimental foundation of our field.

In writing this book, I've been helped by M. Dine, H. Haber, J. Mason, L. Motl, A. Shomer, and K. van den Broek, who've read and commented on all or part of the manuscript. The book would look a lot worse than it does without their input. Chapter 10 was included at the behest of A. Strominger, and I thank him for the suggestion. Chris France, Jared Rice, and Lily Yang helped with the figures. Finally, I'd like to thank my wife Ada, who has been patient throughout all the trauma that writing a book like this involves.

1.2 Why quantum field theory?

Students often come into a class in quantum field theory straight out of a course in non-relativistic quantum mechanics. Their natural inclination is to look for a straight-forward relativistic generalization of that formalism. A fine place to start would seem to be a covariant classical theory of a single relativistic particle, with space-time position variable $x^\mu(\tau)$, written in terms of an arbitrary parametrization τ of the particle's path in space-time.

The first task of a course in field theory is to explain to students why this is not the right way to do things.[1] The argument is straightforward.

Consider a classical machine (an emission source) that has probability amplitude $J_E(x)$ of producing a particle at position x in space-time, and an absorption source, which has amplitude $J_A(x)$ to absorb the particle. Assume that the particle propagates

---

[1] Then, when they get more sophisticated, you can show them how the particle path formalism can be used, with appropriate care.

Fig. 1.1.   **Boosts can reverse causal order for $(x - y)^2 < 0$.**

freely between emission and absorption, and has mass m. The standard rules of quantum mechanics tell us that the amplitude (to leading order in perturbation theory in the sources) for the entire process is (remember our natural units!)

$$A_{AE} = \int d^4x \, d^4y \langle x|e^{-iH(x^0 - y^0)}|y\rangle J_A(x) J_E(y), \tag{1.1}$$

where $|x\rangle$ is the state of the particle at spatial position x. This doesn't look very Lorentz-covariant. To see whether it is, write the relativistic expression for the energy $H = \sqrt{p^2 + m^2} \equiv \omega_p$. Then

$$A_{AE} = \int d^4x \, d^4y \, J_A(x) J_E(y) \int d^3p |\langle 0|p\rangle|^2 e^{-ip(x-y)}. \tag{1.2}$$

The space-time set-up is shown in Figure 1.1. In writing this equation I've used the fact that momentum is the generator of space translations[2] to evaluate position/momentum overlaps in terms of the momentum eigenstate overlap with the state of a particle at the origin. I've also used the fact that (ω_p, p) is a 4-vector to write the exponent as a Lorentz scalar product. So everything is determined by quantum mechanics, translation invariance and the relativistic dispersion relation, up to a function of 3-momentum. We can determine this function up to an overall constant, by insisting that the expression is Lorentz-invariant, if the emission and absorption amplitudes are chosen to transform as scalar functions of space-time. An invariant measure for 4-momentum integration, ensuring that the mass is fixed, is $d^4p \, \delta(p^2 - m^2)$. Since the momentum is then forced to be time-like, the sign of its time component is also Lorentz-invariant (Problem 2.1). So we can write an invariant measure $d^4p \, \delta(p^2 - m^2)\theta(p^0)$ for positive-energy particles of mass m. On doing the integral over p^0 we find $d^3p/(2\omega_p)$. Thus, if we choose the normalization

$$\langle 0|p\rangle = \frac{1}{\sqrt{(2\pi)^3 2\omega_p}}, \tag{1.3}$$

then the propagation amplitude will be Lorentz-invariant. The full absorption and emission amplitude will of course depend on the Lorentz frame because of the coordinate dependence of the sources $J_{E,A}$. It will be covariant if these are chosen to transform like scalar fields.

---

[2] Here I'm using the notion of the infinitesimal generator of a symmetry transformation. If you don't know this concept, take a quick look at Appendix G, or consult one of the many excellent introductions to Lie groups [1–4].

This equation for the momentum-space wave function of "a particle localized at the origin" is not the same as the one we are used to from non-relativistic quantum mechanics. However, if we are in the non-relativistic regime where $|p| \ll m$ then the wave function reduces to $1/m$ times the non-relativistic formula. When relativity is taken into account, the localized particle appears to be spread out over a distance of order its Compton wavelength, $1/m = \hbar/(mc)$.

Our formula for the emission/absorption amplitude is thus covariant, but it poses the following paradox: *it is non-zero when the separation between the emission and absorption points is space-like.* The causal order of two space-like separated points is not Lorentz-invariant (Problem 2.1), so this is a real problem.

The only known solution to this problem is to impose a new physical postulate: every emission source could equally well be an absorption source (and vice versa). We will see the mathematical formulation of this postulate in the next chapter. Given this postulate, we define a total source by $J(x) = J_{\mathrm{E}}(x) + J_{\mathrm{A}}(x)$ and write an amplitude

$$
\begin{aligned}
A_{\mathrm{AE}} &= \int \mathrm{d}^4x \, \mathrm{d}^4y \, J(x) J(y) \int \frac{\mathrm{d}^3p}{2\omega_p (2\pi)^3} [\theta(x^0 - y^0) \mathrm{e}^{-ip(x-y)} \\
&\qquad\qquad\qquad\qquad\qquad + \theta(y^0 - x^0) \mathrm{e}^{ip(x-y)}] \\
&\equiv \int \mathrm{d}^4x \, \mathrm{d}^4y \, J(x) J(y) D_{\mathrm{F}}(x - y),
\end{aligned}
\tag{1.4}
$$

where $\theta(x^0)$ is the Heaviside step function which is 1 for positive x^0 and vanishes for $x^0 < 0$. From now on we will omit the 0 superscript in the argument of these functions. This formula is manifestly Lorentz-covariant when $x - y$ is time-like or null. When the separation is space-like, the momentum integrals multiplying the two different step functions are equal, and we can add them, again getting a Lorentz-invariant amplitude. It is also consistent with causality. In any Lorentz frame, the term with $\theta(x^0 - y^0)$ is interpreted as the amplitude for a positive-energy particle to propagate forward in time, being emitted at y and absorbed at x. The other term has a similar interpretation as emission at x and absorption at y. Different Lorentz observers will disagree about the causal order when $x - y$ is space-like, but they will all agree on the total amplitude for any distribution of sources.

Something interesting happens if we assume that the particle has a conserved Lorentz-invariant charge, like electric charge. In that case, one would have expected to be able to correlate the question of whether emission or absorption occurred to the amount of charge transferred between x and y. Such an absolute definition of emission versus absorption is not consistent with the postulate that saved us from a causality paradox. In order to avoid it we have to make another, quite remarkable, postulate: every charge-carrying particle has an anti-particle of exactly equal mass and opposite charge. If this is true we will not be able to use charge transfer to distinguish between emission of a particle and absorption of an anti-particle. One of the great triumphs of quantum field theory is that this prediction is experimentally verified. The equality of particle and anti-particle masses has been checked to one part in 10^{18} [5].

Now let's consider a slightly more complicated process in which the particle scatters from some external potential before being absorbed. Suppose that the potential is

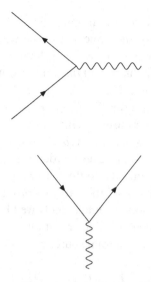

Fig. 1.2.   **Scattering in one frame is production amplitude in another.**

short-ranged, and is turned on for only a brief period, so that we can think of it as being concentrated near a space-time point z. The scattering amplitude will be approximately given by propagation from the emission point to the interaction point z, some interaction amplitude, and then propagation from z to the absorption point. We can draw a space-time diagram like Figure 1.2. We have seen that the propagation amplitudes will be non-zero, even when all three points are at space-like separation from each other. Then, there will be some Lorentz frame in which the causal order is that given in the second drawing in the figure. An observer in this frame sees particles created from the vacuum by the external field! Scattering processes inevitably imply particle-production processes.[3]

We conclude that a theory consistent with special relativity, quantum mechanics, and causality must allow for particle creation when the energetics permits it (in the example of the previous paragraph, the time dependence of the external field supplies the energy necessary to create the particles). This, as we shall see, is equivalent to the statement that a causal, relativistic quantum mechanics must be a theory of quantized local fields. Particle production also gives us a deeper understanding of why the single-particle wave function is spread over a Compton wavelength. To localize a particle more precisely we would have to probe it with higher momenta. Using the relativistic energy–momentum relation, this means that we would be inserting energy larger than the particle mass. This will lead to uncontrollable pair production, rather than localization of a single particle.

Before leaving this introductory section, we can squeeze one more drop of juice from our simple considerations. This has to do with how to interpret the propagation amplitude $D_F(x - y)$ when $x - y$ is space-like, and we are in a Lorentz frame where

---

[3] Indeed, there are quantitative relations, called *crossing symmetries*, between the two kinds of amplitude.

$x^0 = y^0$. Our remarks about particle creation suggest that we should interpret this as the probability amplitude for two particles to be found at time x^0, at relative separation $x - y$. Note that this amplitude is completely symmetric under interchange of x and y, which suggests that the particles are bosons. We thus conclude that spin-zero particles must be bosons. It turns out that this is true, and is a special case of a theorem that says that integer-spin particles are bosons and half-integer spin particles are fermions. What is more, this is not just a mathematical theorem, but an experimental fact about the real world. I should warn you, though, that unlike the other remarks in this section, and despite the fact that it leads to a correct conclusion, the reasoning here is not a cartoon of a rigorous mathematical argument. The interpretation of the equal-time propagator as a two-particle amplitude is of limited utility.

2 Quantum theory of free scalar fields

V. Fock invented an efficient method for dealing with multiparticle states. We will work with delta-function-normalized single-particle states in describing Fock space. This has the advantage that we never have to discuss states with more than one particle in exactly the same state, and various factors involving the number of particles drop out of the formulae. In this section we will continue to work with spinless particles.

Start by defining Fock space as the direct sum $\mathcal{F} = \bigoplus_{k=0}^{\infty} \mathcal{H}_k$, where $\mathcal{H}_k$ is the Hilbert space of k particle states. We will assume that our particles are either bosons or fermions, so these states are either totally symmetric or totally anti-symmetric under particle interchange. In particular, if we work in terms of single-particle momentum eigenstates, $\mathcal{H}_k$ consists of states of the form $|p_1, \ldots, p_k\rangle$, either symmetric or anti-symmetric under permutations. The inner product of two such states is

$$\langle p_1, \ldots, p_k | q_1, \ldots, q_l \rangle = \delta_{kl} \frac{1}{k!} \sum_{\sigma} (-1)^{S\sigma} \delta^3(p_1 - q_{\sigma(1)}) \ldots \delta^3(p_k - q_{\sigma(k)}), \quad (2.1)$$

where the sum is over all permutations σ in the symmetric group, S_k, $(-1)^{\sigma}$ is the sign of the permutation, and the statistics factor, S, is 0 for bosons and 1 for fermions.

The $k = 0$ term in this direct sum is a one-dimensional Hilbert space containing a unique normalized state, called the vacuum state and denoted by $|0\rangle$.

In ordinary quantum mechanics one can contemplate particles that form different representations of the permutation group than bosons or fermions. Although we will not prove it in general, this is impossible in quantum field theory. In one or two spatial dimensions, one can have particles with different statistical properties (braid statistics), but these can always be thought of as bosons or fermions with a particular long-range interaction. In three or more spatial dimensions, only Bose and Fermi statistics are allowed for particles in Lorentz-invariant QFT.

Fock realized that one can organize all the multiparticle states together in a way that simplifies all calculations. Starting with the normalized state $|0\rangle$, which has no particles in it, we introduce a set of commuting, or anti-commuting, operators $a^{\dagger}(p)$ and define

$$|p_1, \ldots, p_k\rangle \equiv a^{\dagger}(p_1) \ldots a^{\dagger}(p_k)|0\rangle. \quad (2.2)$$

These are called the creation operators. The scalar-product formula is reproduced correctly if we postulate the following (anti-)commutation relation between creation operators and their adjoints[1] (called the annihilation operators):

$$[a(p), a^\dagger(q)]_\pm = \delta^3(p - q). \tag{2.3}$$

Fermions are made with anti-commutators, and bosons with commutators.

To get a little practice with Fock space, let's construct the representation of the Poincaré symmetry[2] on the multiparticle Hilbert space. We begin with the energy and momentum. These are diagonal on the single-particle states. The correct Fock-space formula for them is

$$P^\mu = \int d^3p \, p^\mu a^\dagger(p) a(p), \tag{2.4}$$

where $p^0 = \omega_p$. Its easy to verify that this operator does indeed give us the sum of the k single-particle energies and momenta, when acting on a k-particle state. This is because $n_p \equiv a^\dagger(p) a(p)$ acts as the particle number density in momentum space. There is a similar formula for all operators that act on a single particle at a time. For example, the rotation generators are

$$J_{ij} = \int d^3p \, a^\dagger(p) i(p_i \, \partial_j - p_j \, \partial_i) a(p). \tag{2.5}$$

Here $\partial_j = \partial/\partial p^j$.

It is easy to verify that the following formula defines a unitary representation of the Lorentz group on single-particle states:

$$U(\Lambda)|p\rangle = \sqrt{\frac{\omega_{\Lambda p}}{\omega_p}} |\Lambda p\rangle.$$

The reason for the funny factor in this formula is that the Dirac delta function in our definition of the normalization is not covariant because it obeys

$$\int d^3p \, \delta(p) = 1.$$

A Lorentz-invariant measure of integration on positive-energy time-like 4-vectors is

$$\int d^4p \, \theta(p_0) \delta(p^2 - m^2) = \int \frac{d^3p}{2\omega_p}.$$

The factors in the definition of $U(\Lambda)$ make up for this non-covariant choice of normalization.

A general Lorentz transformation is the product of a rotation and a boost, so in order to complete our discussion of Lorentz generators we have to write a formula for

---

[1] This is an extremely important claim. It's easy to prove and every reader should do it. The same remark applies to all of the equations in this subsection. Commutators and anti-commutators of operators are defined by $[A, B]_\pm \equiv AB \pm BA$. Assume that $a(p)|0\rangle = 0$.

[2] Poincaré symmetry is the semi-direct product of Lorentz transformations and translations. Semi-direct means that the Lorentz transformations act on the translations.

the generator of a boost with infinitesimal velocity v. We write it as $v^i J_{0i}$. Under such a boost, $p^i \to p^i + v^i \omega_p$ and $\omega_p \to \omega_p + v^i p_i$. Thus

$$\delta |p\rangle = \left(\frac{p_i v^i}{2\omega_p} + \omega_p v^i \frac{\partial}{\partial p^i} \right) |p\rangle.$$

The Fock-space formula for the boost generator is then

$$v^i J_{0i} = \int d^3 p \left[a^\dagger(p) \left(\frac{p_i v^i}{2\omega_p} + \omega_p v^i \frac{\partial}{\partial p^i} \right) a(p) \right].$$

2.1 Local fields

We now want to model the response of our infinite collection of scalar particles to a localized source $J(x)$. We do this by adding a term to the Hamiltonian (in the Schrödinger picture)

$$H \to H_0 + V(t), \tag{2.6}$$

with

$$V(t) = \int d^3 x \, \phi(x) J(x, t). \tag{2.7}$$

$\phi(x)$ must be built from creation and annihilation operators. It must transform into $\phi(x+a)$ under spatial translations. This is guaranteed by writing $\phi(x) = \int d^3 p \, e^{ipx} \hat{\phi}(p)$, where $\hat{\phi}(p)$ is an operator carrying momentum p.

We want to model a source that creates and annihilates single particles. This statement is meant in the sense of perturbation theory. That is, the amplitude J for the source to create a single particle is small. It can create multiple particles by multiple action of the source, which will be higher-order terms in a power series in J. Thus the field $\phi(x)$ should be linear in creation and annihilation operators:

$$\phi(x) = \int \frac{d^3 p}{\sqrt{(2\pi)^3 2\omega_p}} \left[a(p)\alpha e^{i\mathbf{px}} + a^\dagger(p)\alpha^* e^{-i\mathbf{px}} \right]. \tag{2.8}$$

We have also imposed Hermiticity of the Hamiltonian, assuming that the source function is real.[3] α could be a general complex constant, but we have already defined a normalization for the field, and we can absorb the phase of α into the creation and annihilation operators, so we set $\alpha = 1$.

We now want to study the time development of our system in the presence of the source. We are not really interested in the free motion of the particles, but rather in the question of how the source causes transitions between eigenstates of the free Hamiltonian. Dirac invented a formalism, called the Dirac picture, for studying problems of

---

[3] As we will see, complex sources are appropriate for the more complicated situation of particles that are not their own anti-particles.

this sort. Let $U_S(t, t_0)$ be the time evolution operator of the system in the Schrödinger picture. It satisfies

$$i \, \partial_t U_S = [H_0 + V(t)] U_S \qquad (2.9)$$

and the boundary condition

$$U_S(t_0, t_0) = 1. \qquad (2.10)$$

The Dirac-, or interaction-, picture evolution operator is defined by

$$U_D(t, t_0) = e^{iH_0 t} U_S(t, t_0) e^{-iH_0 t_0} \qquad (2.11)$$

and the same boundary condition. If the interaction $V(t)$ is zero, $U_D = 1$. It describes how H_0 eigenstates evolve under the influence of the perturbation.

Simple manipulations (Problem 2.2) show that the Dirac-picture evolution operator satisfies

$$i \, \partial_t U_D = W(t) U_D, \qquad (2.12)$$

where $W(t) = e^{iH_0 t} V(t) e^{-iH_0 t}$. In words, the formula for $W(t)$ means that it is constructed from the interaction potential in the Schrödinger picture *by replacing every operator by the corresponding Heisenberg picture operator in the unperturbed theory.* For our problem,

$$W(x^0) = \int d^3 x \, J(x, x^0) \phi(x, x^0), \qquad (2.13)$$

where the Heisenberg-picture field is

$$\phi(x, x^0) = \int \frac{d^3 p}{\sqrt{(2\pi)^3 2\omega_p}} [a(p) e^{-ipx} + a^\dagger(p) e^{ipx}]. \qquad (2.14)$$

Note carefully that in the last formula we have Minkowski scalar products $px = p^0 x^0 - px$ ($p^0 \equiv \omega_p$) replacing the three-dimensional scalar products of the Schrödinger picture field. For future reference note also that all of the space and time dependence is contained in the exponentials, which are solutions of the Klein–Gordon equation. Thus the Heisenberg-picture field satisfies the Klein–Gordon equation ($\Box \equiv \partial_0^2 - \nabla^2$)

$$(\Box + m^2)\phi = 0. \qquad (2.15)$$

In fact, the exponentials are the most general solutions of the Klein–Gordon equation, so the formula (2.14) turns solutions of the Klein–Gordon field equations into quantum operators. Willy-nilly we find ourselves studying the theory of quantized fields!

It is easy to solve the evolution equation (2.12) by infinitely iterated integration. This is a formal, perturbative solution, but in the present case it actually sums up to the exact answer:

$$U_D(t, t_0) = \sum_{n=0}^{\infty} (-i)^n \int_{t_0}^{t} dt_1 \int_{t_0}^{t_1} dt_2 \ldots \int_{t_0}^{t_{n-1}} dt_n W(t_1) \ldots W(t_n). \qquad (2.16)$$

Note that the operator order in this formula mirrors the time order: the leftmost operator is at the latest time etc. This suggests a more elegant way of writing the formula. Define the time-ordered product, $TW(t_1)\dots W(t_n)$, to be the product of the operators with the order defined by their time arguments. Thus, e.g.

$$TW(t_1)W(t_2) = \theta(t_1 - t_2)W(t_1)W(t_2) + \theta(t_2 - t_1)W(t_2)W(t_1). \qquad (2.17)$$

Alert readers will note the similarity of this construction to our discussion of causality in the introduction. If, in the formula for U_D, we replace ordinary operator products by time-ordered products, and allow all the ranges of integration to run from t_0 to t, then the only mistake we are making is over-counting the result $n!$ times. A correct formula is then

$$U_D(t, t_0) = \sum_{n=0}^{\infty} \frac{(-i)^n}{n!} \int_{t_0}^{t} dt_1 \dots dt_n \, TW(t_1)\dots W(t_n) \equiv Te^{-i\int_{t_0}^{t} W(t)}. \qquad (2.18)$$

The last form of the equation is a notational shorthand for the infinite sum.

In the case at hand, the formula looks even more elegant, because space and time integrations combine into a Lorentz-invariant measure. We have, in an obvious shorthand,

$$U_D(t, t_0) = Te^{-i\int d^4x \, \phi(x)J(x)}. \qquad (2.19)$$

We have already argued that we should allow the source function $J(x)$ to transform like a scalar field under Lorentz transformations. You will verify, in Problem 2.7, that the field $\phi(x)$ transforms as a scalar as well. The only things in our formula that are not Lorentz-covariant are the (implicit) end points of the time integration and the time-ordering symbol. They both appear because the Hamiltonian formulation of quantum mechanics requires us to choose $H = P^0$ for some particular Lorentz frame.

We can solve the first problem by simply taking the limits of the time integration to $\pm\infty$. We can't expect the answer to what happens to the system as it evolves between two fixed space-like surfaces to be independent of what those surfaces are. The infinite time limit of the Dirac-picture evolution operator is called the scattering operator, or S-matrix. It is reasonable to expect that the S-matrix is Lorentz-covariant. Its only dependence on the Lorentz frame should come from that of the external source $J(x)$.

The formula (2.19) also depends on the Lorentz frame because of the time-ordering operation. When field arguments are at space-like separation, the causal order depends on the Lorentz frame and should not appear in physical amplitudes. We can enforce this by insisting that

$$[\phi(x), \phi(y)] = 0 \qquad \text{if } (x - y)^2 < 0, \qquad (2.20)$$

because then, for space-like separation, the time ordering is superfluous.[4] The requirement that fields commute at space-like separation is called the locality postulate.

---

[4] Actually, one has to be a bit more careful in defining the time-ordered product at coinciding points in order to make sure that the locality postulate is enough to guarantee Lorentz invariance of the S-matrix. The time-ordered products are singular at coinciding points and this issue is properly treated as part of renormalization theory.

Together with some assumptions about the asymptotic behavior of the energy spectrum, this is the defining property of local field theory.

We note that one could have achieved the same end had we insisted that the sources were anti-commuting, or Grassmann, variables and required the fields to anti-commute at space-like distances. Although at first sight this seems bizarre, we will see that it is the right way to treat the fields of particles obeying Fermi statistics. We can now check whether our formula for the field satisfies these requirements. This is the content of Problem 2.5. The result is that, for spinless particles, locality can be imposed only for Bose statistics. This is the first step in the proof of the spin-statistics theorem: in local field theory in four and higher dimensions, all integer-spin particles must be bosons and all half-integer-spin particles fermions. We will not prove this theorem in full generality [6–8], but will see several more examples below. Problem 2.6 generalizes the result to charged particles as well.

A conserved internal symmetry charge is an operator that commutes with the Poincaré generators, which can be written as an integral of a local density, $Q = \int d^3x \, J^0$. We can choose particle states to be eigenstates of Q. A particle with annihilation operator $a(p)$ may have an anti-particle with annihilation operator $b(p)$. We have $[Q, a(p)] = a(p)$ and $[Q, b(p)] = -b(p)$, expressing the fact that the particle and anti-particle have opposite charge. The most general scalar field which creates or annihilates single particles, and annihilates one unit of charge, has the form

$$\phi(x) = \int \frac{d^3p}{\sqrt{(2\pi)^3 2\sqrt{p^2 + m_+^2}}} a(p) e^{ipx} + \alpha \int \frac{d^3p}{\sqrt{(2\pi)^3 2\sqrt{p^2 + m_-^2}}} b^\dagger(p) e^{-ipx},$$

where α is a complex number and $m_\pm$ are the masses of the particle and anti-particle. Problem 2.6 shows that we must choose $\alpha = 1$, $m_+ = m_-$, and Bose statistics to satisfy the locality postulate.

2.2 Problems for Chapter 2

*2.1. Show that a Lorentz transformation, satisfying $\Lambda^T g \Lambda = g$, has det $\Lambda = \pm 1$ and $|\Lambda_0^0| \geq 1$. The Lorentz group thus breaks up into four disconnected components depending on these two choices of sign. Only the $\Lambda_0^0 \geq 1$, det $\Lambda = 1$ component is continuously connected to the identity. These are called *proper orthochronous Lorentz transformations*. Show that we can get to the other components of the group by multiplying proper orthochronous transformations by time reversal, space reflection, or the product of these two. Show that proper orthochronous Lorentz transformations preserve the sign of the time component of time-like and null 4-vectors. Show that, for a space-like vector, an appropriate proper orthochronous Lorentz transformation can change the sign of the time component or set it equal to zero.

*2.2. The Dirac-picture evolution operator is defined by

$$U_D(t, t_0) = e^{iH_0 t} U_S(t, t_0) e^{-iH t_0},$$

where

$$i \, \partial_t U_S = (H_0 + V) U_S$$

is the equation satisfied by the Schrödinger-picture evolution operator. Show that

$$i \, \partial_t U_D = V(t) U_D,$$

where

$$V(t) \equiv e^{iH_0 t} V e^{-iH_0 t}.$$

*2.3. Show that the solution of the Dirac-picture equation is

$$U_D(t, t_0) = T e^{-i \int_{t_0}^{t} V(s) ds}.$$

*2.4. Compute the overlap of the ground-state wave functions of a harmonic oscillator with two different frequencies. A free-bosonic field theory is just a collection of oscillators. Use your calculation to show that the overlap of the ground states for two different values of the mass is zero for any field theory, in any number of dimensions and infinite volume. Show that the overlap is zero even in finite volume if the number of space dimensions is two or greater. This is symptomatic of a more general problem. The states of two field theories, containing the same fields but with different parameters in the Lagrangian, do not live in the same Hilbert space. The formulation of field theory in terms of Green functions and functional integrals avoids this problem.

*2.5. Show that the scalar field

$$\phi(x) = \int \frac{d^3 p}{\sqrt{(2\pi)^3 2\omega_p}} (a(p) e^{-ipx} + a^\dagger(p) e^{ipx})$$

does indeed transform as a scalar, as a consequence of the transformation law of the annihilation operators. Show that it is local if and only if the particles are bosons. Compute the equal-time commutators

$$[\phi(t, x), \phi(t, y)] = 0,$$
$$[\phi(t, x), \dot{\phi}(t, y)] = i\delta^3(x - y).$$

In the language of classical field theory, this means that the field and its time derivative are *canonically conjugate*.

*2.6. Repeat the locality computation for the charged scalar field

$$\phi(x) = \int \frac{d^3 p}{\sqrt{(2\pi)^3 2\omega_p}} (a(p) e^{-ipx} + \eta b^\dagger(p) e^{ipx}).$$

In this case the non-trivial computation is that of the commutator $[\phi(x), \phi^\dagger(y)]$. Show that the mass of the particle and that of the anti-particle (annihilated by a and b, respectively) must be equal and that the complex number η is a pure phase, which can be absorbed into the definition of b.

*2.7. Use the Dirac-picture evolution equations, with $V = -\int \mathrm{d}^3 x\, \mathcal{L}_I$, to show that the interaction-picture evolution operator over infinite times is Lorentz-invariant if $\mathcal{L}_I$ is a local scalar operator.

*2.8. The states of a particle of mass m and spin j at rest transform according to the spin-j representation of the rotation group.

$$U(R)|0,k\rangle = D^{j}_{kl}(R)|0,l\rangle.$$

Define the states with momentum p by

$$|p,k\rangle = \sqrt{\frac{m}{\omega_p}}\, U(L(p))|0,k\rangle.$$

$L(p)$ is a boost that takes the rest-frame momentum into (ω_p, p). Show that the Lorentz transformation law of these states is

$$U(\Lambda)|p,k\rangle = \sqrt{\frac{\omega_{\Lambda p}}{\omega_p}}\, D^{j}_{kl}(R_{\mathrm{W}})|\Lambda p, l\rangle,$$

where $R_{\mathrm{W}} = L^{-1}(\Lambda p)\Lambda L(p)$ is called the Wigner rotation. We can prove that it is a rotation by showing that it leaves the rest frame invariant. Using these transformation laws, show that one can construct fields transforming in the $[2j+1, 1]$ and $[1, 2j+1]$ representations of the Lorentz group (see Chapter 5)[5], which are linear in creation and annihilation operators of these particles. Show that these fields are Hermitian and local only if we choose Fermi statistics for half-integer-spin and Bose statistics for integer-spin particles. Compute the time-ordered two-point functions for these fields and note the ultraviolet behavior of the momentum-space propagator as a function of the spin.

*2.9. Writing an infinitesimal Lorentz transformation as $\Lambda \approx 1 + \mathrm{i}\omega^{\mu\nu} J_{\mu\nu}$, where $J_{\mu\nu}$ is a basis in the space of all η-anti-symmetric 4×4 matrices $(J\eta + \eta J^{\mathrm{T}}) = 0$: η is the Minkowski metric, considered as a matrix). Show that

$$[J_{\mu\nu}, J_{\alpha\beta}] = \mathrm{i}(\eta_{\mu\alpha}J_{\nu\beta} - \eta_{\nu\alpha}J_{\mu\beta} + \eta_{\nu\beta}J_{\mu\alpha} - \eta_{\mu\beta}J_{\nu\alpha}).$$

Write this out for the individual components J_{0i} and $J_{ij} = \epsilon_{ijk}J_k$.

2.10. Show that the stability subgroup of a null momentum[6] is isomorphic to the Euclidean group of translations and rotations in two dimensions. The only finite-dimensional representations of this group have the translation generators set to zero, while the rotation generator has a fixed eigenvalue, the helicity h. Following the method of induced representations we used for massive particles in the previous problem, work out the unitary representation of the Poincaré group on

---

[5] These representations are often denoted by their highest weight, rather than their dimension, and are called $[j,0]$ and $[0,j]$ representations.

[6] Subgroup of the Poincaré group which leaves it invariant.

single massless particle states and on the corresponding creation and annihilation operators. Show that, in order to construct Lorentz-covariant local fields, we must include both signs of helicity, $\pm h$, and the helicity must be quantized in half-integer units (for fermions) and integer units (for bosons). Show that for helicity ± 1 the smallest-dimension field operator transforms like the electromagnetic field strength $F_{\mu\nu}$. Work out the analogous statement for helicity $3/2$ and 2.

*2.11. Calculate the vacuum expectation value of the time-ordered exponential

$$\langle 0| T e^{i \int d^4 x\, \phi(x) J(x)} |0\rangle,$$

for a free field. Do this by writing $\phi(x)$ as a sum of an operator involving only creation operators and another with only annihilation operators. Compute a few terms in the power-series expansion in J, in order to convince yourself that the answer is

$$e^{\frac{i}{2} \int d^4 x \int d^4 y\, J(x)J(y) \int \frac{d^4 p}{(2\pi)^4} \frac{e^{-ip(x-y)}}{p^2 - m^2 + i\epsilon}},$$

which is the exponential of the second-order term. It should be easy for example to prove that all odd terms vanish. We will see a simpler derivation of this result below, in the language of functional integrals. Now consider the case $J(x) = \theta(T - t)\theta(t + T)[\delta^4(x) - \delta^4(x - R)]$, with $T \gg R \gg 1/m$, where m is the mass of the field. Show that in this limit the answer has the form

$$e^{-iTV(R)},$$

where the potential $V(R)$ is the Yukawa potential. We describe this result in the words *static sources for a field experience a force due to exchange of virtual particles.*

Interacting field theory

Our considerations so far give us a description of free relativistic spin-zero particles. Problem 2.10 shows us that exchange of particles between sources leads to what Newton called a force (gradient of a position-dependent potential energy) between the sources. We can think of the mathematical sources of that problem as models for a pair of infinitely heavy particles, separated by a distance R. So forces arise *by the exchange of virtual particles* between other particles. This motivates the idea that the way to introduce interactions between particles is to introduce a perturbation to the Lagrangian density of the Klein–Gordon equation,

$$\mathcal{L} \to \frac{1}{2}[(\partial\phi)^2 - m^2\phi^2] + \mathcal{L}_I, \tag{3.1}$$

where e.g. $\mathcal{L}_I$ has terms with two creation operators and one annihilation operator and vice versa. Then ϕ particles can interact via two operations of $\mathcal{L}_I$. Starting from a two-particle state we can create an extra particle near particle 1, with one operation of $\mathcal{L}_I$, and let that new particle propagate to the vicinity of particle 2, where it is reabsorbed. The condition that these expressions be compatible with Lorentz invariance and causality is, not surprisingly (Problem 2.7), that the interaction Lagrangian $\mathcal{L}_I(x)$ be local:

$$[\mathcal{L}_I(x), \mathcal{L}_I(y)] = 0 \qquad \text{if } (x - y)^2 < 0. \tag{3.2}$$

Thus, causal Lorentz-invariant theories of interacting particles in Minkowski space-time are identical to local quantum field theories with Lagrangians that are not purely quadratic in the fields. The rest of this book is devoted to exploring such theories, mostly in a perturbation expansion when the interactions are weak. We will touch on non-perturbative issues in our discussions of chiral symmetry breaking and confinement in quantum chromodynamics, instantons and solitons, and in our treatment of renormalization.

3.1 Schwinger–Dyson equations and functional integrals

Problem 2.4 shows us that many of the tools of conventional Hilbert-space quantum mechanics become problematic in interacting quantum field theory (QFT). The correct Hilbert-space formulation often depends on the interaction. Wightman and others, following the seminal work of Schwinger and Dyson (SD) [9–11], showed how the Hilbert space could be reconstructed from generalized Green[1] functions. We will follow a variant of the SD approach to the derivation of a set of equations for Green functions, which determine them and lead to a formal solution of the problem of QFT in terms of *Feynman path integrals*, or *functional integrals*. This formulation is manifestly covariant and easily amenable to a variety of approximation schemes.

The generalized *n*-point Green functions are the expressions

$$\langle 0|T\phi(x_1)\dots\phi(x_n)|0\rangle,$$

where ϕ is the Heisenberg field operator. These functions would appear in the Dirac-picture perturbation expansion of the response of a QFT to an external source $H \to H - \int d^3x\, J(x, t)\phi(x, t)$. It is easy to see (Problem 3.2) that in conventional quantum mechanics these would determine the ground-state wave function and the Hamiltonian of the system. In field theory they lead directly to a computation of all particle masses and all scattering amplitudes of those particles, as we will see below in the section on the Lehmann–Symanzik–Zimmermann (LSZ) formula. This is done in a completely covariant manner, without solving the Schrödinger equation or worrying about whether the free and interacting theories live in the same Hilbert space.

The idea of SD was to use the Heisenberg equations of motion and canonical commutation relations (Problem 2.5) to derive a closed set of equations for the Green functions. The generating functional for the Green functions is the vacuum persistence amplitude in the presence of the source,

$$Z[J] \equiv \langle 0|T e^{i\int d^4x\, J(x)\phi(x)}|0\rangle. \tag{3.3}$$

In free scalar field theory it is easy to work this out by operator methods. The action of the source n times can create n particles, which must then be re-annihilated to get a state that has overlap with the vacuum. Our aim will be to find an equation for $Z[J]$, allowing us to solve for it in a general field theory.

If the Lagrangian density is $\mathcal{L} = [\frac{1}{2}(\partial_\mu\phi)^2 - V(\phi)]$, then a naive application of the Heisenberg equations of motion gives

$$\partial^2 \frac{\delta Z}{i\,\delta J(x)} = -\langle 0|T \frac{\partial V(\phi)}{\partial \phi}(x)e^{i\int d^4y\, J(y)\phi(y)}|0\rangle. \tag{3.4}$$

This can be rewritten as

$$\partial^2 \frac{\delta Z}{i\,\delta J(x)} = -\frac{\partial V}{\partial \phi}\left(\frac{\delta}{i\,\delta J(x)}\right)Z[J]. \tag{3.5}$$

In these equations, I have introduced some notation that we will be using throughout the book. Square brackets in $Z[J]$ denote the fact that Z is a *functional* of the function

---

[1] Here I follow a revered teacher who believed that the locution "Green's functions" was an abomination, and at odds with standard practice: Bessel, Legendre, Whittaker, and especially Weierstrass do fine without the possessive. What's so special about Mr. Green?

J. That is, it is a rule that gives us a number for each function. All the functionals we will consider will have power-series expansions of the form

$$Z[J] = \sum \frac{1}{n!} \int Z_n(x_1, \ldots, x_n) J(x_1) \ldots J(x_n),$$

where the integral is over all the indicated coordinates. The functional derivative of such functionals is defined by the formal rule

$$\frac{\delta J(x)}{\delta J(y)} = \delta^4(x - y).$$

The rigorous mathematical meaning of many of our equations will take mathematicians decades to work out. Our attitude is that *all* of the equations of continuum QFT have a formal sense only. They are really defined by the process of regularization and renormalization, which we will study in Chapter 9. The real world is probably not described by a mathematically well-defined continuum QFT. The combination of quantum mechanics and gravity defines a fundamental length scale of order 10^{-33} cm, and QFT almost surely fails at that scale. Our real challenge will be to understand how to define a procedure for making predictions about length scales accessible in the laboratory, which depends as little as possible on the quantum gravitational physics we do not yet understand.

Returning to our formal equation (3.4), the reader should review this carefully to make sure that she/he understands that the time ordering is being applied both to the space-time point x and to the multiple integration variables in the expansion of the exponential. This is what allows us to differentiate operator expressions as if they were ordinary functions. The time ordering takes care of the operator-ordering problems.

Despite this cunning trick, we have made a mistake. The differential operator $\partial^2 = \partial_0^2 - \nabla^2$ contains time derivatives, which act on the Heaviside functions in the time-ordering operation. Consider

$$\partial_0^2 \langle 0|T\phi(x)\phi(y)|0\rangle = \partial_0 [\langle 0|T\partial_0\phi(x)\phi(y)|0\rangle + \delta(x^0 - y^0)\langle 0|[\phi(x), \phi(y)]|0\rangle]. \quad (3.6)$$

The second term comes from differentiating the Heaviside functions ($\partial_0\theta(x - y) = -\partial_0\theta(y - x) = \delta(x^0 - y^0)$).[2] It gives rise to an equal-time commutator, which vanishes by virtue of the canonical commutation relations you worked out in Problem 2.5. However, when we perform the second time derivative, we get a similar term, which involves the equal-time commutator of $\partial_0\phi$ and ϕ. Since these variables are canonically conjugate, this term is proportional to $\delta^3(x - y)$. Thus

$$\partial_0^2 \langle 0|T\phi(x)\phi(y)|0\rangle = \langle 0|T \, \partial_0^2\phi(x)\phi(y)|0\rangle - i\delta^4(x - y). \quad (3.7)$$

When we apply the second derivative operator to $\langle 0|T\phi(x)\phi(y_1)\ldots\phi(y_n)|0\rangle$ we observe a similar phenomenon (Problem 3.1). Differentiation of the Heaviside functions

---

[2] Note that all of the derivatives in this section are x derivatives.

gives rise to equal-time commutators of $\phi(x)$ and $\partial_0\phi(x)$ with each of the fields at y_i:

$$\partial_0^2 \langle 0|T\phi(x)\phi(y_1)\ldots\phi(y_n)|0\rangle = \langle 0|T\,\partial_0^2\phi(x)\phi(y_1)\ldots\phi(y_n)|0\rangle$$
$$- i\sum_j \delta^4(x-y_j)\langle 0|T\phi(y_1)\ldots\phi(y_{j+1})\ldots\phi(y_n)|0\rangle.$$

$$(3.8)$$

If we multiply this equation by $J(y_1)\ldots J(y_n)$, integrate, and use the Heisenberg equation for $\partial_0^2\phi$, we obtain the correct SD equation:

$$\partial^2 \frac{\delta Z}{i\,\delta J(x)} = -\left(\frac{\partial V}{\partial \phi}\left[\frac{\delta}{i\,\delta J(x)}\right] + J(x)\right)Z[J]. \qquad (3.9)$$

This equation may be described succinctly in the following words: Think of the functional differential operator $\delta/i\,\delta J(x)$ as a field and write down the left-hand side of the field equation implied by the Lagrangian $\mathcal{L}+\phi J$, for this field. This gives us a linear functional differential operator $\mathrm{SD}[\delta/i\,\delta J(x)]$, where e.g. $\mathrm{SD}[\phi] = \partial^2\phi + V'(\phi) - J$. The SD equation is $\mathrm{SD}Z[J] = 0$.

3.2 Functional integral solution of the SD equations

In order to understand how to solve the SD equations, it is convenient to think about a regularization of QFT in which space-time is replaced by a finite hypercubic lattice of points. The continuous variable x is replaced by a discrete variable A, with finite range. The field and source functions become finite-dimensional variables ϕ^A and J^A. The differential operator ∂^2 becomes a symmetric matrix K_{AB}. In fact such a regularization is used in numerical solutions of strongly coupled field theories, which are not amenable to other approximation techniques. It converts QFT into a finite problem, which can be solved on a computer.

The discretized SD equations take the form (summation convention for repeated indices):

$$K_{AB}\frac{\partial Z}{i\,\partial J^B} = \left[V'\left(\frac{\partial}{i\,\partial J^A}\right) + J^A\right]Z[J]. \qquad (3.10)$$

Since we have only a finite number of variables, we have replaced functional derivatives by ordinary partial derivatives.

If V is a polynomial of highest order k then this is a set of kth-order partial differential equations (PDEs) in a finite number of variables. The key to solving them is to note that the explicit dependence on J^A is linear. If we Fourier transform a set of linear PDEs, every derivative turns into i times the Fourier transform variable, while every J^A turns into ($-$i times) a derivative w.r.t. the Fourier transform variable. The Fourier transform of the discretized SD equations will be a set of *first*-order PDEs for the Fourier transform. For reasons that will become apparent, we call the Fourier-transform variable ϕ^A, and write

$$Z[J] \equiv \int [\mathrm{d}\phi] e^{\mathrm{i}S[\phi] \,+\, \mathrm{i}J^A\phi^A}, \tag{3.11}$$

where square brackets around the integration measure indicate that it is a multiple integral for which we are contemplating taking the limit of an infinite number of integration variables. Such integrals are called functional integrals. It requires a lot of care to give them a rigorous mathematical definition. Most physicists deal with this problem through the algorithm of renormalization, to which we will turn in Chapter 9.

The Fourier-transformed SD equations (as the reader will kindly verify) are

$$\frac{\partial S}{\partial \phi^A} = K_{AB}\phi^B - \frac{\partial V}{\partial \phi^A},$$

whose solution is

$$S = \frac{1}{2}\phi^A K_{AB}\phi^B - V(\phi) + C, \tag{3.12}$$

where C is a constant independent of ϕ^A. Our equations for Z are homogeneous and do not determine its overall normalization.

Returning to the continuum, we write S as

$$S = \int \mathrm{d}^4x \left[-\frac{1}{2}\phi\,\partial^2\phi - V(\phi) \right]. \tag{3.13}$$

Integrating by parts,

$$S = \int \mathrm{d}^4x \left[\frac{1}{2}(\partial\phi)^2 - V(\phi) \right]. \tag{3.14}$$

The integral defining the Fourier transform of the generating functional is now an integral over some space of functions, or functional integral. For the free-field case, which we will deal with in a moment, such integrals were given a rigorous mathematical definition by Wiener [12]. For interacting field theory they are defined by a process of regularization (turning them into finite integrals) and renormalization to which we will return in Chapter 9. We certainly want our functions to approach constants at infinity in space-time, so that the integration by parts that we did in the action is allowed, but we will not delve further into their properties. In Chapter 10, in discussing gauge theories and low-dimensional scalar field theories we will encounter examples where integrals of total derivatives have to be kept in the action, and we will take care not to drop them where they are important.

If we consider free field theory, where $V(\phi) = \frac{1}{2}m^2\phi^2$, the integral is Gaussian and we can do it explicitly. However, it is an oscillating Gaussian, so we must provide a prescription for evaluating it. One obvious possibility is to make the replacement $m^2 \to m^2 - \mathrm{i}\epsilon$, where ϵ is a small positive number, which is eventually sent to zero.

Consider the Gaussian integral

$$I(J) \equiv \int d^n y \, e^{-\frac{1}{2}K_{ij}y^iy^j + iy^iJ^i}, \tag{3.15}$$

where K is a symmetric matrix with positive definite real part. This can be evaluated by changing variables $y \to K^{-\frac{1}{2}}Y$, evaluating the Jacobian, and doing the resulting single-variable Gaussian integrals. The result is

$$I(J) = e^{-\frac{1}{2}J^i(K^{-1})_{ij}J^j} |\det K|^{-\frac{1}{2}} \left[\sqrt{\frac{\pi}{2}}\right]^n, \tag{3.16}$$

as the knowledgeable reader will know and the diligent reader can verify. In the functional integral limit, we have to invert a differential operator and take its determinant. We will temporarily evade the second of these tasks by noting that the definition of $Z[J]$ in terms of time-ordered products gives us the normalization $Z[0] = 1$. We achieve this by writing $Z[J]$ as the ratio of two functional integrals $I(J)/I(0)$, and note that the functional determinant as well as the pesky factors of $\sqrt{\pi/2}$ cancel out in this ratio.

We are left to evaluate the inverse of the differential operator

$$K = i(\partial^2 + m^2 - i\epsilon). \tag{3.17}$$

That is, we want to find the Green function satisfying

$$i(\partial^2 + m^2 - i\epsilon)D_F(x - y) = \delta^4(x - y). \tag{3.18}$$

This is easily done in momentum space:

$$D_F(x - y) = \int \frac{d^4p}{(2\pi)^4} e^{-ip(x-y)} \frac{i}{p^2 - m^2 + i\epsilon}. \tag{3.19}$$

The integral defining this *Feynman* or *causal* Green function[3] of the Klein–Gordon (KG) operator has poles in the p^0 plane just below the positive real axis and just above the negative real axis. When $x^0 - y^0 > 0$, we evaluate it by closing the contour in the lower half plane. The resulting Green function is an integral over only positive energy waves. Similarly, when the time order is reversed, we close in the upper half plane, and again conclude that only positive energies propagate forward in time. This is the rigorous version of Feynman's idea that "anti-particles are negative-energy particles propagating backwards in time."

The equality

$$D_F(x - y) = \langle 0|T\phi(x)\phi(y)|0\rangle, \tag{3.20}$$

for free Heisenberg fields, may be verified in two illuminating ways. First of all, the free SD equations show that the RHS is a Green function for the KG equation, and the discussion of the previous paragraph shows that the boundary conditions implied by the $i\epsilon$ prescription are precisely those satisfied by the time-ordered vacuum expectation value (VEV). Alternatively, one can compute the RHS by brute force using creation and

---

[3] Also called the Feynman propagator.

annihilation operators. Thus, the $i\epsilon$ trick for making the Gaussian integral converge automatically chooses the right space-time boundary conditions to define the time-ordered Green function of the KG equation.

We can obtain a deeper understanding of this connection by noting that the position of the energy poles in the Feynman Green function allows us to rotate the contour of integration[4] to $p^0 \to ip_4$, if we simultaneously rotate time differences to be imaginary. We realize that D_F is the analytic continuation of the Euclidean Green function,

$$D_E(x) = \int \frac{\mathrm{d}^4 p_E}{(2\pi)^4} \frac{e^{-ipx}}{p_E^2 + m^2}, \qquad (3.21)$$

where the Euclidean scalar product is $p_E^2 = p_4^2 + p^2$. This is the unique Green function of the four-dimensional Helmholtz operator $-\nabla^2 + m^2$, which falls off at Euclidean infinity.

It is easy to see why the correlation functions of QFT should have analytic continuations to imaginary time. Consider (for notational simplicity) the time ordering in which the labels of the points coincide with their time order. Then

$$\langle 0|T\phi(x_1)\ldots\phi(x_n)|0\rangle = \langle 0|\phi(x_1,0)e^{-iH(t_1-t_2)}\ldots e^{-iH(t_{n-1}-t_n)}\phi(x_n,0)|0\rangle, \quad (3.22)$$

where we have used $H|0\rangle = 0$. By inserting complete sets of energy eigenstates, we can rewrite this in the form

$$\int \rho(E_1,\ldots,E_{n-1})e^{-i\sum E_k(t_k-t_{k+1})}, \qquad (3.23)$$

where the spectral function (which also has implicit dependence on the spatial points at which the fields are evaluated) ρ is determined in terms of sums over matrix elements of fields between states of fixed energy. In free-field theory, and every order of perturbation theory around it, it is easy to see that the spectral function is an analytic function of the energy variables. Furthermore, it falls off when the energy variables become large. This is enough to guarantee[5] that these functions have a well-defined analytic continuation to imaginary, or Euclidean, time. Note that, if we take $(t_k - t_{k+1}) \to -i\tau_k$, with $\tau_k > 0$, the exponentials give a Boltzmann-like suppression of high-energy states.[6] Thus, assuming that field matrix elements have at most power-law growth with energy, these functions are well defined and analytic in imaginary time.

---

[4] At this point I am expecting the readers to stop and remember their complex analysis course. If you don't know complex analysis, stop here, learn it, and come back to this point.

[5] In axiomatic approaches to field theory this behavior of the spectral functions is simply postulated. The bound on the growth of the density of states can be derived from the definition of field theory we will give in the chapter on renormalization. A field theory is a perturbation of a conformal field theory by an operator that is negligible at high energy. The density of states of a conformal field theory in volume V is, by virtue of conformal invariance and extensivity, $\rho(E) \sim e^{cVE^{\frac{d-1}{d}}}$.

[6] This is not a coincidence. The Euclidean Green functions with compactified imaginary time compute thermal expectation values of fields.

The Euclidean formulation of field theory[7] leads to mathematical expressions whose properties are simpler to understand. Euclidean methods also turn out to be crucial for understanding field theory at finite temperature and quantum tunneling amplitudes. They are also central to numerical techniques (lattice field theory) for obtaining non-perturbative solutions to field theory. So we will generally take the attitude that a field theory is defined by analytic continuation of a Euclidean functional integral. We simply write the formal expression for the Euclidean action and integrate $e^{-S_E} \phi(x_1) \ldots \phi(x_n)$ over fields in Euclidean space, and then analytically continue the result to get time-ordered products of Lorentzian fields.

In this way, we obtain the definition of the generating functional of Euclidean Green functions, also known as Schwinger functions:

$$Z_E[J] = \frac{\int [d\phi] e^{-S_E[\phi]+i \int \phi J}}{\int [d\phi] e^{-S_E[\phi]}}, \tag{3.24}$$

which has the form of the characteristic function of a probability distribution. Note that in writing this formula we have also analytically continued the source function J to iJ. If we refrain from doing this we obtain the Boltzmann formula for the partition function of a classical statistical system in an external field J. The "potential" of the statistical system is the Euclidean action,

$$S_E \equiv \int d^4 x \left[\frac{1}{2} (\nabla \phi)^2 + V(\phi) \right], \tag{3.25}$$

where we now integrate over four-dimensional Euclidean space, rather than Minkowski space-time. This profound analogy between QFT and classical statistical mechanics has led to an enormously successful cross fertilization between the two fields.

3.3 Perturbation theory

In problems amenable to perturbation theory, the non-quadratic part of the potential is multiplied by a small dimensionless parameter g. The classical action has the form $S = S_0 + g \int V(\phi)$, where S_0 is the action of a free massive field. Therefore, we can write a power-series expansion for $Z[J] \equiv I[J]/I[0]$:

$$I[J] = \sum_{n=0}^{\infty} \frac{g^n}{n!} \int [d\phi] e^{iS_0 + i \int \phi J} \int d^4 x_1 \ldots d^4 x_n (V(x_1) \ldots V((x_n)). \tag{3.26}$$

If V has a power-series expansion in ϕ, then every integral that is required in order to evaluate this formula has the form

$$\int [d\phi] e^{iS_0} \phi(y_1) \ldots \phi(y_m).$$

---

[7] The idea of Euclidean field theory is due to Schwinger, and the rotation of contours in momentum space is called the Wick rotation.

The ϕs in this formula, which come from the same $V(x_i)$, will all have the same value of $y = x^i$, and we will integrate over the x_i variables. When evaluating $Z[J]$, we expand both numerator and denominator. Then it is easy to see that every such integral in the expansion will be accompanied by the denominator I_0, the value of the functional integral at $J = g = 0$. The generating functional for these normalized functional integrals is just $Z_0[J] = \langle 0|Te^{i \int \phi J}|0\rangle$, the free-field vacuum persistence amplitude.

In free-field theory, we can either solve the (linear) SD equation directly or do the Gaussian functional integral to obtain

$$Z_0[J] = e^{-\frac{1}{2} \int d^4x\, d^4y\, J(x)J(y)D_F(x-y)}, \tag{3.27}$$

where $D_F(x)$ is the Feynman Green function or vacuum expectation value of the time-ordered product of two free fields. It follows, by expanding out both the definition and the solution for $Z_0[J]$, that

$$\frac{\int [d\phi]e^{iS_0}\phi(y_1)\dots\phi(y_m)}{\int [d\phi]e^{iS_0}} = \sum_{\text{pairs}} \prod D_F(y_i - y_j). \tag{3.28}$$

This formula is called Wick's theorem. It leads to the following set of rules for perturbation theory, called Feynman rules.

- Write the perturbation as a sum of monomials $V = g \sum (v_k/k!)\phi^k$.
- A term in nth-order perturbation theory in g will have contributions proportional to $v_{k_1} \dots v_{k_n}$. We associate this with a diagram with n vertices. The ith vertex has k_i lines coming out of it. These are called Feynman diagrams or Feynman graphs.
- Each line must be connected either to a line from another vertex or to an external source point. For a contribution to an E-point Green function, there will be E such external lines. Lines connecting two vertices are called internal lines. The number of internal lines I is given by $2I = \sum k_i - E$.
- Write a factor $D_F(y_i - y_j)$ for an internal line connecting two space-time points.
- Write a factor $D_F(y_i - x_a)J(x_a)$ for an external line.
- Integrate over all internal and external points. If one wants to directly compute individual Green functions, rather than the generating functional, omit the factors of $J(x_a)$ and the integral over the external points.
- We can easily generalize to interaction vertices like $\phi^2(\partial_\mu \phi)^2$, just letting the derivatives act on the appropriate argument of D_F. Similarly, it is easy to generalize these rules to theories of multiple scalar fields.
- A given Feynman diagram in nth order comes with a factor $(ig)^n v_{k_1} \dots v_{k_n} S_G$, where the combinatorial factor S_G is a combination of the inverse factorials from the definition of v_k and *the number of times a given graph can be obtained in the sum over all possible pairings (also called all possible contractions)*. S_G is called the symmetry number of the graph, because it can be shown that it is equal to the inverse of the order of the group of geometrical symmetries of the graph. I have always found that the art of figuring out the geometrical symmetries is harder than reproducing S_G by counting the number of contractions which give the graph. It's a matter of taste. An algorithm for doing the counting can be found in [13].

- By Fourier transforming everything in sight, one can derive an equivalent set of rules in momentum space. These are somewhat easier to write down, because the Fourier transform $i/(p^2 - m^2 + i\epsilon)$ of D_F is so simple. The rules involve integrating over the momenta of closed loops in the diagram, with measure $d^4p/(2\pi)^4$ (equivalently, integrating over all internal line momenta, with $2\pi^4\delta^4(\sum q_i)$ enforcing momentum conservation at each vertex; one momentum δ function of the sum over external momenta is left). Derivatives in the interaction turn into factors of momentum, but one must take care to make sure that one has the right sum of momenta in these factors.
- In Euclidean space, the rules are even easier. We replace the factor of i in $(ig)^n$ by a minus sign, and drop the i from each propagator. The Euclidean propagator is just $1/(p_E^2 + m^2)$.

It's extremely important for the reader to work out the derivation of these rules, following the outline we have supplied above.

The Feynman rules for a large class of field theories, along with standard diagrammatic conventions, are collected in Appendix D. There one can also find several examples of working out the combinatoric factors for individual graphs. This is probably a good point in the exposition for students to stop and do a few exercises to convince themselves that they know how to set up the computation of Feynman graphs. If you go too far in this exercise you'll encounter the disturbing fact that most of the loop graphs are infinite. We won't learn how to tackle this problem, and what it means, until Chapter 9.

3.4 Connected and 1-P(article) I(rreducible) Green functions

The free generating functional has the form $e^{\frac{1}{2}\int J(x)J(y)D_F(x-y)} \equiv e^{iW_0[J]}$, i.e. it is an exponential of something simple. W_0 has a nice graphical interpretation. If we write out all of the graphs contributing to $Z_0[J]$, only the connected graphs contribute to W_0. The Feynman rules show us that this is a general property of the interacting theory as well. The sum over all graphs is the exponential of the sum over connected graphs. This follows from the symmetry-number Feynman rule. If I have a disconnected graph with k identical disconnected pieces, there is an obvious S_k geometrical symmetry. The order of the symmetric group S_k is $k!$. Thus, in general, we define $Z[J] \equiv e^{iW[J]}$ and W is the generating functional of connected Green functions:

$$Z_0[J] = \sum \frac{1}{n!} \begin{matrix} J \\ J \end{matrix} \underline{\qquad} \begin{matrix} J \\ J \end{matrix} = e^{iW_0[J]}.$$

Wick's theorem and connected diagrams

The Feynman rules for W are the same as those for Z, but summing only over connected diagrams. In particular, since all contributions from corrections to the denominator $I[0]$ are disconnected, we never have to worry about them. *Aficionados* of statistical mechanics will recognize the relation between Z and W to be essentially the same as that between the partition function and the Helmholtz free energy.[8]

Often, one can conveniently reorganize perturbation theory into a semi-classical expansion of the functional integral. This happens whenever, by rescaling the fields, we can write the action as $S = (1/g^2)s[\phi]$, where g is a small dimensionless parameter, and s contains only positive powers of g (typically only g^0). One then recognizes that one can do the integral by the stationary-phase, or steepest-descents, method. That is, one looks for stationary points of the logarithm of the integrand, and expands the integral around them. The first step requires us to solve

$$\frac{\delta s}{\delta \phi(x)} + J(x) = 0, \qquad (3.29)$$

which is just the classical field equation in the presence of the source. If $\phi(x, J]^9$ is the classical solution then the connected generating functional is given by

$$W_c[J] = \frac{1}{g^2}\left(S[\phi[J]] + \int J\phi[J]\right). \qquad (3.30)$$

In words, *in the classical approximation, the connected generating functional is the Legendre transform of the classical action.*

If the functional $s[\phi]$ is g-independent, then the semi-classical expansion has a topological interpretation in terms of Feynman graphs. If we expand around a stationary point $\phi = \phi_0(x)$ of the $J = 0$ functional integral, the first correction to the classical action is

$$S_{(2)}[\phi_0, \delta\phi] \equiv \frac{1}{g^2}\int d^4x\, d^4y\, \delta\phi(x)\delta\phi(y)\frac{\delta^2 s}{\delta\phi(x)\delta\phi(y)}\bigg|_{\phi=\phi_0}. \qquad (3.31)$$

If we define $\Delta = g\,\delta\phi$, we can write $(1/g^2)s[\phi] = (1/g^2)s[\phi_0] + S_2[\Delta] + \sum g^{k-2}V_k[\Delta]$. Viewed as a field theory for the generating functional of Δ Green functions, this looks like our perturbation problem, except with a link between the power of the field and the power of the perturbation parameter g.

Now consider a contribution to an E-point function at order g^n. This means that the number of lines, $V = \sum k_i$, emanating from vertices v_{k_i} in the diagram must sum up to

$$\sum(k_i - 2) = \sum k_i - 2V = n,$$

where V is the number of vertices. The number of internal lines, I, is

$$I = \frac{1}{2}\sum k_i - E,$$

---

[8] Up to factors of i and -1. In Euclidean QFT, the parallel is exact.

[9] The peculiar bracket structure in $\phi(x, J]$ is supposed to convey the fact that it is both a function of x and a functional of J.

and the number of loops for a connected diagram

$$L = I - V + 1 = \frac{1}{2}\sum k_i - E - V + 1.$$

Thus

$$L = \frac{n}{2} - E + 1.$$

For a fixed number of external lines, the expansion in powers of g^2 is an expansion in the number of loops. Note that if we look at correlation functions of the original fluctuations $\delta\phi$, rather than Δ, the powers of g associated with external lines disappear from these formulae. An order-one fluctuation of Δ is an $o(g)$ fluctuation of the original fields. Solving the classical field equations with a source J of order one sums up all tree diagrams, with any number of external legs.

These considerations motivate the introduction of a new generating functional $\Gamma[\phi]$, which is related to the exact $W[J]$ by a Legendre transform

$$W[J] = \Gamma[\phi[J]] + \int J\phi[J], \tag{3.32}$$

where $\phi(x, J]$ is a solution of

$$\frac{\delta W}{\delta J(x)} = \phi(x, J]. \tag{3.33}$$

The expansion coefficients of Γ,

$$\Gamma[\phi] = \sum \frac{1}{n!}\int \Gamma_n(x_1 \ldots x_n)\phi(x_1)\ldots\phi(x_n)\mathrm{d}^4 x_1 \ldots \mathrm{d}^4 x_n, \tag{3.34}$$

are called one-particle irreducible (1PI) Green functions. Connected Green functions are constructed from tree diagrams with 1PI functions as vertices and propagators given by the full connected two-point function.

Diagrammatically, 1PI functions are constructed as the sum of all diagrams that cannot be cut into disconnected parts by cutting a single propagator line. At tree level, these just give the vertices constructed from the classical action. Γ is often called the *quantum effective action*. We will drop the word "effective," in order to reserve it for another use in the discussion of renormalization. In Chapter 9 we will learn that the cut-off-dependent *effective action* defined by the renormalization group is equal to the *quantum action* defined here only in the limit that the momentum cut-off scale is taken to zero, and only if there are no massless states in the theory.

3.5 Legendre's trees

The graphical assertions of the previous two paragraphs were that the Legendre transform relation

$$W[J] = \int \phi(x)J(x) + \Gamma[\phi] \tag{3.35}$$

(where $\phi(x)$ is identified with $\delta W/\delta J(x)$ and $J(x)$ with $-\delta\Gamma/\delta\phi(x)$) generates an expansion of the (connected) Green functions obtained from the expansion of W in terms of tree diagrams whose vertices are the 1PI Green functions and whose branches are the full propagator

$$\frac{\delta^2 W}{\delta J(x) \delta J(y)}.$$

In leading order in the semi-classical expansion we have $\Gamma = S/g^2$.

The proof of this relation is simple. We start from the obvious identity

$$\frac{\delta^2 W}{\delta J(x) \delta J(y)} = \frac{\delta \phi(y)}{\delta J(x)} = \left(\frac{\delta J(y)}{\delta \phi(x)} \right)^{-1} = -\frac{\delta^2 \Gamma}{\delta \phi(x) \delta \phi(y)}.$$

The inverse in the second equality is meant in the sense of integral operators,

$$\int \mathrm{d}^4 z \, K(x, z) K^{-1}(z, y) = \delta^4(x - y),$$

which is the continuous analog of matrix inversion. We write this as $W_2 = -\Gamma_2^{-1}$, where Γ_2 is the integral operator made from the 1PI two-point function.

Differentiate this identity with respect to $J(z)$,

$$\frac{\delta^3 W}{\delta J(x) \delta J(y) \delta J(z)} = -\frac{\delta \Gamma_2^{-1}(x, y)}{J(z)}, \qquad (3.36)$$

and use the continuous analog of the matrix differentiation formula $\mathrm{d}K^{-1} = -K^{-1} \, \mathrm{d}K \, K^{-1}$ to write this as

$$W_3(x, y, z) = \int \mathrm{d}^4 w_1 \, \mathrm{d}^4 w_2 \, W_2(x, w_1) W_2(y, w_2) \frac{\delta \Gamma_2(w_1, w_2)}{\delta J(z)}. \qquad (3.37)$$

Using the chain rule for differentiation and the relation $\phi = \delta W / \delta J$, this becomes

$$W_3(x, y, z) = \int \mathrm{d}^4 w_1 \, \mathrm{d}^4 w_2 \, \mathrm{d}^4 w_3 \, W_2(x, w_1) W_2(y, w_2) W_2(z, w_3) \Gamma_3(w_1, w_2, w_3). \qquad (3.38)$$

This is the formula implied by the three-point position-space Feynman diagram of Figure 3.1.

It is now all over except for the shouting, which mathematicians call induction. When we differentiate W_3 w.r.t. $J(s)$, the derivatives act either on the propagators or on Γ_3. The latter contribution gives the first term in the four-point diagram of Figure 3.1. For the former, we simply rerun the previous derivation to get the second four-point term. Similarly, for W_n, either differentiation acts on the Γ_k, with $3 \leq k \leq n$, and generates Γ_{k+1} and an extra external W_2 leg, or it acts on one of the propagators, adding an extra three-point vertex and external propagator. In words, W_n is generated from all possible tree diagrams.

This result is very powerful. Below we will show that the exact momentum-space two-point function, $W_2(p)$ (when $J = 0$ the two-point function is translation-invariant and its Fourier transform depends only on one variable), has a pole at the mass of any stable particle that can be created from the vacuum by the quantum field ϕ. The

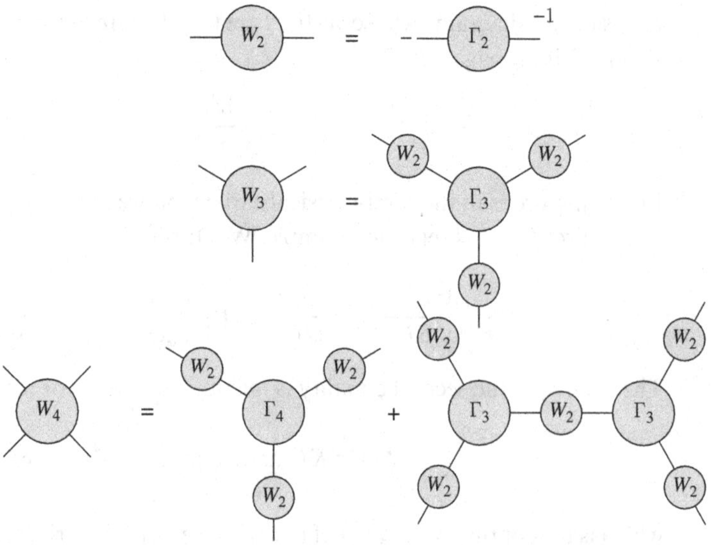

Fig. 3.1.   Legendre transform $W[J] = \Gamma[\phi] + \int \phi J$ makes trees.

1PI expansion of connected Green functions then tells us that *every* W_n has a pole on each of its external legs. We will see that the residue of this multiple pole, after proper normalization, contains the S-matrix for all scattering processes in which the total number of ingoing and outgoing particles is n.

Before concluding this subsection, we want to remind the reader that our notation has conflated the quantum field operator ϕ, the functional integration variable ϕ, and the argument of the quantum action ϕ. These are distinct concepts, but it is typographically insane to use different versions of the same Greek letter to separate them. The reader who has truly understood the discussions above should have no trouble distinguishing the meaning of ϕ from its context.

3.6 The Källen–Lehmann spectral representation

For $x^0 > 0$, we can write the two-point function of the interacting Heisenberg field $\phi(x)$ as

$$\langle 0|T\phi(x)\phi(0)|0\rangle = \int d^4p \sum_n \delta^4(p - p_n)e^{-ipx}|\langle 0|\phi(0)|n\rangle|^2. \tag{3.39}$$

We have inserted a complete set of states between the two fields, and used the action of the translation generator on the Heisenberg fields and on the states. p_n is the momentum of the state $|n\rangle$. As we will discuss in more detail below, we assume that the theory has a complete set of *scattering states*, i.e. states that are composed of multiple stable particles traveling freely at large space-like separation from each other. The energies and momenta of such states are determined by the usual free-particle relations. We will assume further that the theory has a mass gap. That is, apart from the non-degenerate vacuum state, the lowest value of P^2 is a non-zero number equal to the mass of the lightest stable particle in the theory.

For theories with massless particles, the general form of the Källen–Lehmann [14,15] representation will remain unchanged, but the identification of particle masses with poles is slightly more subtle, and depends in detail on the form of the massless particle interactions. Let m be the mass of the lightest stable particle that can be created from the vacuum by a single action of ϕ. Then it follows from Lorentz invariance that

$$\langle 0|\phi(0)|p\rangle = \sqrt{\frac{Z}{2\omega_p(2\pi)^3}}. \tag{3.40}$$

The constant Z is called the *on-shell wave-function renormalization constant of* ϕ.

In any field theory, with or without a mass gap, we will have only states of time-like or null momentum, with positive energy. So we can rewrite the two-point function as

$$\begin{aligned}
\langle 0|T\phi(x)\phi(0)|0\rangle = \int_0^\infty \mathrm{d}\mu^2 \int \mathrm{d}^4p\,\theta(p^0)\delta(p^2-\mu^2) \\
\times \sum_n \delta^4(p-p_n)\mathrm{e}^{-ipx}|\langle 0|\phi(0)|n\rangle|^2.
\end{aligned} \tag{3.41}$$

We recognize that this is

$$\langle 0|T\phi(x)\phi(0)|0\rangle = \int_0^\infty \mathrm{d}\mu^2\,\rho(\mu^2)D_F(x;\mu^2), \tag{3.42}$$

where D_F is the free-field two-point function and

$$\sum_n \delta^4(p-p_n)|\langle 0|\phi(0)|n\rangle|^2 \equiv \rho(p^2). \tag{3.43}$$

Indeed, since $U^\dagger(\Lambda)\phi(0)U(\Lambda) = \phi(0)$, for every Lorentz transformation Λ, the positive function $\rho(p)$ is Lorentz-invariant, and depends only on p^2.

In a theory with a mass gap, the lower limit on the integral is m^2, coming from the one-particle state, and then there is a gap until the lightest multiparticle state that can be created by the action of ϕ. Thus there is a contribution

$$\rho(\mu^2) = Z\delta(\mu^2 - m^2) + C(\mu^2),$$

where the continuum contribution C is positive and has support starting at the invariant mass squared of the lowest multiparticle state.

It follows that the exact momentum-space two-point function $W_2(p)$ has a pole at the stable particle mass, with residue Z. Note that nothing in this derivation requires that the field ϕ be the fundamental variable that appears in the Lagrangian, or that perturbation theory be applicable. Any field with a non-zero Z will do. In cases where perturbation theory is applicable, the fundamental fields will have poles in their Green functions. However, in quantum chromodynamics (QCD), the theory of the strong interactions, perturbation theory is inapplicable in the vicinity of hadron mass scales. The fundamental quark and gluon fields are not associated with stable particles, and hadrons are instead created by multi-linear functions of the quark fields. In lattice gauge theory, the functional integral of QCD is done numerically, and hadron masses are computed by finding poles in the two-point functions of quark–anti-quark bilinears and quark trilinears.

3.7 The scattering matrix and the LSZ formula

Most of the real experimental data to which QFT can be applied are the results of scattering experiments. It is an experimental fact that there exist (approximately) stable single-particle states in the world, as well as states of multiple particles at large relative space-like distances, which behave, to a very good approximation, like free particles. Interactions fall off at large spatial separation, and the *cluster property* of QFT provides a neat mathematical explanation of this [16–17]. The cluster property states that, at large space-like separation, the connected parts of Green functions fall to zero. In perturbation theory, this follows from the falloff of a single Feynman propagator. If all particles are massive, the falloff is exponential, whereas if there is no mass gap we expect power-law falloff. In this case, a variety of behaviors is possible, and there is not always a scattering theory.

The idea of scattering theory is to derive formulae for idealized amplitudes in which some number of widely separated particles come in from past infinity, interact, and go off to become a state of (possibly different) space-like separated particles in the asymptotic future. The central quantity one computes is the *scattering* or *S*-matrix: the amplitude

$$\langle \text{out}\, f_1 \ldots f_n | g_1 \ldots g_m\, \text{in} \rangle$$

for m particles with asymptotic wave functions g_i to turn into n particles with asymptotic wave functions f_i. The g_i and f_j run over complete sets of single free-particle states.

Non-relativistic quantum mechanics has a well-developed scattering theory [18–19]. In that theory one proves that the S-matrix for potential scattering is a *partial isometry*, a unitary transformation on a subspace (the scattering states) of the Hilbert space orthogonal to all bound states of the particles. The *in* and *out* states are two different bases for this subspace and the S-matrix is the transformation relating them. In QFT all bound states can be simply thought of as additional particles. So the entire Hilbert space will be spanned by scattering states, as long as we treat bound states as separate particles.

The basic assumption of scattering theory is thus that the Hilbert space has two complete bases of states, each of which is isomorphic to the Fock space of some collection of free particles. A list of the masses and spins of the particles completely specifies the spectrum of the Hamiltonian. We have noted in Problem 2.3 how difficult it is to discuss the eigenstates of an interacting quantum field theory, and claimed that the introduction of Green functions side-steps all these difficulties. The Lehmann–Symanzik–Zimmermann (LSZ) formula [20] shows us how to find the masses and spins of all single-particle states, and to compute the S-matrix, in terms of Green functions. To derive it, we assume the existence of a single-particle state of mass m and spin 0[10]. We also assume the existence of a local field $\Phi(x)$, such that

$$\langle 0|\Phi(x)|p\rangle = \sqrt{\frac{Z}{(2\pi)^3 2\omega_p}}\, e^{-ipx},$$

with $Z \neq 0$. The functional form of the matrix element follows from Lorentz invariance (prove it if you don't believe me). Φ might be proportional to the Lagrangian field of an interacting field theory, or it might be a complicated function of such fields. We will let $Z[J]$ denote the generating functional of Green functions of Φ.

Consider a source of the following form: $J = \sum J_{in}^i + \sum J_{out}^j$. Each component of this composite source is defined as follows:

$$\int J_{in}^i \Phi(x) = \frac{i}{\sqrt{Z}} \epsilon_{in}^i \lim_{t\to\infty} \int d^3x (\Phi\, \partial_0 \phi_{in}^i - \phi_{in}^{i*}\, \partial_0 \Phi).$$

There is an analogous formula for the outgoing source. Here ϵ_{in}^i is infinitesimal and all formulae are to be interpreted by expansion to lowest order in all the $\epsilon_{in/out}^i$. ϕ_{in}^i is a normalizable, positive-energy solution of the KG equation. If Φ were a free field of mass m, this formula would just define the creation operator for a single-particle state with wave function ϕ_{in}^i, the spatial integral would be time-independent and the limit superfluous. For any local field Φ, the source creates a state localized around an infinitely distant spatial point as $t \to \infty$. Furthermore, if we consider the matrix element of $\langle \eta| \int J\Phi|\Psi\rangle$ between states $|\Psi\rangle$ and $|\eta\rangle$ of fixed 4-momentum, then, unless $(P_\Psi^\mu - P_\eta^\mu)^2 = m^2$, the limit will vanish. If we choose all of the incoming and outgoing wave functions[11] to be localized around different asymptotic directions as $t \to \pm\infty$, then this operator acts just like an *in* creation operator even in the interacting theory[12]. Similarly, the out part of the source acts like a sum of annihilation operators. Since $Z[J] = \langle 0|T\exp(-i\int J\Phi)|0\rangle$, all the annihilation operators sit to the left of all of the creation operators. Thus, it is plausible that the generating functional with this source is nothing but the generating functional for the scattering matrix. That is, the coefficient of $\epsilon_{in}^1 \cdots \epsilon_{in}^m \epsilon_{out}^1 \cdots \epsilon_{out}^n$ in the expansion of $Z[J]$ for this source[13] is the scattering-matrix element

$$\langle \text{out}\, f_1 \ldots f_n \mid g_1 \ldots g_m\, \text{in}\rangle,$$

---

[10] The extension to other spins is straightforward and we will not go into the details, but simply use the appropriate formulae when necessary.

[11] The outgoing wave functions are negative-energy solutions of the KG equation, and we take the limit $t \to -\infty$, instead of $+\infty$.

[12] This is the *assumption* that particle interactions fall off at large distances. It can be motivated/proven using certain plausible axioms about QFT [21–22].

[13] We emphasize that other terms in the expansion of the composite source have no particular use. They correspond to creating or annihilating multiple particles in exactly the same scattering state.

where the single-particle states f_i and g_j are defined by

$$\phi_{in}^j = \int \frac{d^3p}{\sqrt{2\omega_p(2\pi)^3}} e^{(-i\omega_p t - px)} g_j(p)$$

and

$$\phi_{out}^i = \int \frac{d^3p}{\sqrt{2\omega_p(2\pi)^3}} e^{(i\omega_p t - px)} f_i(p).$$

Conventional time-dependent perturbation theory offers an alternative definition of the scattering matrix as the infinite-T limit of the Dirac-picture evolution operator $U_D(T, -T)$, and it can be verified in all examples that this coincides with our definition in terms of $Z[J]$. More importantly, the LSZ formula can be massaged into a form that allows us to find particle masses and scattering amplitudes directly from Green functions. We first write

$$\int J_{in}^i \Phi = \int dt\, i\epsilon_{in}^i\, \partial_0 \left[\int d^3x (\Phi\, \partial_0 \phi_{in}^i - \phi_{in}^i\, \partial_0 \Phi) \right].$$

This is not quite right. The boundary term at $t = -\infty$ gives what we want, but the term at $t = \infty$ contains a creation operator for a state localized in the future. However, if all of the final states are orthogonal to all of the initial states, this term just acts to the left on the vacuum, and vanishes. In the case where this is not true (the case of partially forward scattering) the LSZ formula we are about to write must be slightly modified. In practice, this modification is never of any consequence. Part of the forward-scattering amplitude is due to disconnected processes happening independently at large space-like separation. The rest can be obtained by analytic continuation from non-forward scattering. We now act with the outer time derivative on the KG scalar product, and exploit its Wronskian form and the fact that ϕ_{in}^i satisfies the KG equation to write

$$\int J_{in}^i \Phi = i \int d^4x [(\nabla^2 - m^2)\phi_{in}^i \Phi - \partial_0^2 \Phi\, \phi_{in}^i] = i \int d^4x\, \phi_{in}^i [\Box + m^2]\Phi.$$

We can do a similar manipulation for all of the in and out sources. For outgoing particles, we want to pick up the creation operator rather than the annihilation operator, so we complex conjugate the source equation, and end up with the complex conjugate of the positive-energy outgoing wave function. The result is the LSZ formula for the S-matrix

$$\langle out\, f_1 \ldots f_n | g_1 \ldots g_m\, in \rangle = \left(\frac{i}{\sqrt{Z}} \right)^{m+n} \int \prod_k d^4x_k \phi_{in}^k(x_k) \prod_j d^4y_j (\phi_{out}^j)^*(y_j)$$

$$\times \prod_k [\Box_{x_k}^2 + m^2] \prod_j [\Box_{y_j} + m^2]$$

$$\times \langle 0 | T\Phi(x_1) \ldots \Phi(x_m)\Phi(y_1) \ldots \Phi(y_m) | 0 \rangle.$$

The product of differential operators includes one KG operator for each integration variable. It is conventional in scattering theory to replace the normalizable initial- and final-state wave packets by plane waves. As we shall see, this causes some trivial infinities to show up in cross sections, but they are easily dealt with. Recall again that the field Φ in the LSZ formula *might* be an elementary field, ϕ in the Lagrangian, but need not be. All that is required is that it have a finite amplitude to create single-particle states, which is verified by finding a pole in its two-point function.

The LSZ formula has two remarkable properties. The first is that it is symmetric between in and out states except for the fact that the out wave functions are negative-energy solutions and the in wave functions positive-energy solutions of the KG equation. This leads one to surmise that different scattering amplitudes (e.g. electron–electron scattering and electron–positron annihilation) are just analytic continuations of each other. These *crossing symmetry* relations were discussed in a heuristic manner in the introduction. The hard part in proving them is the proof of analyticity [23].

Secondly, in momentum space, the formula reads

$$\langle \text{out } q_1 \ldots q_n | p_1 \ldots p_m \text{ in}\rangle = \left(\frac{i}{\sqrt{Z}}\right)^{m+n} \prod \left[\sqrt{\frac{1}{2\omega_p(2\pi)^3}}(p_i^2 - m^2)\right]$$
$$\times \prod \left[\sqrt{\frac{1}{2\omega_q(2\pi)^3}}(q_j^2 - m^2)\right]$$
$$\times G_{n+m}(-q_1 \ldots - q_n; p_1 \ldots p_m).$$

G_{n+m} is the Fourier transform of the time-ordered product of $n + m$ Φ fields. Note that for the outgoing states q_i is the physical, positive-energy, 4-momentum, but the Fourier transform is evaluated at negative-energy outgoing momenta. Because of translation invariance, the Fourier transform contains a delta function of the sum of all its arguments, and because of the previous remark, this is just energy–momentum conservation $\delta(\sum p_i - \sum q_j)$. When we square the amplitude to get a cross section, one of these momentum-space delta functions gives us $\delta^4(0)$, which is interpreted as the space-time volume. In a translation-invariant system, we want to compute the probability per unit volume per unit time for a plane wave to interact. With normalizable wave packets, this infinity does not occur.

The most remarkable part of this formula is that all the momenta are on mass shell and the formula contains what appears to be a product of zeros. One concludes that the S-matrix element will be zero unless the Green function has a single pole at the mass shell on each external leg. The connection between full connected Green functions and 1PI Green functions shows how easy it is for this to happen. The full Green functions are evaluated as tree diagrams whose vertices are 1PI functions and whose internal and external lines are full propagators. In momentum space, each external leg has a factor $W_2(p)$, the full connected two-point function. The Lehmann representation shows that each of these Green functions has a simple pole, giving precisely the multiple-pole structure that we need, in order for the LSZ formula to predict a finite S-matrix.

3.8 Problems for Chapter 3

3.1. Consider a quantum mechanics problem with a single variable x (the generalization is easy) and a normalizable ground state, denoted by $|0\rangle$, with energy 0. Assume that you are given the Green functions

$$\langle 0|Tx(t_1)\ldots x(t_n)|0\rangle. \qquad (3.44)$$

Choose a time ordering. By inserting complete sets of intermediate states, show that knowledge of these functions allows you to read off all of the eigenvalues of the Hamiltonian. Show that you can also calculate

$$\langle 0|x^m p^n|0\rangle$$

for any m and n from your knowledge of these Green functions. Argue that this allows you to calculate the wave function $\psi_0(x)$ of the ground state in the x representation (up to an overall constant phase). Use the Schrödinger equation to determine the potential in terms of ψ_0. Thus, knowledge of the Green functions is equivalent to knowledge of the functional form of the Hamiltonian, as well as its eigenspectrum.

*3.2. This problem is a repetition of Problem 2.11, by functional methods. Compute the generating functional for a free field of mass m in the presence of a source

$$J(t, x) = q_1 \delta^3(x) + q_2 \delta^3(x - R), \qquad |t| < T,$$

$$J(t, x) = 0, \qquad |t| > T.$$

This source causes a static disturbance of the field, at two spatial points, over a time interval $2T$. When $T \gg R \gg 1/m$, this amplitude should have the form $e^{-2V(R)T}$ where $V(R)$ is the lowest energy of states with two localized disturbances. Show that this is the case and that, apart from an additive constant, V is the Yukawa potential. Thus, particle exchange is responsible for forces between static disturbances. This motivates the claim that all particle interactions can be understood in terms of particle creation, annihilation, and exchange.

3.3. Carefully derive the source term in the SD equations by looking at the action of the second derivative ∂_0^2 on the Heaviside functions in the definition of a time-ordered product.

3.4. Compute all diagrams for four-point functions at the tree level in the theory with Lagrangian

$$\mathcal{L} = \frac{1}{2}[(\partial_\mu \phi)^2 - m_0^2 \phi^2] - \frac{\lambda_3}{3!}\phi^3 - \frac{\lambda_4}{4!}\phi^4.$$

Do the computations in Euclidean space and analytically continue to Minkowski space in order to understand the relation between Euclidean and Lorentzian Feynman rules. Of what order in the loop approximation parameter g are $\lambda_{3,4}$? Argue that, at tree level, the particle mass is m_0 and the wave-function renormalization $Z = 1$. Use the LSZ formula to compute the two-to-two scattering amplitude, for particles of momenta $p_{1,2}$ to scatter into particles of momenta $p_{3,4}$. Introduce the Mandelstam variables $s \equiv (p_1 + p_2)^2$, $t \equiv (p_1 - p_3)^2$, $u \equiv (p_1 - p_4)^2$, and prove that $s + t + u = 4m_0^2$. Evaluate s, t, u in terms of center-of-mass energy and scattering angle. The center of mass is the Lorentz frame where $p_1 + p_2 = 0$. Show that the amplitude is symmetric under interchange of s, t, u and interpret this as a crossing symmetry relation.

*3.5. Compute the combinatoric factors for all diagrams, through two loop orders and eight external legs for the Lagrangian in Problem 3.4. Do this both by figuring out the geometrical symmetries of the diagrams and by combining the inverse factorials from perturbation theory with the number of contractions that give a particular graph. A useful notation for counting contractions is to write out powers of the field, as in $\phi^4(x_1) \to \phi_1 \phi_1 \phi_1 \phi_1$ and draw under-brackets $\phi_i \phi_j$, or draw over-brackets $\overbrace{\phi_i \phi_j}$ between fields that are contracted.

*3.6. Prove that, in a general theory of one scalar field with action $(1/g^2)[\frac{1}{2}(\partial\phi)^2 - V(\phi)]$, if we make the decomposition $\phi = \phi_{\rm cl} + g\chi$ discussed in the text, then the general diagram for an E-point connected Green function has the power g^{2L-2+E}, where the Feynman rules have a power g^{k-2} for a k-point vertex, and *NO* powers of g for the external lines. L is the number of loops in the diagram.

4 Particles of spin 1, and gauge invariance

4.1 Massive spinning particles

We now want to extend our discussion of particles and fields to particles with spin. The reader has been asked to do this on her/his own in Problem 2.6. This section can be viewed as an extended answer to parts of that exercise. The first step is to understand the Lorentz transformation properties of single-particle states. We start with massive particles in their rest frame. The stability subgroup of the rest-frame momentum is SU (2), so single-particle states are classified as irreducible representations of this group, which is the same as the usual non-relativistic definition of spin. A particle of spin j has $(2j + 1)$ states, $|0, m\rangle$ identified with eigenvalues m between $[-j, j]$, of the operator J_3 representing infinitesimal rotations around the 3-axis.

To generalize our discussion of spinless particles, we *define* the states of a massive spinning particle, with momentum p and spin component m, as

$$|p, m\rangle \equiv \sqrt{\frac{m}{\omega_p}} U(L(p))|0, m\rangle.$$

Here $L(p)$ is defined as a boost along the 3-direction, with velocity $|p|/\omega_p$, followed by a clockwise rotation of the 3-direction into the direction of p. It transforms the rest frame to the frame where the particle has momentum p. Then, using the fact that $U(\Lambda)$ is supposed to be a representation of the Lorentz group, we have

$$U(\Lambda)|p, m\rangle = \sqrt{\frac{m}{\omega_p}} U(L(\Lambda p)) U(L^{-1}(\Lambda p)\Lambda L(p))|0, m\rangle.$$

The transformation $R_W(\Lambda, p) \equiv L^{-1}(\Lambda p)\Lambda L(p)$, called the Wigner rotation, takes the rest frame into itself, and hence belongs to SU(2). We therefore know how it acts on the rest-frame states. Thus

$$U(\Lambda)|pm\rangle = \sqrt{\frac{m}{\omega_p}} U(L(\Lambda p))\mathcal{D}_{mk}(R_W)|0, k\rangle = \sqrt{\frac{\omega_{\Lambda p}}{\omega_p}}\mathcal{D}_{mk}(R_W)|\Lambda p, k\rangle, \qquad (4.1)$$

where $\mathcal{D}$ is the usual $(2j + 1)$-dimensional representation of the Wigner rotation. It's easy to verify that this defines a unitary representation of the Poincaré group on the space of states of a massive spinning particle. The creation and annihilation operators for these particles will thus carry a label $a_m(p)$, transforming in the spin-j representation of SU (2).

4.2 Massless particles with helicity

When we try to generalize these considerations to massless particles, we run into a surprise. We can always transform a null momentum to a frame where $p = (E, 0, 0, E)$. This is invariant under rotations around the 3-direction. The combination of transverse boost and rotation generators $T_i = K_i + \epsilon_{ij} J^j$ also leaves the null momentum invariant. Together with the transverse rotation these form the two-dimensional Euclidean group. This group has infinite-dimensional "continuous spin" representations unless $T_i = 0$. There are no such infinitely degenerate massless particles in nature, and one cannot make a local field theory which includes them.[1] The remaining representations are characterized by the eigenvalue of J_3, called the helicity, which must be quantized in half-integer units in order to have a local field theory.[2] Furthermore, local field theory requires that a helicity value h be accompanied by $-h$. Indeed, if this were not the case we could trace the causal order of emission and absorption processes between space-like separated points by following the helicity flow. However, the causal order can be changed by doing boosts along the momentum of the massless particle, which preserve helicity. On a technical level we find that we need both helicities (and helicity quantization) in order to construct a local field.

A general null momentum $(|p|, p)$ can be obtained from the canonical one $(E, 0, 0, E)$ by performing a boost of rapidity $\ln(|p|/E)$ along the 3-direction, followed by a rotation of the 3-direction into the direction $p/|p|$. Call the product of these two transformations $\mathcal{L}(p)$. By analogy with the massive case, we can derive the Lorentz transformation rule for massless single-particle states. It is

$$U(\Lambda)|p, h\rangle = \frac{\omega_{\Lambda p}}{\omega_p} \, U(\mathcal{L}^{-1}(\Lambda p)\Lambda\mathcal{L}(p))|p, h\rangle = e^{ih\Theta(\Lambda, p)}|p, h\rangle. \tag{4.2}$$

The second equality follows because $\mathcal{L}^{-1}(\Lambda p)\Lambda\mathcal{L}(p)$ is in the stability subgroup of $(|p|, p)$ and so acts on the states like a rotation around the direction of motion. $\Theta(\Lambda, p)$ is the angle of that rotation.

In Problem 4.3, you will show that the construction of local fields from the creation and annihilation operators of massless particles requires that h be quantized in half-integer multiples and that a particle of helicity $-h$ exists for every particle of helicity h. For particles that participate in parity-conserving interactions, it is conventional to call the state with helicity $-h$ another spin state of the same particle, whereas for neutrinos, which have only parity-violating interactions, it is called the anti-neutrino.

Note that for $|h| > \frac{1}{2}$ there is a discontinuity in the number of degrees of freedom of a massive particle of spin h and a massless particle with helicity h. The "missing" degrees of freedom are called *longitudinal*. A field theory of massive particles of spin h, which has a smooth massless limit, must acquire some sort of symmetry principle in the massless limit, which decouples the unwanted longitudinal states. By dimensional

---

[1] T. Banks (1973), *Continuous Spin Representations and Locality* (unpublished).

[2] You will prove the assertions on this page in the problems at the end of the chapter.

analysis, going to large momentum must be related to the zero-mass limit, so a theory with good high-energy behavior must have a similar decoupling principle. We will see that this requirement is the origin of gauge invariance, both for Maxwell's field and for much of the rest of the *standard model* of particle physics. To that end, we proceed to the field theory of massive spin-1 particles, after which we will take the massless limit.

4.3 Field theory for massive spin-1 particles

The analysis we did of the states of massive spinning particles implies that the creation operators of massive spin-1 particles satisfy

$$U^\dagger(\Lambda)a_i^\dagger(p)U(\Lambda) = \sqrt{\frac{\omega_{\Lambda p}}{\omega_p}} R_i^j(W(\Lambda, p))a_j^\dagger(\Lambda p),$$

where R is the usual 3×3 matrix representation of rotations.

As in the scalar case we want to find a local field linear in creation and annihilation operators, which is a model for a device that can create maximally localized states of a single particle. The covariant field we construct from the creation operators must have at least three components, each of which transforms as a vector under rotations. The stability subgroup of a single space-time point, in the Poincaré group, is the Lorentz group, so the fields at a point should transform as a representation of this group. The particle momentum is Fourier conjugate to its position, so the components of the field are related to internal degrees of freedom of the particle. We have described particles with a finite number of degrees of freedom per momentum state, so we should have fields that transform in finite-dimensional representations of the Lorentz group. None of these representations is unitary, but we are not talking about the action of the Lorentz group on the Hilbert space. If that unitary action is denoted $U(\Lambda)$ then we want the field A_K to transform as

$$U^\dagger(\Lambda)A_K(x)U(\Lambda) = S_K^L(\Lambda)A_L(\Lambda^{-1}x),$$

where $S(\Lambda)$ is the finite-dimensional representation. S is not a unitary matrix, but $U(\Lambda)$ is an infinite-dimensional unitary operator, which we constructed in the previous section.

The only three-dimensional representations of the Lorentz group are the (complex) self-dual and anti-self-dual tensors

$$B_{\mu\nu} = \pm\frac{1}{2}\epsilon_{\mu\nu\lambda\kappa}B^{\lambda\kappa}.$$

These are mapped into each other by reflection. In order to construct a Hermitian Hamiltonian we would need both of these fields, so we really have six real components. The smallest representation we can use to describe spin-1 particles is the 4-vector B_μ, and this can be reduced to three components by the covariant constraint

$$\partial_\mu B^\mu = 0.$$

Thus we may guess at a formula

$$B_\mu(x) = \int \frac{d^3p}{\sqrt{2\omega_p(2\pi)^3}}[e^i_\mu(p)a_i(p)e^{-ipx} + \text{h.c.}],$$

where $p^\mu e_\mu(p) = 0$.

There are three solutions to the latter equation. We normalize the space component of e_μ by $e^j_i(p = 0) = \delta^j_i$, and allow it to transform as a 4-vector, which defines it for all values of p. In the exercises, the reader will verify that this choice leads to a Lorentz 4-vector transformation law for the field $B_\mu(x)$. We can build an anti-symmetric tensor

$$B_{\mu\nu} = \partial_\nu B_\mu - \partial_\mu B_\nu,$$

so we see that the six-component anti-symmetric tensor is a derived field. Since B_μ was constructed to satisfy the Klein–Gordon equation and the transversality condition $\partial_\mu B^\mu = 0$, this field satisfies

$$\partial_\nu B^{\mu\nu} + \mu^2 B^\mu = 0,$$

which is called the Proca [24] equation. Note that, conversely, the Proca equation implies the transversality condition.

A Lagrangian that leads to the Proca equation is

$$\mathcal{L}_P = -\frac{1}{4}B^2_{\mu\nu} + \frac{\mu^2}{2}B^2_\mu.$$

To canonically quantize this Lagrangian (see the problems) we must use the constraint equation to eliminate B_0, which does not have a canonical conjugate. The procedure is slightly painful, but ends up giving the obvious answer: the generating functional for Green functions $Z[J_\mu]$ is given by a functional integral with this Lagrangian coupled to a source via $\delta\mathcal{L} = B_\mu J^\mu$. The generating functional for free connected Green functions is obtained by doing a Gaussian integral:

$$W_0[J_\mu] = -\frac{i}{2}\int d^4x\, d^4y\, J_\mu(x)J_\nu(y)\frac{d^4p}{(2\pi)^4}e^{-ip(x-y)}\frac{\eta_{\mu\nu} - p_\mu p_\nu/\mu^2}{p^2 - \mu^2 + i\epsilon}.$$

The reader should be able to see from this expression's symmetry that it would be inconsistent to quantize these fields as fermions. As in the case of spin-zero particles, the connected two-point function of the free Proca field is just the Green function (in the sense of partial differential equations) of the Proca equation, with Feynman boundary conditions.

It should by now be obvious how to derive Feynman diagrams for any perturbation around this free-field Lagrangian. It would be a good idea for the reader to work out the diagrams for two-, three- and four-point functions up to one loop (without doing the loop integrals) for the interaction $\mathcal{L}_I = igB_\mu(\phi^* \partial^\mu\phi - \phi \partial^\mu\phi^*)$ between the Proca field and a charged scalar.

The propagator of a massive vector boson[3] will be denoted by

$$\sim\!\!\sim\!\!\sim\!\!\sim \qquad \frac{-\mathrm{i}\left(\eta_{\mu\nu} - \frac{p_\mu p_\nu}{\mu^2}\right)}{p^2 - \mu^2 + \mathrm{i}\epsilon} \qquad \frac{\left(\delta_{\mu\nu} - \frac{p_\mu p_\nu}{\mu^2}\right)}{p^2 + \mu^2}$$

Massive photon propagator in Minkowski and Euclidean signature

One look at the vector boson propagator tells us that something dramatic happens for spin 1 and zero mass. Indeed, the longitudinal part of the propagator blows up in this limit. The only way we can get a consistent limiting expression is to insist that the source satisfy $\partial_\mu J^\mu = 0$. *A massless vector meson must be coupled to a conserved current.* This remark becomes even more interesting when we realize that the $\mu = 0$ limit of the Proca equation coupled to a source is just Maxwell's equation for the electromagnetic field.

4.3.1 The Stueckelberg formalism

We can obtain more insight by introducing a redundant parametrization of the space of field variables, known as the Stueckelberg formalism. We introduce another vector field A_μ, as well as a scalar, θ, via the formula

$$B_\mu = A_\mu - \partial_\mu\theta.$$

The Lagrangian is obtained by just substituting this combination into the Proca Lagrangian. The well-known gauge invariance of Maxwell's field strength tensor with respect to gauge transformations of the vector potential implies that

$$B_{\mu\nu} = \partial_\nu A_\mu - \partial_\mu A_\nu \equiv F_{\mu\nu}.$$

The full Lagrangian now has a gauge invariance under

$$A_\mu \to A_\mu + \partial_\mu\Omega,$$
$$\theta \to \theta + \Omega,$$

even for non-zero mass. If we couple a current to A_μ, then it must be conserved, in order to preserve gauge invariance (one must integrate by parts to show this, which means that either the gauge transformation or the current must vanish sufficiently rapidly at infinity). If it is conserved, then $\int J^\mu A_\mu = \int J^\mu B_\mu$. For such conserved sources, the longitudinal part of the B_μ propagator cancels out, and there are no divergences when $\mu \to 0$.

From the point of view of free-particle physics, Maxwell's gauge invariance is thus the result of the discontinuity of the number of degrees of freedom of a spin-1 particle

---

[3] Here and henceforth, we record the Feynman rules in both Minkowski and Euclidean space.

in the massless limit. The Stueckelberg formalism incorporates the massless split of the degrees of freedom into a spin-zero particle and a massless helicity ± 1 particle into the massive theory by introducing a redundant degree of freedom. It is a hint of the Higgs mechanism, which we will discuss in Chapters 6 and 8. The connection between gauge invariance and a redundancy of degrees of freedom will reappear again and again in our discussions.

4.4 Problems for Chapter 4

4.1. Find the form, in an arbitrary frame, of the three transverse polarization vectors $\epsilon_i^\mu(p)$ which satisfy $\epsilon_i^0 = 0$ and $\epsilon_i^j = \delta_i^j$ in the rest frame of a massive spin-1 particle. Using the transformation law of the creation operators $a_i(p)$, show that the field $B_\mu(x)$ transforms like a vector field.

4.2. Canonically quantize the Proca Lagrangian. Note that the canonical conjugate to B_0 vanishes and that the Euler Lagrange equation for B_0 allows us to write it at any fixed time as a function of the variables B_i. Then quantize the spatial components using standard procedures. Compute the Green function $\langle 0|TB_\mu(x)B_\nu(0)|0\rangle$ using canonical commutators, and show that it is covariant despite the asymmetric treatment of different Lorentz components. Show that this function is the Green function of the Proca equation with Feynman boundary conditions.

4.3. Write down a covariant local field transforming in the $(2j + 1, 0)$- or $(0, 2j + 1)$-dimensional representation of the Lorentz group, built as a linear combination of creation and annihilation operators of helicity $\pm j$ massless particles. The two representations are complex conjugates of each other. Show that, if we have only one sign of the helicity, no local field is possible. Note that the helicity j must be quantized in order for this construction to work.

4.4. Prove that the formula (4.1) defines a unitary representation of the Poincaré group.

5 Spin-$\frac{1}{2}$ particles and Fermi statistics

For spin-$\frac{1}{2}$ particles, we will introduce a change of pace and start directly from field theory. We have to ask ourselves what kind of fields could create spin-$\frac{1}{2}$ particles. This leads to the question of what kinds of fields there are, which is answered by saying that fields at a fixed point form finite-dimensional representations of the Lorentz group. This guarantees that the only infinity in the number of states of a single particle comes from the different values of momentum it can carry.

We're lucky to live in four dimensions, where the analysis of finite-dimensional representations is particularly easy. In particular, the kind reader will verify that the combinations of Lorentz generators[1]

$$\epsilon_{ijk} J_{ij} \pm i J_{0k}$$

form two commuting copies of the algebra of SU (2). We can use our complete knowledge of the representations of SU (2) to list all of the finite-dimensional representations of SO (1, 3). The finite-dimensional representations of SU (2) are all equivalent to unitary representations, so we can always use a basis in which the two SU (2) generators are Hermitian. Note that, using this construction, the rotation generators will be Hermitian and the boost generators anti-Hermitian. This was to be expected. We cannot have a finite-dimensional unitary representation of the non-compact Lorentz group.[2] Thus, the general Lorentz-covariant field carries two integer labels $N \equiv [n_L, n_R]$ corresponding to the dimensions of the representations of the two SU (2) groups. Recall that even dimensions correspond to half-integer and odd to integer spin. Note that the two kinds of integer are interchanged by space reflection. Thus, fields of the form $[1, n_R]$ and $[n_L, 1]$ have a handedness, and are called right and left chiral fields.

The rotation generators are the sum of the left and right SU (2) generators, so we can get spin $\frac{1}{2}$ with either $[1, 2]$ or $[2, 1]$. These are called right- and left-handed Weyl spinors, respectively. In fact the $[1, 2]$ representation is complex,[3] and its complex conjugate is

---

[1] We are picking a particular rotation subgroup of SO (1, 3) in order to exhibit the invariant isomorphism between groups.

[2] It is important to remember that this is the representation of the Lorentz group on field labels. The representation on the Hilbert space is infinite-dimensional and unitary.

[3] SO (1, 3) is locally isomorphic to SL (2, C) and the $[1, 2]$ is the fundamental representation of the latter.

the [2, 1]. To see this note that the rotation and boost generators in the [1, 2] are $J_i = \sigma_i/2$, $J_{0i} = -i\sigma_i/2$. The representation of rotations is pseudo-real, $\sigma_i^* = -\sigma_2\sigma_i\sigma_2$, but the boost operator has an extra sign change under conjugation, so that we get the [2, 1] representation. We can describe a spin-$\frac{1}{2}$ particle either in terms of a left-handed Weyl fermion field v, in the [2, 1] representation, and its complex conjugate, or equivalently, in terms of the four-component field[4]

$$\psi = \begin{pmatrix} v \\ i\sigma_2 v^* \end{pmatrix},$$

which is a Dirac spinor field, satisfying the Majorana condition

$$\psi^* = \begin{pmatrix} 0 & -i\sigma_2 \\ i\sigma_2 & 0 \end{pmatrix}\psi.$$

The equivalence between Weyl and Majorana descriptions of a neutral spin-$\frac{1}{2}$ particle was not always recognized, and there is a lot of confusion in the literature, as late as the 1970s. A general Dirac spinor has the form $\psi_1 + i\psi_2$, with ψ_i satisfying the Majorana condition.

Following van der Waerden, we describe left-handed Weyl fields as two-component fields v_a, whereas right-handed fields are written as $\psi_{\dot{a}}$. Each of these representations has a Lorentz-invariant product

$$v_a^{(1)} v_b^{(2)} \epsilon^{ab} \quad \psi_{\dot{a}}^{(1)} \psi_{\dot{b}}^{(2)} \epsilon^{\dot{a}\dot{b}}.$$

We use the Levi-Civita ϵ symbols[5] to raise spinor indices $v^a = \epsilon^{ab} v_b$ (note that we contract on the second index). We also introduce ϵ_{ab} by

$$\epsilon_{ab}\epsilon^{bc} = \delta_a^c,$$

and similarly for dotted indices.

The product $[2, 1] \otimes [1, 2]$ of the two Weyl representations of opposite chirality is the [2, 2] or 4-vector representation. The matrices $\sigma^\mu_{a\dot{b}} = (1, \sigma)_{a\dot{b}}$ are the Clebsch–Gordan coefficients relating the tensor product of the two opposite-chirality Weyl spinor representations to the conventional basis in the 4-vector representation. These matrices map the right-handed spinor representation into the left-handed spinor. The corresponding map in the opposite direction is given by $\bar{\sigma}^\mu_{\dot{b}a} = (1, -\sigma)_{\dot{b}a}$. These symbols allow us to write the Lagrangian

$$\mathcal{L}_{\text{Weyl}} = \frac{1}{2}i(v^*)^{\dot{b}}(\bar{\sigma}^\mu)_{\dot{b}a}\partial_\mu v^a + \text{h.c.} \tag{5.1}$$

---

[4] In this section we will use the symbol $*$ to denote Hermitian conjugation of the operators in each component of a spinor field. This is to distinguish it from $\dagger$, which could be taken to imply turning row indices into column indices as well as Hermitian conjugation. Later on, most of our equations will be written in terms of complex Dirac spinors, and we will revert to the usual symbol for Hermitian conjugation.

[5] $\epsilon^{21} = -\epsilon^{12} = 1$, and the same for dotted indices.

The two terms are equal, up to a total divergence, and it is conventional to simply drop both the factor of one half and the Hermitian conjugate term in the Weyl Lagrangian. If we vary it with respect to v^* we get

$$i\bar{\sigma}^\mu \, \partial_\mu v = 0,$$

or in momentum space

$$p_\mu \bar{\sigma}^\mu v(p) = 0.$$

This tells us that positive-energy solutions have negative helicity. The opposite correlation is valid for right-handed spinors.

The Weyl matrices satisfy

$$\sigma_\mu \bar{\sigma}_\nu = \eta_{\mu\nu} + \bar{\eta}^a_{\mu\nu} \sigma_a,$$

$$\bar{\sigma}_\mu \sigma_\nu = \eta_{\mu\nu} + \eta^a_{\mu\nu} \sigma_a.$$

σ ($\bar{\sigma}$) maps the right (left)-handed spinor into the left (right)-handed one, so the first product maps the left-handed spinor to itself and the second maps the right-handed spinor to itself. The product of the [1, 2] with itself is [1, 1] $\oplus$ [1, 3], and the second of these is the anti-self-dual second-rank anti-symmetric tensor $\epsilon_{\mu\nu\alpha\beta} T^{\alpha\beta} = -2T_{\mu\nu}$. $\bar{\eta}$ is the Lorentzian continuation of a standard basis for Euclidean anti-self-dual tensors, introduced by 't Hooft [25], which we will study in the last chapter. η is the analog for self-dual tensors. In Lorentzian signature

$$\eta^a_{\mu\nu} = (\delta_{\nu 0}\delta^a_\mu - \delta_{\mu 0}\delta^a_\nu) - i\epsilon^a_{\mu\nu},$$

$$\bar{\eta}^a_{\mu\nu} = (\delta_{\mu 0}\delta^a_\nu - \delta_{\nu 0}\delta^a_\mu) - i\epsilon^a_{\mu\nu}.$$

The three-index Levi-Civita symbol appearing in these equations is the usual rotation-invariant symbol in three dimensions, with indices raised by the Euclidean 3-metric.

The Weyl equation implies that $p^2 = 0$ and that the solution with positive energy has negative helicity (is left-handed) while the negative-energy solution has positive helicity. The positive- and negative-*helicity* solutions are constrained two-component spinors $v_\pm(p)$, with only one free component.

The general solution of the Weyl equation is

$$v(x) = \int \frac{\mathrm{d}^3 p}{\sqrt{(2\pi)^3 |p|}} [a(p)\mathrm{e}^{-ipx} v_-(p) + b^\dagger(p)\mathrm{e}^{ipx} v_+(p)].$$

Note that it is inconsistent to insist that a and b be the same operator, because the field v is complex. Quantization of a and b as either bosons or fermions is consistent with the requirement that the Heisenberg equations derived from the Hamiltonian[6] be the

---

[6] We are here anticipating Noether's theorem from Chapter 6. The point is that time-translation symmetry allows us to construct the Hamiltonian from the Lagrangian without using the canonical formalism. Canonical commutators *or* anti-commutators can be *derived* by insisting that Heisenberg's equations be consistent with the Euler–Lagrange equations.

Weyl equation. The covariant Feynman Green function of the Weyl equation, which can be constructed without thinking about how the fields are quantized, is

$$S_F(x) = \int \frac{d^4p}{(2\pi)^4} e^{-ipx} \frac{\sigma_\mu p^\mu}{p^2 - m^2 + i\epsilon}.$$

It is odd under interchange of spin indices combined with $x \to -x$ (one should write things out in terms of the four real components of the Majorana spinor to see the anti-symmetry). Thus, it is compatible only with Fermi statistics. Correspondingly (see Problem 5.1) the Weyl field cannot have local commutation relations for either choice of statistics, but has local anti-commutation relations for Fermi statistics. Thus, spin-$\frac{1}{2}$ particles *must* be fermions. The proof of the spin-statistics theorem follows the same pattern for all spins [6–8], by analyzing the symmetry properties of the Feynman Green function.

It is worth noting that we cannot allow fields with local anti-commutation relations in the Lagrangian density, which must commute with itself at space-like separation, but that even functions of them are allowed. Indeed, Lorentz invariance also requires that we have only even functions of half-integer spin fields in the Lagrangian, so the spin-statistics connection replaces these two constraints by one. It can be stated in symmetry terms. Let $(-1)^F$ be the operator which is -1 on all states created from the vacuum by products of fields containing an odd number of fermion operators and $+1$ on the other states. Then the spin-statistics connection is the equation

$$(-1)^F = e^{2\pi \hat{n} J}.$$

($\hat{n}$ is any unit 3-vector and J is the generator of angular momentum in any Lorentz frame). The operator $(-1)^F$ commutes with all observables (operators that can be viewed as infinitesimal changes of the Lagrangian). It should be emphasized that the proof of the spin-statistics theorem follows from assuming that the Hilbert space of the theory consists only of positive-norm particle states. Later on we will encounter fermionic scalar fields, called Faddeev–Popov ghosts, which live in a "Hilbert" space of indefinite norm. A gauge equivalence principle will forbid them from being produced in the scattering of physical particles, but they are extremely useful at intermediate stages of calculation.

5.0.1 Chiral symmetry

The Weyl Lagrangian has a symmetry under $\nu \to e^{i\alpha}\nu$, which gives rise via Noether's theorem (see Chapter 7) to a conservation law. At the level of free particles this conserved quantity is simply helicity. This symmetry is called a chiral symmetry because it acts differently on particles of different helicities. It cannot remain conserved for a massive particle. Indeed, if we try to add a non-derivative term to the Weyl Lagrangian, in order to generate a mass, then it must have the form

$$\mathcal{L}_m = m\epsilon^{\alpha\beta}\nu_\alpha \nu_\beta + \text{h.c.},$$

where m is a complex number. Note that this term would be identically zero if the fields were c-numbers. Instead, in the next section we will show that they are anti-commuting (Grassmann) numbers. In the presence of this term, the Weyl equation becomes

$$i\bar{\sigma}^\mu \, \partial_\mu v + m^* v^* = 0.$$

By multiplying this equation by $\sigma^\mu \partial_\mu$ and using the complex-conjugate equation, we obtain

$$(\partial^2 + mm^*)v = 0,$$

indicating that the field creates and annihilates massive particles, with mass $|m|$.

The appearance of both v and v^* in the equation suggests that things will look more elegant if we introduce the four-component Majorana field ψ. In terms of ψ, the massive field equation takes the form

$$[i\gamma^\mu \, \partial_\mu \psi - mP_- - m^* P_+]\psi = 0.$$

Here we have introduced the Dirac matrices,

$$\gamma^\mu \equiv \begin{pmatrix} 0 & \sigma^\mu \\ \bar{\sigma}^\mu & 0 \end{pmatrix}.$$

$P_\pm$ are the projectors on spinors of fixed chirality: P_- projects on the first two components of a Dirac spinor. Note that

$$\gamma_5 = P_+ - P_- = \begin{pmatrix} -1 & 0 \\ 0 & 1 \end{pmatrix}$$

anti-commutes with all the γ^μ and that $P_\mp = \frac{1}{2}(1 \mp \gamma_5)$. We define $A \equiv A_\mu \gamma^\mu$ for any 4-vector A_μ. In the problems and Appendix C, you will find many properties of the Dirac matrices. They mostly follow from the anti-commutation relation

$$[\gamma^\mu, \gamma^\nu]_+ = 2\eta^{\mu\nu},$$

which the kind and diligent reader will easily verify.

We can do a chiral transformation on v or ψ to make the mass of a single free fermion real. The Lagrangian for the Majorana field with real mass is

$$\mathcal{L} = \bar{\psi}(i \, \partial\!\!\!/ - m)\psi, \qquad (5.2)$$

where[7] $\bar{\psi} = \psi^\dagger \gamma^0$. This is called the Dirac Lagrangian. The general solution of its equations of motion does not satisfy the Majorana condition, but can be decomposed into a pair of fields, which do. This is analogous to the breakup of a complex scalar field into real and imaginary parts. In fact (Problem 5.2) we can change basis in such a way that the Dirac matrices are all imaginary. In this basis the Dirac equation has real solutions and the Majorana condition is just $\psi^* = \psi$.

---

[7] The reader should prove that the γ^0 is necessary for the Lorentz invariance of $\mathcal{L}$.

5.1 Dirac, Majorana, and Weyl fields: discrete symmetries

The Lorentz group has four disconnected components, which can be obtained by appending the time-reversal operation T and space-inversion operation P to the proper orthochronous Lorentz group. Our study of field theory suggests the existence of another operation called *charge conjugation*, C, which takes any particle into its anti-particle. The time has come to see how these symmetries are or are not implemented in quantum field theory. Reflection symmetry reverses spatial positions and momenta, while preserving angular momenta. The standard approach to this subject is to work out the most general constraints on particle states and then find further constraints that follow from local field theory. We will work directly with the fields.

If these operations correspond to symmetries there must be operators $U(P)$, $U(T)$, and $U(C)$ in quantum field theory, which implement these operations on the Hilbert space and commute with the scattering matrix. In the case of P and C these are unitary operators. $U(T)$ must be anti-unitary, because $U^\dagger(T)e^{-iP_0 t}U(T) = e^{iP_0 t}$. If $U(T)$ were a unitary linear operator, commuting with the Hamiltonian, then the Hamiltonian could not be bounded from below. Note also that, by virtue of its definition, $U(T)$ must map in states to out states and vice versa. Wigner pointed out long ago that the solution to this problem was to choose $U(T)$ to be the product of a unitary transformation and the non-linear but idempotent operation of complex conjugation of the coefficients in the expansion of states in the Hilbert space in *some* orthonormal basis.[8] Such operators are called anti-unitary. They still satisfy $U^\dagger(T)U(T) = 1$, but $U^\dagger(T)cU(T) = c^*$ for a general c-number. From now on, we will designate the unitary transformations $U(T, C, P)$ by the simpler notation T, C, P.

The most general space-reflection transformation (also called a parity transformation) on a set of n scalar fields must have the form

$$P^{-1}\phi_A(t, x)P = O_A^B \phi_B(t, -x). \tag{5.3}$$

P is the unitary representative of the parity transformation on the Hilbert space of our field theory, and O is an $O(n)$ matrix, the most general internal symmetry (see Chapter 7) of free massless scalar field theory. The square of this parity transformation is a purely internal symmetry, which acts on the multiplet of scalars via the matrix O^2. The matrix O is the product of a reflection in n-dimensional space and rotations by angles θ_i in some set of orthogonal 2-planes. If any of the rotation angles is irrational, O^2 would generate an infinite discrete group. A theory with polynomial Lagrangian could be invariant under this group only if it were invariant under the continuous one-parameter subgroup of rotations with arbitrary θ angles. In that case we could find a product of an internal symmetry and our parity transformation, which was just simple reflection of all coordinates, with at most an internal reflection.

---

[8] Of course any two orthonormal bases are related by a unitary transformation, so the choice of basis here is irrelevant.

If the angles are rational, O generates a Z_k group for some k. If k is odd, then O^{-1} is in the group generated by O^2, and we can again find a parity transformation with no internal rotation. If k is even, then the group generated by O is $Z_2 \times Z_{k/2}$, and we can eliminate the internal $Z_{k/2}$ transformation. If we diagonalize the Z_2 action, we find a set of fields that transform as

$$P^{-1}\phi_A(t, x)P = \epsilon^B \phi_B(t, -x), \tag{5.4}$$

where each ϵ^B is ± 1. Fields with $\epsilon^B = 1$ are called scalars, the others, *pseudo-scalars*.

For vector fields, the story is even simpler. Geometrically, a 4-vector could transform like an ordinary or an axial vector under parity, depending on whether its space component changes sign under reflection. We may ask whether there exists a possibility of appending internal symmetry transformations to the definition of parity, as we did for scalars. However, with one exception that we will describe below, all internal symmetry transformations for vector fields are gauge invariances (see Chapter 8), and the Lagrangian for such fields contains a term of the form

$$\partial_\mu A_\nu^a f^{abc} A_\mu^b A_\nu^c,$$

where f^{abc} are the structure constants of the group (see Appendix E for an explanation of the terminology). When the structure constants are non-vanishing, this term allows only the ordinary vector transformation law for the fields.

There are two exceptions to this rule. Abelian (U(1)) gauge fields can be pseudo-vectors, because the structure constants vanish and Maxwell's Lagrangian is invariant under reflection of the fields. If we want to construct a parity-invariant Lagrangian, we must couple these fields only to pseudo-vector currents. The second exception occurs if we have a discrete Z_2 automorphism of the gauge group, which can be appended to the definition of the parity transformation. The simplest examples involve a group G $\times$ G where the Z_2 exchanges the groups. Hypothetical models of this type for an extension of the standard model of electro-weak interactions based on the case G = SU(2) have been studied in great detail [26–29].

For spin-$\frac{1}{2}$, the story of parity is more complicated, because the simple [2, 1] Weyl representation is not mapped to itself by reflection. Before studying this case, we introduce the notion of *charge conjugation*.

5.1.1 Charge conjugation

The basic idea of charge conjugation is a symmetry that interchanges particles and anti-particles, without transforming space-time variables. Particles are distinguished from anti-particles by their internal symmetry charges. So charge conjugation must satisfy

$$C^{-1}UC = U^\dagger,$$

where U is the unitary operator implementing any global symmetry. This will be true, in particular, for the global residuum of any gauge invariance (Chapter 8). For scalar fields,

$$C^{-1}\phi_A(x)C = \mathcal{C}_A^B\phi_B(x),$$

and $\mathcal{C}$ must satisfy

$$\mathcal{C}O = O^{-1}\mathcal{C}, \tag{5.5}$$

for the matrices O of all internal symmetry rotations. We must also have $\mathcal{C}^2 = 1$.

At this point in the exposition, the reader should study Appendix E on Lie groups and Lie algebras. This is the mathematical tool for studying symmetries that depend on continuous parameters. In that appendix, she/he will encounter the notion of *infinitesimal generators* of transformations on fields

$$\chi_M \rightarrow \chi_M + i\omega_n(t^n)_M^N\chi_N.$$

The parameters ω_a are infinitesimal real numbers and the t^a are linearly independent Hermitian matrices, which satisfy

$$[t^m, t^n] = iF^{mnk}t^k.$$

Any set of matrices with these commutation relations defines a finite-dimensional unitary representation of the continuous group of transformations whose parameters are ω_a. We will denote by G_{global} the group of all global symmetries of our Lagrangian. It acts on the space of scalar fields via unitary or orthogonal matrices. This set of matrices is called the representation R_S of the group G_{global} on the space of scalar fields. The complex-conjugate matrices also represent the same group and we can ask whether our representation is equivalent to its complex conjugate. That is, is there a unitary or orthogonal transformation on the internal labels of the scalar fields that obeys

$$U^\dagger O(g)U = O^*(g),$$

for every group element $g \in G_{global}$. If not, we say the representation is complex, and there is an inequivalent representation $\bar{R}_S$. For any group represented by unitary or orthogonal matrices in the representation R_S, the expression

$$\bar{\phi}^A\phi_A,$$

where $\bar{\phi}$ transforms in $\bar{R}_S$, is invariant. If $\bar{R}_S = R_S$ (in the sense of unitary equivalence), then this implies that one can make a bilinear invariant out of the components of ϕ_A itself. If this invariant is symmetric then R_S is said to be real, because one can choose a basis in which the action of all the group matrices is real. If it is anti-symmetric, we say R_S is *pseudo-real*.

In Problem 5.16, you will show that the existence of a charge-conjugation operator implies that the representation of G on scalar fields is real or pseudo-real. A famous theorem in group theory, the Peter–Weyl theorem, says that any finite-dimensional unitary representation of a group is a direct sum of a finite number of irreducible unitary representations. In an irreducible representation, anything that commutes with all the group generators is proportional to the identity matrix (Schur's lemma). The irreducible components of R_S are not necessarily themselves real or pseudo-real, but complex representations must come in conjugate pairs $R + \bar{R}$ that are interchanged by the charge-conjugation matrix $\mathcal{C}$. The form of $\mathcal{C}$, up to conjugation by an element of G_{global}, is mostly determined by group theory. However, if there are several different fields transforming in the same irreducible representation of G_{global} then the action of $\mathcal{C}$ is undetermined in that subspace. Given a set of mass terms and interactions, there will be at most one choice of the matrix $\mathcal{C}$, which leaves the Lagrangian invariant. Note also that $\mathcal{C}$ takes fields in pseudo-real representations into their complex conjugates.

The charge-conjugation properties of vector fields A_μ^a are more constrained, because, as we shall see in Chapter 8, they are all gauge fields, associated with a gauge equivalence group $G \subset G_{\text{global}}$. Call the generators of G T^A. The structure constants of the Lie algebra of G are defined by

$$[T^A, T^B] = \mathrm{i} f^{ABC} T^C.$$

The constants f^{ABC} are real and totally anti-symmetric. The Jacobi identity for double commutators is equivalent to the statement that the matrices

$$(T_{\text{adj}}^A)^{BC} = \mathrm{i} f^{BAC}$$

form a representation of the same Lie algebra, called the adjoint representation. Gauge fields always transform in the adjoint representation of G, and are invariant under continuous global symmetries that are not in G. It is a real representation of the group, obtained by writing a general element as $O(g) = \mathrm{e}^{\mathrm{i}\omega^A T_{\text{adj}}^A}$. The matrices $O(g)$ are all rotation matrices. A Cartan subalgebra of the Lie algebra is a maximal independent set of commuting generators H^i. Given the Cartan subalgebra, we can form linear combinations of the other independent generators, such that

$$[H_i, E_r] = r_i E_r.$$

The number, r, of elements in the Cartan subalgebra is called the rank of the group, and the r-dimensional vectors r_i are called the roots of the Lie algebra. The generators H_i can be thought of as generators of rotations in orthogonal planes, in the vector space on which the matrices T_{adj}^A act.[9] Thus, their eigenvalues, the roots, come in pairs of equal magnitude and opposite sign. The charge-conjugation transformation on the adjoint representation flips the sign of all the Cartan generators and exchanges positive with negative roots. Thus, in the familiar example of SU(2), charge conjugation flips the sign of the gauge field associated with the third component of isospin and exchanges the two fields with $I_3 = \pm 1$.

We are finally ready to talk about the charge-conjugation properties of Weyl fermions. The discussion parallels that of scalar fields, except for the case of pseudo-real representations. For scalars, charge conjugation turned a scalar in a pseudo-real representation into its complex conjugate. However, the complex conjugate of a left-handed

---

[9] Both the matrices and the space itself are often called the adjoint representation by physicists.

Weyl field is right-handed. Complex conjugation of Weyl fields does not commute with the action of the Lorentz group. Thus, the only way in which we can have a charge-conjugation-invariant theory with Weyl fermions in a pseudo-real representation is to have two copies of the pseudo-real representation. We can always choose one of them to transform under G as the complex conjugate of the other and define charge conjugation to exchange them.

5.1.2 CP transformations for Weyl fields

We have now learned enough to understand that the correct name for the space-reflection transformation on Weyl fields is CP, the combined symmetry of parity and charge conjugation. A space reflection *must* take Weyl fields into their complex conjugates. As such it *must* also change all internal symmetry transformations into their complex conjugates, and therefore also performs what we have called charge conjugation in the previous section. The most general CP transformation for M Weyl fields is

$$(CP)^{-1}\psi_i^a(x,t)CP = U_{ij}(i\sigma_2)^{a\dot{a}}\psi_{j\dot{a}}^\dagger(-x,t),$$

where U is an element of the U(M) internal symmetry of the free Weyl Lagrangian. As for scalars, the transformation of ψ_i by U^2 must be part of the internal symmetry group of the full interacting Lagrangian, if CP is a symmetry. We can use this freedom to redefine the CP transformation law.

The most general Lagrangian involving spin-$\frac{1}{2}$ fermions coupled to vector and scalar fields (the most general renormalizable couplings) has the form

$$\psi^\dagger i\bar{\sigma}^\mu(\partial_\mu - A_\mu(x))\psi + \psi M(x)\psi + \text{h.c.} \tag{5.6}$$

ψ is a vector of left-handed Weyl fermion fields. Here the matrix-valued vector potential A_μ must be composed of matrices that transform in the adjoint representation of a compact Lie group G, which *must* be an exact gauge symmetry of the Lagrangian (Chapter 8). The matrix-valued scalar $M(x)$ is a general compact symmetric matrix in the internal space (the fermion bilinear that couples to it is anti-symmetrized in spin to make a Lorentz scalar). These couplings between a single scalar field and fermion bilinears are called *Yukawa couplings*. The rest of the Lagrangian, which we have not written, has to be invariant under the action of G on M ($M \to (U^T)^{-1}MU^{-1}$) induced by the action on ψ. There may also be a global symmetry group, larger than G. The fermion gauge field couplings automatically preserve CP (this restricts the matrix U to commute with the action of G on ψ_i). It can be shown [30] that the rest of the Lagrangian is CP-invariant if and only if there is a basis of fields for which all of the Yukawa couplings are real. However, we will see in Chapter 8 that there are other terms depending only on the gauge fields, which have dimension 4 and violate CP. We will also see that transformations on the fermions of the form $\psi_i \to e^{i\beta}\psi_i$, which we might want to do to make the Yukawa couplings real, can change the values of these CP-violating pure gauge terms. Thus, the real criterion for CP invariance is that the couplings are real in the same basis as that in which the CP-violating gauge terms vanish.

Of course, there are operators of higher mass dimension, which can be added to the Lagrangian, and it is hard to believe that one would ever find a basis in which *all* coefficients in the Lagrangian were real. These higher-dimension terms are, by dimensional analysis, always multiplied by inverse powers of a mass scale M. At energy scales small compared with M they will be negligible (Chapter 9), and CP invariance might be approximately valid. Experiment seems to show only small violations of CP invariance in all physical processes except those involving the heaviest generation of quarks. If there were no experimental evidence for CP violation, we might have *imposed* CP invariance as a fundamental property of nature. This is not possible. It is therefore important to understand why the violation of CP in nature is so small at low energy.

In the semi-classical approximation, we expand around a constant value for M, which gives rise to masses for some of the fermion fields. The propagators for the massive fields will involve correlators of ψ_i with itself and with $\psi_i^\dagger$. This suggests the utility of a four-component notation. The nature of the four component fields depends on the structure of the mass matrix. Let us assume that all the fields acquire mass, since that appears to be the case in the real world.

In the mass terms, each component of ψ_i can be paired either with itself or with some other component.[10] In the latter case, we can make a four-component complex Dirac field, and the kinetic terms preserve a U(1) symmetry under which this field is multiplied by a phase. The mass term is called a Dirac mass. In the standard model of particle physics, the corresponding approximately conserved quantum number (Chapter 7) is called a quark or lepton flavor. A particular linear combination of these flavors is called electric charge, and it appears to be exactly conserved. The neutral leptons, or neutrinos, do not carry this charge. The experiments that show they are massive are not yet sensitive enough to tell us whether we have six Weyl fields, or only three, which are paired with themselves.

A mass term that pairs a Weyl field with itself is called a Majorana mass, and the corresponding particle is called a Majorana particle. The only internal quantum number such a particle can carry is Z_2-valued, which is compatible with the particle being its own anti-particle. Some authors insist that a Majorana field/particle has such a Z_2 symmetry, but in general the quantum number might have to be carried by other fermions as well. One can construct Lagrangians whose only conserved fermion number is $(-1)^F = e^{2\pi i J_3}$.

The charged standard-model fermions are best described by complex Dirac fields with a diagonal mass matrix. In the approximation that neutrino masses vanish, which is valid for most purposes, it is convenient to append a right-handed neutrino to the standard model. The neutrino is described by a four-component complex field, but its right-handed components simply do not interact. This is achieved by putting left-hand projection operators in the interaction Lagrangian.

---

[10] The mass matrix is a symmetric complex matrix, which can be brought to canonical form [31].

5.1.3 Time reversal and TCP

The anti-unitary time-reversal operator T maps left-handed fields into themselves. This is because both momentum and angular momentum change sign under time reversal, so helicity is invariant. The most general time-reversal transformation for Weyl fermions is

$$T^\dagger \psi_i(x, t) T = \mathcal{T}_i^j i\sigma_2 \psi_j(x, -t),$$

where $\mathcal{T}$ is a U(M) matrix. Again, $\mathcal{T}^2$ must be an ordinary internal symmetry of the Lagrangian and we can use this to simplify the time-reversal transformation.

We can try to use the freedom in the definition of the various transformations to make a given Lagrangian invariant. Note in this connection that the various matrices that act on the spinors and scalars (the gauge field matrices are strongly constrained by gauge invariance, and in simple cases there is no freedom at all) in the definitions of T, C, and P all commute with each other. This is because the symmetries commute geometrically, and these matrices express their action on the algebra of observables, not on the states of the theory. The action on the states need only be a representation up to a phase of the geometrical group.

We have stated that the condition for CP invariance is simply that there be a basis of fields in which all couplings are real. The conditions for individual C and P invariances are stronger. If there is no action of P on the gauge group itself (which can occur only if there is a Z_2 outer automorphism of the group) then this can be satisfied only if the scalar representation of the gauge and global symmetry groups is real or pseudo-real, while the Weyl fermions lie in a real representation.

It is remarkable, by contrast, that the combined transformation CPT is a symmetry of every local Lorentz-invariant quantum field theory (see a rigorous proof in [11]). We can get a rough understanding of this by referring to the Euclidean formulation of field theory. The geometrical operation of reflecting all coordinates has determinant 1 in four dimensions, and is part of the connected component of the group SO(4). Thus, any field theory that can be obtained by Wick rotation of a Euclidean field theory will be invariant under some kind of time- and space-reflecting transformation. It remains to understand why this transformation must involve charge conjugation. When we rotate back to Minkowski space, the transformation changes positive into negative frequencies, and thus exchanges creation operators with annihilation operators. A local charged field contains the creation operator of an anti-particle and the annihilation operator of the particle. Thus, on continuing to Euclidean space, doing a rotation by π in two orthogonal planes, and continuing back to Minkowski space, we have reversed both space and time and changed particles into anti-particles. This is the CPT transformation. We remark in passing that the transformation PT alone turns left-handed into right-handed fields, and cannot be a symmetry unless the spectrum of left-handed fields is in a real representation of the symmetry group. Like most arguments in this book, this discussion is a bit "quick and dirty," but can be turned into a rigorous proof of the CPT theorem.

5.2 The functional formalism for fermion fields

Although we have worked explicitly only with a single scalar field, the functional formalism immediately generalizes to any kind of local field theory. However, the formalism we have considered so far integrates over c-number functions. It turns out that this is appropriate for fields associated with bosonic particles. For fermion fields we start from anti-commutation relations for creation and annihilation operators and it turns out that fields cannot commute at space-like separation. The best we can do is to force them to anti-commute at space-like separation. We will base our approach to the functional quantization of fermion fields on this observation.

We have already seen that particles of spin zero and 1 cannot be fermions and particles of spin $\frac{1}{2}$ cannot be bosons. This generalizes to the spin-statistics theorem: integer-spin particles are bosons and half-integer-spin particles are fermions. In this section we will not make any commitment to the character of spin indices carried by the field. However, we will work mostly with complex fermion fields, so we bias our notation to deal with that case. We will touch briefly on the properties of real fermions when we discuss Pfaffians.

In order to define covariant time-ordered products for fermion fields, we must change the definition of time ordering to include a factor of (-1) every time we interchange a pair of fermions in the process of reordering the fields from their order on the page to their proper time order. For example,

$$T\psi(x)\bar{\psi}(y) = \theta(x - y)\psi(x)\bar{\psi}(y) - \theta(y - x)\bar{\psi}(y)\psi(x).$$

Note also that, if ψ were a Dirac field, then, as a consequence of this definition and the canonical anti-commutation relations (Problem 5.1),

$$\gamma^{\mu}\, \partial_{\mu} T(\psi(x)\bar{\psi}(y)) = T(\gamma^{\mu}\, \partial_{\mu}\psi(x)\bar{\psi}(y)) + \delta^{4}(x - y).$$

If we consider a fermionic analog of the coupling to an external source,

$$T\mathrm{e}^{\mathrm{i}\int [\bar{\eta}(x)\psi(x) + \bar{\psi}(x)\eta(x)]},$$

then if the sources are c-numbers all but the first term in the series vanish. This is a conflict between the symmetry of a product of ordinary numbers $\eta(x)\eta(y) = \eta(y)\eta(x)$ and the anti-symmetry of the fermionic time-ordered product. It can be cured by changing the multiplication rule of source functions to read $\eta(x)\eta(y) = -\bar{\eta}(y)\eta(x)$. We will first consider the physical meaning of such a rule and then elaborate on its mathematical implementation.

A source function is a mathematical idealization of a large classical machine, which probes the dynamics of quantum field theory. If such a machine creates a single fermion, without violating the conservation law of $(-1)^{\mathrm{F}}$ in the entire Universe, then another fermion must be created inside the machine. The requirement that the machine be classical is, in this case, simply the requirement that there be so many different fermionic excitations in the machine that the Pauli exclusion principle is irrelevant. As a mathematical model, think of η as being constructed from an average over a large number of canonical fermion fields

$$\eta(x) = \frac{1}{N} \sum \psi_i.$$

In the large-N limit, such average fields anti-commute with all other fermionic fields, including their own Hermitian conjugates.

In the nineteenth century, Grassmann invented the algebra of anti-commuting numbers generated by a finite set of generators $[\eta_a, \eta_b]_+ = 0$. These had the properties usually attributed to *infinitesimals*, dx_a in calculus. In modern mathematics, Grassmann numbers are used to describe differential forms. The Grassmann sources of field theory are made by combining an infinite-dimensional Grassmann algebra with a basis in function space: $\eta(x) = \sum \eta_a f_a(x)$. For Dirac spinors, the $f_a(x)$ are just a basis of ordinary spinor-valued functions on space-time.

The square of any individual Grassmann generator is zero, so functions on a finite-dimensional Grassmann algebra are all polynomials, and functions on an infinite-dimensional algebra are all power series. Differentiation is easily defined by the usual algebraic rules, supplemented by the proviso that $\partial/\partial\eta_a$ anti-commutes with all Grassmann numbers. Integration is defined by insisting that it be a linear functional, invariant under translation. For a single Grassmann variable, the most general function is $a + b\eta$, where a, b are complex numbers:

$$\int d\eta(a + b\eta) = a \int d\eta 1 + b \int d\eta \, \eta.$$

Invariance under $\eta \to \eta + \psi$ implies that

$$\int d\eta 1 = 0,$$

and we normalize $\int d\eta \, \eta = 1$. Multiple integration is defined by iteration.

It is easy to see that this leads to the rule

$$\int d^n\eta \, f(\eta) = f_{a_1 \ldots a_n},$$

where the right-hand side is the coefficient of the term $f_{a_1 \ldots a_n} \eta_{a_1} \ldots \eta_{a_n}$ in the function $f(\eta)$. In particular, if A is an $N \times N$ anti-symmetric matrix, then

$$\int d^n\eta \, e^{\frac{1}{2}A^{ab}\eta_a\eta_b} = \epsilon_{a_1 \ldots a_N} A_{a_1 a_2} \ldots A_{a_{N-1} a_N} \equiv \mathrm{Pf}[A],$$

if N is even, and vanishes if N is odd. This combination of matrix elements of an even-dimensional anti-symmetric matrix is called the Pfaffian. We can define the Pfaffian of an odd-dimensional anti-symmetric matrix to be zero.

Similarly, if $\bar{\eta}_a$ and η_a are two independent sets of Grassmann variables, then

$$\int d^N\bar{\eta} \, d^N\eta \, e^{\bar{\eta}M\eta} = \det M,$$

for any complex matrix.

Using these formulae, and the invariance of Grassmann integration under shift symmetry, it is easy to prove that Wick's theorem is valid for Gaussian fermionic integrals (Problem 5.9). The only thing we have to be careful of is the minus signs, which arise between terms with different contractions.

5.3 Feynman rules for Dirac fermions

In this book, we will give Feynman rules only for complex spin-$\frac{1}{2}$ fields satisfying the Dirac equation. Weyl fermions are to be treated as Dirac fermions whose interactions satisfy a constraint such that their right-handed components are free fields. That is, the correct couplings will never let the right-handed components propagate in internal lines, or scatter. Furthermore, we will discuss only Lagrangians quadratic in fermion fields, which have a U(1) fermion-number symmetry. These will have the form

$$\mathcal{L}_D = \bar{\psi} D \psi,$$

where $D = i\slashed{\partial} + M$ and M is a $4N \times 4N$ matrix, which depends on the bosonic fields in the theory. N is the number of independent complex Dirac fields, which we call the number of species. The result of doing the path integral

$$\int [\mathrm{d}\bar{\psi} \ \mathrm{d}\psi] e^{i \int \mathrm{d}^4 x \mathcal{L}_D} \psi_{i_1}(x_1) \bar{\psi}_{k_1}(y_1) \dots \psi_{i_n}(x_n) \bar{\psi}_{k_n}(y_n) \tag{5.7}$$

is

$$\det \Big[S_{i_m k_l}(x_m, y_l) \Big] \mathrm{Det} \ D. \tag{5.8}$$

In these equations i_m and k_l are composite indices combining the spin and species indices. S is the Green function for $(1/i)D$, with Feynman boundary conditions (or the Euclidean Green function regular at infinity). The lower-case det just means the determinant of the $4Nn \times 4Nn$ matrix of propagators, while the upper-case Det is the functional determinant of D.

In the particular case of free-field theory, where M is a constant proportional to the unit matrix, the functional determinant is just a number[11] and factors out of the answers for connected correlation functions, since they are ratios of two functional integrals. The propagator $S(x, y)$ depends only on $x - y$, and its Fourier transform is

$$S(p) = \frac{i}{\slashed{p} - m + i\epsilon}. \tag{5.9}$$

m is the mass of the free field. The propagators are oriented, carrying an arrow which shows the flow of the conserved U(1) particle–anti-particle number (which, for electrons, is minus the electron charge because of an unfortunate choice of conventions in the nineteenth century). The momentum p which appears in the propagator is the momentum flowing in the direction of the particle-number arrow.

---

[11] It is infinite, but that is properly treated under the heading of renormalization.

The Feynman diagram for a free Dirac propagator is

$$p \longrightarrow \qquad \frac{i}{\not{p} - m + i\epsilon} \qquad \frac{1}{\not{p} + im}$$

Dirac propagator in Minkowski and Euclidean space

5.4 Problems for Chapter 5

*5.1. Write the general solution of the Dirac equation,

$$(i\,\not{\partial} - m)\psi,$$

in terms of Fourier coefficients $a_i(p)u_i(p)$ for positive-frequency solutions and $b_i^\dagger(p)v_i(p)$ for negative frequencies. $(\not{p} - m)u_i = 0$ and $(\not{p} + m)v_i = 0$. There are two linearly independent solutions of each equation (see Appendix C). Show that $a_i^\dagger$ and $b_i^\dagger$ transform like the creation operators of massive spin-$\frac{1}{2}$ particles, if the field ψ transforms in the $[2, 1] \oplus [1, 2]$ representation of the Lorentz group. Show that the commutation or anti-commutation relation

$$[\psi_a(x, t), \psi_b^\dagger(y, t)]_\pm = \delta_{ab}\delta^3(x - y)$$

implies bosonic or fermionic commutation relations for the creation and annihilation operators. Show that only the fermionic choice leads to a local field, which anti-commutes with itself and its conjugate, at space-like separation.

5.2. Starting from the Weyl representation of the Dirac matrices, find a change of basis that makes all the matrices imaginary (Majorana representation) and another one in which γ^0 is $\sigma_3 \otimes 1$ and $\gamma_\mu^\dagger = \gamma^0 \gamma_\mu \gamma^0$.

*5.3. The most general action of the discrete symmetries T, C, and P on mass eigenstates of spin-$\frac{1}{2}$ particles is (the ηs are complex phases)

$$T^\dagger a^\dagger(p, s)T = \eta_T a(-p, -s),$$
$$P^\dagger a^\dagger(p, s)P = \eta_P a^\dagger(-p, s),$$
$$C^\dagger a^\dagger(p, s)C = \eta_C b^\dagger(p, s).$$

Recall that the T operator is anti-unitary, while the others are unitary. If the anti-particle operator b is not the same as that of the particle (i.e. the particle carries a charge), then there are in principle independent phases $\bar{\eta}_{T,P}$ for the anti-particle. Find the connections between these phases implied by a homogeneous transformation law for local fields, and write down the transformation laws for Dirac fields in the Weyl representation.

*5.4. The Dirac anti-commutation relations imply that independent elements of the algebra generated by the Dirac matrices are the antisymmetrized products $\gamma^{\mu_1 \cdots \mu_k}$, where k runs from 0 to 4. Rewrite the $k = 3, 4$ cases in terms of the matrices γ_μ and γ_5. Use the results of the previous problem to characterize the T, C, and P transformation laws of all Dirac bilinears $\bar\psi\gamma^{\mu_1\cdots\mu_k}\psi$.

*5.5. Given a set of Weyl fields v_a^i, the most general non-derivative, bilinear term one can add to the Lagrangian has the form

$$M_{ij}\epsilon^{ab}v_a^i v_b^j + \text{h.c.}$$

Anti-commutativity of the fields and anti-symmetry of the Levi-Civita symbol imply that M is a symmetric complex matrix. Suppose that there is a U(1) transformation acting on the v_a^i, which commutes with M. Then the bilinear can be written as a sum over subsets of fields with charges $\pm q$ under U(1), $\sum_q v_a^I(q)v_b^J(-q)m_{IJ}(q)$. The part of this sum with $q = 0$ is called a Majorana mass term, while the rest of the bilinear is called a Dirac mass term. $m_{IJ}(q)$ is a matrix acting only on the subspace of fields with charge q. Show that by doing independent unitary transformations on the charge q and $-q$ fields (for $q \neq 0$) we can transform each $m_{IJ}(q)$ into a diagonal matrix with real positive entries. Thus, the fermion kinetic term for Dirac fermion masses actually has a U(1)$^{\text{No. of charges}}$ symmetry and preserves T, C, and P. In the context of the strong interactions these U(1) charges are called *quark flavors*. They are not exactly preserved by weak interactions, but this is a small effect. Flavor is an example of an accidental approximate symmetry. What can you say about the Majorana mass matrix with regard to putting it in a canonical form, and its transformation under discrete Lorentz transformations?

5.6. Derive the Källen–Lehmann spectral representation for a Dirac field $\psi_\alpha(x)$ and for a conserved vector current operator $J_\mu(x)$ satisfying $\partial_\mu J^\mu = 0$.

5.7. Solve the Dirac Green function equation

$$(i\slashed{\partial} - e\slashed{A} - m)S(x, y) = \delta^4(x - y)$$

in a constant background electromagnetic field $A_\mu(x) = F_{\mu\nu}x^\nu$

*5.8. Show that $\gamma_5 = i\gamma^0\gamma^1\gamma^2\gamma^3$ and that $[\gamma^\mu, \gamma^\nu]_+ = 2\eta^{\mu\nu}$. Prove that $\gamma_5^2 = 1$ and that $[\gamma_5, \gamma^\mu]_+ = 0$.

*5.9. Use the rules of Grassmann integration to prove the analog of Wick's theorem for fermions.

*5.10. Using the transformation $X_\pm = (\bar\eta \pm \eta)/2$, in the Grassmann integral formula for the determinant of M, for the case in which M is anti-symmetric, relate the determinant and Pfaffian formulae and prove that $\det A = (\text{Pf } A)^2$.

*5.11. Derive the Feynman rules for the Yukawa interaction

$$\Delta\mathcal{L} = \phi(g_S\bar\psi\psi + g_P\bar\psi\gamma_5\psi)$$

between a spin-zero field and a Dirac fermion.

*5.12. Prove all of the trace and contraction identities in Appendix C, and generalize them to at least one higher power of Dirac matrices. Evaluate all of the traces when an additional γ_5 is inserted, along with the Dirac matrices γ^μ.

*5.13. Consider an interaction $\bar\psi\Gamma^A\psi g_A\phi^A + \frac{1}{2}(\phi^A)^2$, where Γ^A runs over the 16 independent anti-symmetrized Dirac products, ϕ^A is a tensor field of the appropriate kind to make a Lorentz-invariant product, and g_A is the same for each component of an irreducible Lorentz representation. The fields ϕ^A have no kinetic terms. Solve for them algebraically, and show that you get the most general four-fermion interaction for the ψ field, which preserves the symmetry $\psi \to e^{i\alpha}\psi$. This trick can be generalized to higher powers of ψ and to interactions that don't preserve the symmetry. Using this trick, one can formally do the fermion functional integral in terms of the Green function for the Dirac operator $i\,\partial\!\!\!/ - \phi^A\bar\psi\Gamma^A\psi$ and the determinant of this operator. If ψ were a boson, we would get the inverse determinant instead of the determinant. Show that this replacement implies the extra Feynman rule of (-1) for each closed fermion loop.

*5.14. Prove the Gordon identity

$$2m\bar u(k)\gamma^\mu u(p) = \bar u(k)[(p+k)^\mu + \gamma^{\mu\nu}(k-p)_\nu]u(p)$$

and the analogous identity for the anti-particle spinors $v(p)$.

*5.15. Solve the momentum-space Dirac equations $(p\!\!\!/ - m)u(p,s) = 0 = (p\!\!\!/ + m)v(p,s)$, in both the Dirac and the Weyl bases for the Dirac matrices (see Appendix C).

*5.16. Show that charge-conjugation symmetry implies that the representation of the internal symmetry group G is real or pseudo-real.

Massive quantum electrodynamics

In this chapter we will do some perturbative calculations in a theory called *massive QED*. It is quantum electrodynamics with the photon replaced by a massive vector field. Actually, it is the simplest example of a theory with a Higgs mechanism, since we have seen that the massive vector can be written as a gauge theory by introducing an additional scalar degree of freedom. Indeed, we can write a theory of electrodynamics of spinor and scalar fields in terms of the Lagrangian

$$g^2 \mathcal{L} = |D_\mu \phi|^2 - \frac{1}{4} F_{\mu\nu} F^{\mu\nu} + \bar{\psi} \gamma^\mu (iD_\mu - m)\psi + \kappa (|\phi|^2 - v^2)^2. \quad (6.1)$$

If we write $\phi = \rho e^{i\theta}$, and treat θ as the Stueckelberg field in the gauge formalism for the massive vector field, then we can eliminate θ in favor of a massive vector field B_μ with a mass that depends on the field ρ. ρ is called the Higgs field. In the formal limit $\kappa \to \infty$, fluctuations around $\rho = v$ are infinitely costly, and we get the theory called massive QED. It turns out to be a perfectly good quantum field theory (it is renormalizable in perturbation theory), in the sense that, if g^2 is small and we introduce an ultraviolet cut-off Λ to make sure the theory is well defined (see the chapter on renormalization), then as long as $\Lambda < e^{c/g^2} E$, where c is a number of order 1 that we will learn to compute later, the predictions of the theory at energy scales $E \ll \Lambda$ are insensitive to the value of Λ and to the precise way in which we implement the cut-off.

We introduce massive QED for several reasons. First, it allows us to do computations in QED without worrying about gauge invariance. This is because the massive theory has no gauge invariance, and can be quantized in a straightforward manner. The Heisenberg equations of motion and canonical commutation relations lead, in a familiar fashion, to the path-integral formula for the generating functional of Euclidean Green functions

$$Z[J^\mu, \eta, \bar{\eta}] = \frac{\int [dB_\mu \, d\psi \, d\bar{\psi} \,] e^{-S + \int d^4 x (B_\mu J^\mu + \bar{\eta}\psi + \bar{\psi}\eta)}}{\int [dB_\mu \, d\psi \, d\bar{\psi} \,] e^{-S}}. \quad (6.2)$$

This leads to Euclidean Feynman rules with the propagators for B_μ:

$$\langle B_\mu(p) B_\nu(-p) \rangle = \frac{\delta_{\mu\nu} - p_\mu p_\nu / \mu^2}{p^2 + \mu^2}. \quad (6.3)$$

The Euclidean fermion propagator is

$$\frac{1}{\not{p} + im}.$$

As usual we have written the propagators of fields that are rescaled to make the quadratic terms in the Lagrangian e-independent. The gauge boson mass is $\mu = ev$.

The interaction vertex corresponds to the following amputated[1] Feynman diagram:

$$i e \gamma^\mu \qquad\qquad -e\gamma^\mu$$

Fermion gauge vertex

and has the value $-e\gamma^\mu$, which goes to $-ie\gamma^\mu$ in Lorentzian signature. It is a non-trivial, but true, statement that, when we restrict attention to sources satisfying $\partial_\mu J^\mu = 0$, this generating functional has a finite limit as $\mu \to 0$ and the vector boson becomes massless. In this limit the particle states split into two different representations of the Poincaré group, a massless scalar (the erstwhile longitudinal component of the massive spin-1 particle) and a massless particle with helicity ± 1. The restriction to conserved sources decouples the massless scalar. A particular class of conserved currents, of the form $J^\mu = \partial_\nu M^{\mu\nu}$, with $M^{\mu\nu} = -M^{\nu\mu}$, generates Green functions of the electromagnetic field strength $F_{\mu\nu}$ in the $\mu \to 0$ limit.

When we continue to Lorentzian signature and compute scattering amplitudes we see several interesting effects in the massless limit. First of all the amplitudes for producing longitudinal gauge bosons from initial states that had no longitudinal gauge bosons in them goes to zero in this limit. Secondly, amplitudes involving scattering of fermions diverge in perturbation theory. This divergence is primarily due to the fact that in the $\mu \to 0$ limit one can emit an arbitrary number of vector bosons with a finite cost in energy. Indeed, in this limit the probability of emitting any finite number of bosons goes to zero.[2] This is the *infrared catastrophe* of Maxwell's QED. As a catastrophe, it rates pretty low on the Richter scale. It simply means that *any* scattering of charged particles is accompanied by a low-energy burst of classical *bremsstrahlung* radiation, which more or less follows the trajectories of the outgoing charged particles. If we restrict the energy ϵ in this radiation, rather than the number of particles (which is always infinite in the $\mu \to 0$ limit) we get a finite answer. The advantage of massive QED is that we can derive this finite answer by summing up finite amplitudes in perturbation theory and then taking μ to zero, rather than formally resumming a series of infinite terms.

We do not have space in this text to actually do a bremsstrahlung calculation, and will leave it to the problem set. Instead we will do two tree-level computations of fermion annihilation processes, and a calculation through one loop level of low-energy scattering in an external magnetic field. It will be convenient to add a second fermion to the theory, with the same charge as the original one, and a mass $M \gg m$. We will call the light fermions electrons and positrons, and the heavy ones muons and anti-muons, in honor of famous characters who have appeared in the *Real World*.[3]

---

[1] This adjective implies, as usual, that "the legs have been cut-off."

[2] To understand how a zero can look infinite in perturbation theory, consider the expression $\mathrm{e}^{-\mathrm{e}^2 (\ln \mu)^2}$.

[3] A well-known piece of international cinema, of little artistic value, but enormous popular appeal.

6.1 Free the longitudinal gauge bosons!

Our first calculation will be the tree-level amplitude for annihilation of electron and positron into spin-1 particles. There are two Feynman diagrams (Figure 6.1), related by the Bose statistics of the gauge bosons.

At tree level, we can short circuit much of the LSZ formula. The poles in external lines sit at the bare Lagrangian masses, which are thus equal to the physical masses at this order. Wave-function renormalization constants are all equal to one. The general rule for external-line wave functions is that for incoming lines they are the coefficients of annihilation operators, while for outgoing lines they are the coefficients of creation operators, in the free fields which create and annihilate scattering states. Thus we have

$$u(p, s) - \text{incoming fermion,}$$

$$\bar{u}(p, s) - \text{outgoing fermion,}$$

$$v(p, s) - \text{outgoing anti-fermion,}$$

$$\bar{v}(p, s) - \text{incoming anti-fermion.}$$

The amplitude for the first diagram is thus

$$\mathcal{M} = (-\mathrm{i}e)^2 \bar{v}(p_2, s_2) \gamma^\mu \frac{\mathrm{i}}{\not{p} - m + \mathrm{i}\epsilon} \gamma^\nu u(p_1, s_1)(\epsilon^*)^\mu(k_1, a_1)(\epsilon^*)^\nu(k_2, a_2). \qquad (6.4)$$

The momentum in the internal fermion line is

$$p^\mu = (k_1 - p_2)^\mu.$$

The amplitude for the second diagram is the same expression with the two massive-photon polarization vectors interchanged, and $p^\mu = k_2 - p_2$. The two diagrams are to be added, which shows that the Feynman rules implement Bose statistics.

We have written the *invariant amplitude*, $\mathcal{M}$. The full S-matrix element for any process is gotten by multiplying the invariant amplitude by

$$S = \prod \frac{1}{\sqrt{(2\pi)^3 2\omega(p_i)}} (2\pi)^4 \delta^4 \left(\sum_{\text{in}} p_i - \sum_{\text{out}} p_i \right) \mathcal{M}.$$

The product of energy factors comes from our non-relativistic state normalization, and runs over all initial and final particles. The momentum-conservation delta function is written here with the convention that all momenta have positive energy.

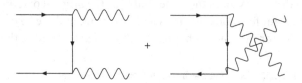

Fig. 6.1.   Gauge-boson production in e$^+$, e$^-$ annihilation.

It is instructive to examine this amplitude for the case of longitudinally polarized massive photons. The longitudinal polarization vector is

$$\epsilon^\mu(k, L) = \frac{1}{\mu}\left(|k|, \mathbf{k}\frac{\omega(k)}{|k|}\right).$$

Since it blows up at $\mu = 0$, production of these particles would seem to rule out the possibility of taking the massless limit. Furthermore, at high energies the dimensionless polarization vector must blow up like $|k|$. At the very least we seem to be faced by a breakdown of perturbation theory in both the massless and the high-energy limit.

The observation that saves the day is that, at $|k| \gg \mu$, we have

$$\epsilon^\mu(k, L) = \frac{k^\mu}{\mu} + o(\mu/|k|).$$

Now consider the production amplitude for ϵ_1 longitudinal and center-of-mass energy much bigger than μ. It is, in leading order,

$$-\mathrm{i}e^2 \bar{v}\left(\not{k}_1 \frac{1}{\not{k}_1 - \not{p}_2 - m}\not{\epsilon}_2 + \not{\epsilon}_2\not{k}_1 \frac{1}{\not{k}_2 - \not{p}_2 - m}\not{k}_1\right)u. \tag{6.5}$$

Now use the identities

$$\not{k}_2 - \not{p}_2 - m = \not{p}_1 - \not{k}_1 - m,$$
$$\not{k}_1 = \not{k}_1 - \not{p}_2 - m + \not{p}_2 + m,$$
$$\not{k}_1 = -(\not{p}_1 - \not{k}_1 - m) + \not{p}_1 - m,$$
$$(\not{p}_1 - m)u = \bar{v}(\not{p}_2 + m) = 0.$$

The terms coming from the two possible substitutions for $\not{k}_1$ each consist of one term that cancels out the fermion propagator and another that vanishes because the external spinors satisfy the Dirac equation. The terms with canceled propagators from the two diagrams have equal magnitude and opposite sign. Thus, the divergent contribution to the longitudinal production amplitude, in the zero-mass or high-energy limit, cancels out exactly. This cancelation is the heart of the proof of renormalizability of Higgs models.

6.2 Heavy-fermion production in electron–positron annihilation

In this section we will calculate the leading-order contribution, in massive QED, to the annihilation cross section for a light charged particle (electrons and positrons) into a heavier one. We will call the heavier particles muons, but the calculation is valid for τ leptons as well. As a consequence of asymptotic freedom, it can also be used for calculating the $e^+e^- \to$ hadrons cross section above the QCD confinement scale (see Chapter 8 for a definition). In that case, the rigorous use of the calculation involves an analytic continuation from Euclidean space [32], which we will not delve into. Since we

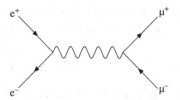

Fig. 6.2.   Heavy-fermion production in e^+, e^- annihilation.

work with massive photons, our calculation can also be adapted to calculate the part of the full cross section that comes from Z-boson, rather than photon, exchange. We need only substitute the correct couplings to the Z boson, and set μ equal to m_Z in the calculation below. We will learn how to calculate the Z couplings in Chapter 8. Given the format of this book, we cannot possibly do justice to the full extent of the physics of this process. The reader is urged to consult the wonderful section in Chapter 5 of Peskin and Schroeder [33] for a detailed description of it.

There is only one diagram (Figure 6.2) in leading order, and its value is

$$\mathcal{M} = (-\mathrm{i}e)^2 \bar{u}(k_-, r_-)\gamma^\kappa v(k_+, r_+)\frac{-\mathrm{i}(\eta_{\kappa\lambda} - q_\kappa q_\lambda/\mu^2)}{q^2 - \mu^2 + \mathrm{i}\epsilon}\bar{v}(p_+, s_+)\gamma^\lambda u(p_-, s_-). \quad (6.6)$$

Here $q = p_+ + p_- = k_+ + k_-$, and the labels indicate incoming and outgoing fermion momenta and spins in what I hope is an obvious manner. The labels are related to the scattering energy and angle in the center-of-mass frame by Figure 6.3.

We note the identity

$$\bar{v}(p_+, s_+)(p_+ + p_-)_\lambda \gamma^\lambda u(p_-, s_-) = (-m_\mathrm{e} + m_\mathrm{e})\bar{v}u(p_-, s_-) = 0,$$

which eliminates the longitudinal part of the gauge-boson propagator from the amplitude.

We will now calculate the *spin-averaged* total cross section for this process where we sum over initial and average over final spins. This corresponds to the simplest experimental situation, in which the initial beams are unpolarized and no spin measurements

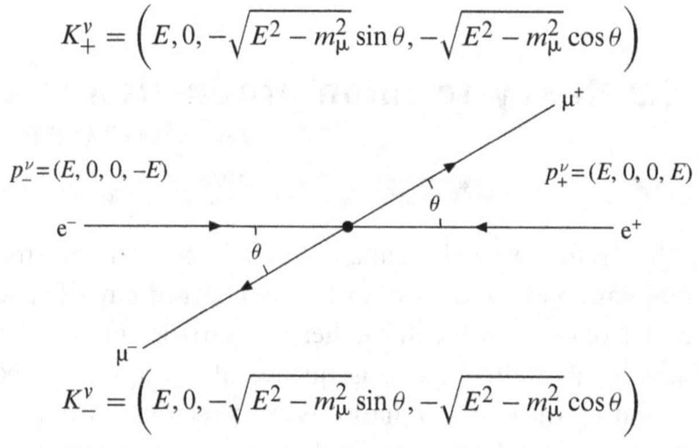

$$K_+^\nu = \left(E, 0, -\sqrt{E^2 - m_\mu^2}\sin\theta, -\sqrt{E^2 - m_\mu^2}\cos\theta\right)$$

$p_-^\nu = (E, 0, 0, -E)$ μ^+ $p_+^\nu = (E, 0, 0, E)$

e^- θ e^+

μ^-

$$K_-^\nu = \left(E, 0, -\sqrt{E^2 - m_\mu^2}\sin\theta, -\sqrt{E^2 - m_\mu^2}\cos\theta\right)$$

Fig. 6.3. μ^+, μ^- Pair production by annihilation of massless e^+, e^- in the center-of-mass frame.

are made on the final particles. Polarized beams and polarization detectors are powerful experimental tools in e^+e^- annihilation. The reader is urged again to consult Peskin and Schroeder [33] for a detailed description of polarization amplitudes. At this point the reader should also turn to the appendix on Diracology. He/she is urged to take all of the results stated there as exercises, even those whose proofs are given. Mastery of this technology is an important part of the repertoire of any respectable field theorist.

Using the polarization sum identities of Diracology, the spin-averaged squared amplitude is

$$\frac{1}{4} \sum_{r_\pm, s_\pm} |\mathcal{M}|^2 = \frac{e^4}{4|(q^2 - \mu^2 + i\epsilon)|^2} \, \text{tr}\Big[(\not{p}_+ - m_e)\gamma^\lambda (\not{p}_- + m_e)\gamma^\kappa \Big]$$

$$\times \text{tr}\Big[(\not{k}_- + m_\mu)\gamma_\lambda (\not{k}_+ - m_\mu)\gamma_\kappa \Big]. \tag{6.7}$$

Now we use the trace formulae of the appendix to write this as

$$\frac{1}{4} \sum_{r_\pm, s_\pm} |\mathcal{M}|^2 = \frac{e^4}{4|(q^2 - \mu^2 + i\epsilon)|^2}$$

$$\times \Big[(p_- k_-)(p_+ k_+) + (p_- k_+)(p_+ k_-) + m_\mu^2 (p_- p_+) \Big]. \tag{6.8}$$

We have dropped a term proportional to m_e^2 (which the reader should calculate), since these formulae are always used at center-of-mass energy above twice the muon mass.

We evaluate the scattering cross section in the center-of-mass frame (which is not quite the lab frame in an e^+e^- collider, because of bremsstrahlung). There, according to Figure 6.3,

$$(p_+ + p_-)^2 = 4E^2; \qquad (p_+ p_-) = 2E^2;$$

$$(p_- k_-) = (p_+ k_+) = E^2 - E\sqrt{E^2 - m_\mu^2} \cos\theta;$$

$$(p_- k_+) = (p_+ k_-) = E^2 + E\sqrt{E^2 - m_\mu^2} \cos\theta.$$

We have again made the approximation that the electrons are massless and travel at the speed of light. E is the energy and absolute value of the momentum of the electron or positron. The square root in the above formulae is the absolute value of the muon or anti-muon momentum.

We now consult the appendix on cross sections, using the two-to-two formula with $E_A = E_B = E$, and $|v_A - v_B| = 2$ to write

$$\frac{d\sigma}{d\Omega} = \frac{\sqrt{1 - m_\mu^2/E^2}}{1024\pi^2} \sum |\mathcal{M}|^2, \tag{6.9}$$

$$= \frac{\alpha^2}{(4E^2 - \mu^2)^2} \sqrt{1 - \frac{m_\mu^2}{E^2}} \Big[(E^2 + m_\mu^2) + (E^2 - m_\mu^2) \cos^2\theta \Big]. \tag{6.10}$$

We have introduced the fine-structure constant $\alpha \equiv e^2/4\pi \approx 1/137$.

There are three interesting things to note about this cross section. The first is the square-root turn on of the cross section near threshold and the second is the angular distribution of the reaction products. Both the square root and the angular distribution are characteristic of spin-$\frac{1}{2}$ particles. Observations of thresholds like this in e^+e^- annihilation signal new quarks and leptons and give us one of the pieces of evidence that they all have spin $\frac{1}{2}$. If we are ever lucky enough to see a supersymmetric partner of a quark or lepton with spin zero, we will see something quite different. The electromagnetic current for these particles carries a factor of momentum, so the turn on of the cross section above threshold is slower by an extra square root. The angular distribution is also different. You should do Problem 6.4, to appreciate these differences.

Finally, if we calculate the total cross section

$$\sigma_{\text{tot}} = \frac{16\alpha^2\pi}{3(4E^2 - \mu^2)^2}\sqrt{1 - \frac{m_\mu^2}{E^2}}\left(E^2 + \frac{1}{2}m_\mu^2\right) \tag{6.11}$$

and take the limit $E \gg$ all mass scales, then we get the scale-invariant result

$$\sigma_{\text{tot}} = \frac{\pi\alpha^2}{3E^2}. \tag{6.12}$$

The fact that this is scale-invariant seems obvious, but it is a clue to the very essence of what a quantum field theory is. You'll have to hold your breath till the chapter on renormalization to find out.

6.3 Interaction with heavy fermions: particle paths and external fields

If the muon mass were extremely heavy, we would not be able to create muons in low-energy processes involving electrons, photons, and muons. Furthermore, in this limit, muons move without much disturbance from their interactions (momentum transfer much less than the muon mass). Thus we can control the trajectories of any muons and anti-muons in the initial state, and restrict our attention to trajectories where no annihilation processes are possible. In this section we will discuss a set of approximations that are useful for describing this limit.

In the path-integral formalism for QED, we can do the integral over the muon field exactly, with the other fields held fixed, in terms of the Green function of the muon Dirac operator in an external photon field. If we are trying to evaluate a Green function with $2N$ muon fields, appropriate for situations in which there are N muons or anti-muons in the initial state of a scattering process, then the result is

$$\det[S(x_i, y_j)]\text{Det}[i\,\partial\!\!\!/ - e A\!\!\!/ - M_\mu]. \tag{6.13}$$

$S(x, y)$ is the solution of

$$[i\,\partial\!\!\!/ - e A\!\!\!/(x) - M_\mu]S(x, y) = i\delta^4(x - y),$$

with Feynman boundary conditions. The lower-case det refers to the ordinary determinant of the $N \times N$ matrix of propagators, corresponding to all possible contractions of initial and final points. The upper-case Det refers to the functional determinant of the Dirac operator. The functional determinant is negligible as long as all external fields and momenta are small compared with M_μ.

To evaluate the propagator S we write

$$S = (i\not{D} - M_\mu)^{-1} = (i\not{D} + M_\mu)\left(-D^2 - M_\mu^2 + \frac{i}{2}\sigma^{\mu\nu}F_{\mu\nu}\right)^{-1}, \qquad (6.14)$$

where D_μ is the covariant derivative. To evaluate the inverse of the second-order operator, we think of it as the Hamiltonian for a quantum particle with four position coordinates[4] and a spin, and write the inverse in terms of a mixture of path-integral formalism for the positional mechanics and a time-ordered formula for the spin dynamics:

$$S = (i\not{D} + M_\mu)\int_0^\infty ds\, e^{-isM_\mu^2}[dx^\mu(s)]e^{i\int_0^s d\tau\left[\left(\frac{dx}{2d\tau}\right)^2 + A_\mu\frac{dx^\mu}{d\tau}\right]}Te^{\frac{i}{2}\int d\tau\,\sigma^{\mu\nu}F_{\mu\nu}(x(\tau))}. \tag{6.15}$$

The path integral is over paths satisfying $x(0) = x$, $x(s) = y$. On introducing the dimensionless variables $sM_\mu^2 \equiv u$, $\tau M_\mu^2 = v$, it is easy to see that, for large M_μ, the particle action is large, and is dominated by the free-particle kinetic term, which is of order M_μ^2. The term involving the line integral of the vector potential is of order one, while the interaction with the spin is down by $1/M_\mu^2$.

Thus we can saturate the path integral by the straight-line path between x and y. The result, to leading order in M_μ, is

$$S(x,y) = \int_0^\infty ds\, e^{-isM_\mu^2 + i\frac{(x-y)^2}{4s}}e^{\frac{i}{s}\int_0^s d\tau A_L(x+\frac{\tau}{s}(y-x))}. \tag{6.16}$$

$A_L \equiv (x - y)^\mu A_\mu$, and the last exponential is just the line integral of the vector potential along the straight-line particle path and, in particular, it is independent of the parameter s. Thus, the effect of integrating out the massive fermion is to add a source term to the A_μ Lagrangian. If we shift the A_μ field to eliminate the source, it is shifted exactly by the classical (massive) electromagnetic field produced by the current of the heavy point particle. Obviously we can repeat this procedure for all the fermion propagators. To leading order in the mass, the effect of a collection of massive particles on the rest of the theory is simply to shift the background value of A_μ from zero to some other classical field. This justifies the study of such classical external-field problems.

6.4 The magnetic moment of a weakly coupled charged particle

One of the most interesting calculations we can do with the external field method is of the correction to the magnetic moment of a particle whose strongest interaction is

---

[4] This actually looks a little more conventional in Euclidean space, but we will not bother to go through the analytic continuations.

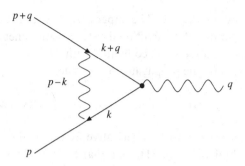

Fig. 6.4.   One-loop vertex correction.

electromagnetic. The classic result of Dirac is that the gyromagnetic ratio of a point particle of spin $\frac{1}{2}$ is 2. What we will show is that purely electromagnetic interactions correct that result. In fact, higher-order calculations and modern experimental techniques have combined to make this calculation/measurement for the electron and the muon one of the most precise agreements we have between theoretical physics and the real world. The current state of the art is sensitive to the effects of virtual strongly interacting particles and of the weak interactions. It can even be used to put interesting bounds on physics at quite high energy scales. The electron and muon dipole moments are known to 13 and 10 significant digits experimentally. For the electron there is complete agreement between theory and experiment at that level of accuracy. Recent (November 2006) results indicate a possibly significant deviation for the muon, which might be an indication of physics beyond the standard model. We will calculate only the leading correction to the anomalous magnetic moment. A general reference for precision QED results is [34].

We will employ the external field methods we have just learned, specialized to a field of the form $F_{ij} = \epsilon_{ijk}B^k$, where B^k is the constant value of the magnetic field, in the rest frame of the particle. The electric field is assumed zero in this frame. We will do the computation to leading order in the corrections to the result for the Dirac equation in an external field. The reader should first do Problem 6.5, which shows that the prediction of the Dirac theory is a gyromagnetic ratio of 2.

To first order in the external field, the invariant amplitude is given by

$$(-ie_0) \int d^4q \, A^\mu(q)\bar{u}(p+q)\Gamma_\mu(p,q)u(p).$$

We have suppressed the spin indices on the incoming and outgoing spinors. We will evaluate the 1PI vertex Γ_μ to one-loop order (Figure 6.4), taking into account that to this order the bare mass parameter m_0 is equal to the physical mass m. Thus, the diagram is given by

$$\bar{u}\Gamma_\mu u = \frac{(-i)(-ie_0)^2}{(2\pi)^4}\bar{u}(p+q) \int d^4k \, \frac{\eta_{\lambda\kappa} - (k-p)_\lambda(k-p)_\kappa/\mu^2}{(k-p)^2 - \mu^2 + i\epsilon}$$
$$\times \frac{\gamma^\lambda[(\slashed{k}+\slashed{q}+m)\gamma_\mu(\slashed{k}+m)]\gamma^\kappa}{[(k+q)^2 - m^2 + i\epsilon][k^2 - m^2 + i\epsilon]}.$$

Let us first deal with the term which apparently diverges as $\mu \to 0$. Using the identities $\bar{u}(p+q)(\not{p} - \not{k}) = \bar{u}(p+q)(m - \not{k} - \not{q})$ and $(\not{p} - \not{k})u(p) = (m - \not{k})u(p)$, we see that the denominators in the two fermion propagators are canceled out by the numerators, and we are left with a term proportional to

$$\frac{e_0^2}{\mu^2} \bar{u} \gamma_\mu u \int \frac{\mathrm{d}^4 k}{k^2 - \mu^2 + i\epsilon}.$$

This term can be absorbed into a rescaling of the fermion fields because it has the same form as the tree-level vertex. The precise value of this *renormalization* depends on the method we use for cutting off ultraviolet divergences in the theory, a subject to which we will return in Chapter 9. There we will distinguish between cut-off procedures that preserve gauge invariance and those which don't. With a gauge-invariant regulator, like the analytic continuation of Feynman diagrams in the space-time dimension[5] the apparently divergent integral in the previous expression is actually finite and proportional to μ^2. We have to be careful about the regulation procedure because, as we have noted, massive electrodynamics is really a gauge theory in its Higgs phase. The upshot of this discussion is that we can ignore the longitudinal term in the gauge-boson propagator, and replace it by

$$-i\frac{\eta_{\mu\nu}}{p^2 - \mu^2 + i\epsilon}$$

in this diagram.

Before continuing the computation let us stop to ask what we expect to get. The current matrix element is $\bar{u}(p+q)\Gamma_\mu u(p)$, and the vertex function Γ_μ can be expanded in terms of the linearly independent Dirac matrices $1, \gamma_5, \gamma_\mu, \gamma_\mu \gamma_5, \gamma_{\mu\nu}$. The two expressions with γ_5, as well as anything else involving the Levi-Civita symbol, cannot appear, because massive electrodynamics conserves parity. We must combine the matrices with the 4-vectors p, q to make a 4-vector, and, when we sandwich Γ_μ between on-shell spinors, we can use

$$(\not{p} - m)u(p) = (\not{p} + \not{q} - m)u(p+q) = 0$$

and the conjugates of these equations to prove that

$$\bar{u}(p+q)\not{q}u(p) = 0$$

and

$$\bar{u}(p+q)(2p^\mu + q^\mu)u(p) = \bar{u}(p+q)[2m\gamma^\mu + \gamma^{\mu\nu}q_\nu]u(p).$$

The last identity, named after Gordon, was proved in the exercises to Chapter 5. Using these identities, it is easy to see that anything in Γ_μ can be reduced to

$$F_1(q^2, m^2)\gamma_\mu - F_2(q^2, m^2)\frac{\gamma_{\mu\nu}q^\nu}{2m} + F_3(q^2, m^2)q_\mu,$$

---

[5] This is called *dimensional regularization* and abbreviated DR.

when evaluated between on-shell spinors. The form factors F_i depend on the Lorentz invariants q^2, $p^2 = m^2$ and $2pq = -q^2$.

The current matrix element is conserved, which means $q^\mu \bar{u} \Gamma_\mu u = 0$. This is satisfied for arbitrary $F_{1,2}$ and implies that $F_3 = 0$. In our calculation, we will seek to reduce all expressions in Γ_μ (after loop integration) to linear combinations of $(2p + q)^\mu, \gamma^\mu$, and q^μ. The diligent reader will prove that, once this is done, the coefficient of q^μ vanishes. The Gordon identity will allow us to identify the individual $F_{1,2}$. An extremely important point of the analysis is that, for $q^2 = 0$, which is the only thing our constant external field probes, F_1 serves merely to *renormalize* the value of the electric coupling e_0. Thus, it cannot change the gyromagnetic ratio of 2 implied by the Dirac equation for any coupling. The change in the gyromagnetic ratio is due entirely to the second form factor. F_1 is called the electric and F_2 the magnetic form factor.

At this point in our studies, this is fortunate. A direct calculation of $F_1(0)$ (though not its derivatives w.r.t. q^2) would lead to infinite integrals, whose meaning I am not yet prepared to explain. We will find, however, that F_2 is completely finite and unambiguous. This can be seen by dimensional analysis. The integral defining Γ_μ contains four powers of loop momentum in the denominator. It can, at most, give rise to a logarithmic divergence. If we differentiate it w.r.t. q, we get a finite integral. Only the first term in the Taylor expansion of Γ_μ diverges. The term containing F_2 vanishes at $q = 0$ and will therefore be finite. This sort of dimensional analysis will be the key to the understanding of renormalization, which we will achieve in Chapter 9.

For those readers who wish to practice their skills by calculating the entire vertex function, I will introduce a method to render all integrals finite. It consists in changing the denominator in the photon propagator to $(p^2 - \mu^2 + i\epsilon)K(p^2/\Lambda^2)$, where K is a smooth function, which is 1 for $|p^2| < \Lambda^2$ and blows up at least exponentially rapidly above that. We will argue in Chapter 9 that, as long as Λ is much larger than the electron and photon masses and the external momenta, the answers for physical amplitudes will be independent of the detailed shape of K.

I will also employ an identity, invented by Feynman (its derivation follows in the problems), which states that

$$\frac{1}{A_1 A_2 A_3} = \int_0^1 \mathrm{d}x \, \mathrm{d}y \, \mathrm{d}z \, \delta(1 - x - y - z) \frac{2}{D^3},$$

where

$$D = xA_1 + yA_2 + zA_3.$$

For our purposes, the three A_i are the Feynman denominators, and

$$D = k^2 + 2k(yq - zp) + yq^2 + zp^2 - (x + y)m^2 + i\epsilon.$$

If we introduce $r = k + yq - zp$, then

$$D = r^2 - U + i\epsilon,$$

with

$$U = -xyq^2 + (1 - z)^2 m^2 - z\mu^2.$$

This is extremely useful, because the numerator contains terms of zeroth, first, and second order in r_μ. The linear terms vanish upon integration, by virtue of Lorentz invariance, as long as we are careful to regulate the divergences with a Lorentz-invariant prescription like DR, or our modified photon propagator. The quadratic terms are also simplified by Lorentz invariance:

$$\int d^4r \, \frac{r_\mu r_\nu}{D^3} = \frac{\eta_{\mu\nu}}{4} \int d^4r \left(\frac{1}{D^2} + \frac{U}{D^3} \right) = \int d^4r \, \frac{D+U}{D^3}.$$

Using the anti-commutation relations, we can rewrite the numerator of the diagram as

$$\text{Num} = \gamma^\lambda [\slashed{q}\gamma_\mu(\slashed{k}+m) + (m^2 - k^2)\gamma_\mu + 2k_\mu(\slashed{k}+m)]\gamma_\lambda.$$

We use the contraction identities from Appendix C,

$$\gamma_\lambda \gamma_\mu \gamma^\lambda = -2\gamma_\mu,$$

$$\gamma^\lambda \gamma_\mu \gamma_\nu \gamma_\lambda = 4\eta_{\mu\nu},$$

$$\gamma^\lambda \gamma_\mu \gamma_\nu \gamma_\alpha \gamma_\lambda = -4\eta_{\nu\alpha}\gamma_\mu + 2\gamma_\nu \gamma_\alpha \gamma_\mu,$$

and the fact that $\bar{u}(p+q)\slashed{q}u(p) = 0$, to rewrite this as

$$\text{Num} = 4mq_\mu + 4k_\mu(2m - \slashed{k}) + \gamma_\mu[\slashed{k}\slashed{q} + 2(k^2 - m^2)].$$

We substitute $k = r + zp - yq$ and use the relations for integrals over r to express this as

$$\text{Num} = \gamma_\mu \left[(1-x)(1-y)q^2 + (1 - 2z + z^2) - \frac{1}{2}(D+U) \right]$$
$$+ (2p+q)_\mu mz(z-1) + q_\mu m(z-2)(x-y).$$

We have again used the fact that the numerator is sandwiched between on-shell spinors. Note in particular that

$$\bar{u}(p+q)\gamma_\mu \slashed{q}u(p) = 2\bar{u}(p+q)(p_\mu + q_\mu - \gamma_\mu m)u(p).$$

We have also reintroduced $x = 1 - y - z$, in order to make it clear that the term proportional to q_μ is odd under interchange of x and y. It therefore vanishes when integrated. We now use the Gordon identity to find that the magnetic form factor is given by

$$\frac{ie^2}{2\pi^4}\bar{u}(p+q)\frac{-\gamma^{\mu\nu}q_\nu}{2m}u(p)\int d^4r \int_0^1 dx \, dy \, dz \, \delta(1 - x - y - z)\frac{m^2 z(1-z)}{D^3}. \quad (6.17)$$

Note that we have left off the effect of the regulator for the photon propagator, as well as the photon mass. In fact, we will find that the answer for the magnetic form factor is completely finite. In Chapter 9 we will learn the reason for the UV finiteness: expressed as corrections to the quantum Lagrangian for the Dirac field, the terms in the Taylor expansion of the magnetic form factor around $q=0$ are all operators of dimension higher than four. In a theory whose couplings have non-negative mass dimension (which we will learn to call a renormalizable theory at the Gaussian fixed point), such terms are independent of the cut-off.

The r integral is done by analytic continuation to Euclidean space, following the $i\epsilon$ prescription

$$\int d^4 r_E \frac{1}{(r_E^2 + U)^3} = \frac{1}{U}(2\pi^2) \int_0^\infty \frac{t^3 \, dt}{(t^2 + 1)^3}.$$

The factor $2\pi^2$ is the 3-volume of the unit sphere embedded in four dimensions. The Lorentzian signature integral is $-i$ times the Euclidean one. The i comes from $r^0 = i r_E^4$, and the minus sign from $r^2 - U = -(r_E^2 + U)$. The result is

$$\frac{e^2}{4\pi^2} \bar{u}(p+q) \frac{-\gamma^{\mu\nu} q_\nu}{2m} u(p) \int_0^1 dx \, dy \, dz \, \delta(1 - x - y - z) \frac{m^2 z(1-z)}{(1-z)^2 m^2 - xyq^2}. \quad (6.18)$$

At this point, we have set $\mu^2 = 0$, since we will obtain a finite result in this limit. The electric form factor has an infrared divergence when $\mu = 0$, reflecting the zero probability for scattering without any emission of massless photons, but, to this order in perturbation theory, the magnetic form factor is IR finite. For $q^2 = 0$ it is easy to do the remaining integrals. They give $F_2(0) = \alpha/(2\pi)$, where $\alpha = e^2/(4\pi)$ is the fine-structure constant. In the exercises, you will verify that $F_2(0) = (g-2)/2$, also called the anomalous magnetic momentum of the electron (g is the gyromagnetic ratio). This result was first obtained by Schwinger in 1948. Since then, experimental and theoretical determinations have competed with each other in precision, with no definite discrepancy yet having been found for the electron magnetic moment. Possible deviations for the muon moment, if they exist, are probably indications for physics beyond the standard model. We will learn more about the standard model of particle physics in Chapter 8.

6.5 Problems for Chapter 6

*6.1 Use the Feynman rules of QED to compute the amplitude for Compton scattering (scattering of a photon by a charged particle) to leading order in perturbation theory.

*6.2 Compute the scattering of an electron by a heavier spin-$\frac{1}{2}$ charged particle at leading order in perturbation theory. Show that it is related by crossing symmetry to the annihilation amplitudes discussed in the text. Compute the spin-averaged differential cross section. Take the limit where the heavy-particle mass goes to infinity and show that you obtain Rutherford's formula for scattering of a charged particle from a nucleus.

6.3 Set up an expansion for the functional determinant of $i\gamma^\mu \partial_\mu - M - V(x)$, in the limit of large mass M. $V(x)$ is a smoothly varying external field, which may be a matrix in spinor space as well as in the external index space of a collection of Weyl fields. For simplicity, insist that $V = iA_\mu(x)\gamma^\mu + i\phi(x)$, where A_μ and ϕ

are Hermitian matrices in internal space. Work in Euclidean space (although the result is also valid for Minkowski signature).

*6.4. Compute the differential cross section for electron–positron annihilation into a particle–anti-particle pair of spin-zero bosons with charge Q.

*6.5. Compute the magnetic moment for a Dirac particle, with no interaction corrections. Show that the gyromagnetic ratio is 2. Find the general relation between the gyromagnetic ratio g and the magnetic form factor $F_2(0)$.

*6.6. Let A_i, $i = 1, \ldots, n$, be positive real numbers. Start from the obvious identity (Schwinger)

$$\frac{1}{A_i} = \int_0^\infty ds_i \, e^{-s_i A_i},$$

and prove the Feynman identity used in the text by making the change of variables $s_i = x_i s$, with $\sum x_i = 1$.

*6.7. Show that the one-loop amplitude for scattering an electron in a background field contains an infrared divergence, behaving like $\ln \mu$ as the photon mass is taken to zero. Consider the inclusive cross section for scattering in the field, including the possibility of emitting a finite-energy photon of energy E. Show that, if E is kept finite as $\mu \to 0$, the inclusive cross section for either no photon or one photon emitted in the scattering process is finite. The generalization of this, to all orders in perturbation theory, is that we must include the possibility of any number of photons, of total energy less than E. Faddeev and Kulish [35] addressed the problem of defining IR finite *amplitudes* rather than just inclusive cross sections.

7 Symmetries, Ward identities, and Nambu–Goldstone bosons

Emmy Noether, one of the first great female mathematical physicists, proved a theorem central to the study of symmetries in classical and quantum physics. Noether's theorem was proved in the context of classical mechanics, but the path-integral formalism allows us to immediately generalize it to the quantum theory.

Noether's theorem applies to groups of transformations that depend on a continuous parameter, also known as Lie groups.[1] In general, the group will have a number of independent continuous parameters (e.g. the Euler angles of the rotation group), but Noether's theorem concentrates on one parameter at a time.

The classical Noether theorem states that any one-parameter group of global symmetries of the action leads to a conservation law. In order to discuss Noether's theorem in a general way we introduce a field vector Φ^i that contains all components of all possible elementary fields in our theory. Using the condensed notation of the field vector, we write the infinitesimal variation of the fields under the symmetry as $\epsilon \, \delta_G \Phi^i(x)$, where ϵ is the (x-independent) infinitesimal group parameter and $\delta_G \Phi^i$ is the variation of the field vector under the symmetry transformation. Generally it will be a function of fields and their first derivatives, for an action that depends on at most first derivatives of the field. The statement of invariance of the action is

$$\delta_G S \equiv \int \frac{\delta S}{\delta \Phi^i(x)} \, \epsilon \, \delta_G \Phi^i(x) = 0.$$

Note that this holds whether or not the fields satisfy the equations of motion, but only for constant ϵ.

If ϵ is allowed to vary over space-time, the action is not usually invariant (unless we have a gauge symmetry, but, as we shall see, that situation is completely different). For an action depending only on first derivatives of the fields, the change in the action will be a linear functional of $\partial_\mu \epsilon$:

$$\delta_G S = \int \partial_\mu \epsilon \, J_N^\mu = - \int \epsilon \, \partial_\mu J_N^\mu.$$

The first of these equalities defines the *Noether current* $J_N^\mu(x)$, while the second follows from the first if ϵ is chosen to vanish rapidly enough at infinity to justify integration by parts. The equation for variation of the action shows us that, *if the fields satisfy the*

---

[1] Before reading this section, the reader should consult Appendix E, and perhaps some of the references in [1–4].

classical equations of motion, then the variation vanishes even for variable ϵ. Comparison with the last form of the Noether formula tells us that, when the fields are on shell, the current must be conserved. This is Emmy Noether's celebrated theorem.

To derive the quantum version of this theorem in the path-integral formalism, one performs a change of variables in the numerator path integral for $Z[J]$, which has the form of a space-time dependent symmetry transformation: *viz.* $\Phi^i \rightarrow \Phi^i + \epsilon(x)\delta_G\Phi^i$. One assumes that the measure of integration $[d\Phi]$ is invariant under this field redefinition.[2] Using the fact that a change of variables does not change the value of an integral, we obtain, to first order in ϵ,

$$0 = \int [d\Phi] e^{iS + \int \Phi^i J_i} \left[\int d^4x \, \epsilon \, \partial_\mu J_N^\mu + \delta_G \Phi^i J_i \right].$$

We will generally deal with theories in which the action of the symmetry group is linear in Φ, $\delta_G\Phi^i = T_j^i\Phi^j$. The most common case of non-linear action is that of Nambu–Goldstone bosons, where the space of fields is a curved sub-manifold of a linear space on which the group acts linearly. This case can be subsumed under linear actions by coupling sources to the original linear fields and incorporating the sub-manifold constraint into the measure of the path integral. For linear action, we can write

$$\int T_j^i \Phi^j J_i = \int \Phi^i (T^{\mathrm{T}})_i^j J_j,$$

which is the effect on the generating functional of an infinitesimal linear change of variables, $\delta J = T^{\mathrm{T}}J$. If, when ϵ is constant, we can drop the term $\int \partial_\mu J^\mu$ (because it is the integral of a total divergence), then we derive the quantum version of Noether's theorem: $Z[J]$ is invariant under transformations of J inverse to those of Φ. The same is obviously true for the connected generating functional $W[J]$, while the 1PI functional $\Gamma[\Phi]$ is invariant under the original transformations on Φ. We could have derived this statement simply by making our change of variables with constant ω_a. We have carried out the more general transformation in order to understand how the derivation could fail. As we will see, the failure, when it occurs, has to do with the fact that the integral of the divergence of the current doesn't vanish.

To see how this result is used, let's study the special case of scalar fields with an $O(n)$ internal rotation symmetry. The statement that $W[J]$ is invariant under rotations of J is equivalent to the statement that the Green functions

$$\langle \phi^{a_1}(x_1) \ldots \phi^{a_k}(x_k) \rangle$$

are constructed from invariant tensors of the $O(n)$ group. These tensors are the Kronecker δ_{ab} and the Levi-Civita symbol $\epsilon_{a_1 \ldots a_n}$. In particular, the one-point function vanishes, and the two-point function satisfies

$$\langle \phi^a(x)\phi^b(y) \rangle = \delta^{ab} G(x - y).$$

---

[2] In cases where it is not, we have what is called a quantum anomaly in the symmetry, or an anomaly for short.

On applying the Lehmann spectral representation, we conclude that the particles created from the vacuum by different components of the ϕ^a field all have the same mass: we have an n-fold-degenerate particle multiplet, as a consequence of the O(n) symmetry. In the exercises, the diligent reader will employ the symmetry to constrain the properties of particle scattering amplitudes.

7.1 Space-time symmetries

The treatment of space-time symmetries is roughly similar to that of internal symmetries, but the differences are illuminating. Let us start from space-time translations. The first obvious difference is that the Lagrangian density is no longer invariant. Under translations,

$$\mathcal{L}(x) \to \mathcal{L}(x + a),$$

or, infinitesimally,

$$\delta\mathcal{L} = a^\nu \, \partial_\nu \mathcal{L}.$$

As a consequence, when we (using the field vector notation) make an infinitesimal, space-time-dependent translation $x^\mu \to x^\mu + f^\mu(x)$, we find

$$\delta S = \int \mathrm{d}^4 x \left[\frac{\partial\mathcal{L}}{\partial\Phi^i} f^\nu(x)\partial_\nu\Phi^i(x) + \frac{\partial\mathcal{L}}{\partial\partial_\mu\Phi^i} \partial^\mu(f^\nu(x)\partial_\nu\Phi^i(x)) \right].$$

Thus,

$$\delta S = \int \mathrm{d}^4 x \, \partial_\mu f_\nu \left[\frac{\partial\mathcal{L}}{\partial\partial_\mu\Phi^i} \partial_\nu\Phi^i(x) - \eta^{\mu\nu}\mathcal{L} \right],$$

$$= \int \mathrm{d}^4 x \, \partial_\mu f_\nu T^{\mu\nu}(x).$$

Thus, in classical mechanics, when the equations of motion are satisfied, the *stress-energy tensor* or *energy-momentum tensor* $T^{\mu\nu}(x)$ satisfies

$$\partial_\mu T^{\mu\nu} = 0,$$

so that

$$P^\nu = \int \mathrm{d}^3 x \, T^{0\nu}$$

is time-independent. P^ν is the energy momentum, also identified in the quantum theory as the operator that generates infinitesimal translations in space-time on the Hilbert space of states. The quantum analog of stress-energy conservation is the Ward identity

$$\partial_\mu^x \langle 0| T(T^{\mu\nu}(x)\Phi^{i_1}(x_1)\ldots\Phi^{i_n}(x_n)) |0\rangle$$
$$= -\mathrm{i}\sum \delta^4(x - x_k)\langle 0| T(\phi^{i_1}(x_1)\ldots\partial^\nu\Phi^{i_k}(x_k)\ldots\Phi^{i_n}(x_n)|0\rangle.$$

As before, this is derived by implementing a change of variables in the path integral, replacing the fields by their values after a space-time-dependent translation, and then noting that the change of variables does not change the value of the functional integral.

The fact that an infinitesimal space-time-dependent translation is actually a general infinitesimal coordinate transformation suggests that the conserved currents associated with Lorentz, or other, space-time symmetries will also be associated with the stress tensor. In fact, this is true, but only after we take into account the fact that Noether's definition of the stress tensor is somewhat ambiguous. In fact, it is easy to verify (see the problems) that the change

$$T^{\mu\nu} \to T_{\mu\nu} + \partial_\lambda M^{\mu\nu\lambda},$$

where $M^{\mu\nu\lambda} = -M^{\lambda\nu\mu}$, changes neither the fact that the stress tensor is conserved nor the value of the translation generator, P^ν.

The easiest way to resolve this ambiguity is to consider the quantum field theory we are studying on a general space-time geometry, rather than sticking to flat Minkowski space. This is of course a good thing to do, because Einstein's theory of gravitation tells us that gravitation is nothing but curved space-time geometry, with the geometry determined by the matter in it. (To quote J. A. Wheeler: "Space-time tells matter how to move, and matter tells space-time how to curve.") The basic idea of curved space-time geometry is that the Minkowski metric $\eta_{\mu\nu}$, which determines infinitesimal space-time intervals between points, is replaced by a general symmetric tensor $g_{\mu\nu}(x)$, which is an invertible matrix with signature $(+, -, -, -)$. It is easy to generalize the action for scalar and Maxwell fields simply by replacing $\eta_{\mu\nu}$ by $g_{\mu\nu}(x)$ and the volume element by

$$d^4x \to d^4x \sqrt{-g},$$

where g is the determinant of the metric. Spinor fields require a more involved discussion, which we will take up after we have understood non-abelian gauge theory and vector bundles.

The scalar Lagrangian is

$$\frac{1}{2}\sqrt{-g}(g^{\mu\nu}\,\partial_\mu\phi\,\partial_\nu\phi - 2V(\phi)),$$

while the Maxwell Lagrangian is

$$-\frac{1}{4}\sqrt{-g}(g^{\mu\lambda}g^{\nu\kappa}F_{\mu\nu}F_{\lambda\kappa}).$$

It is easy to see that, if we make a general coordinate transformation on the fields, *as well as the transformation*

$$g_{\mu\nu}(x) \to g_{\lambda\kappa}(y(x))\frac{\partial x^\lambda}{\partial y^\mu}\frac{\partial x^\kappa}{\partial y^\nu},$$

then the action is invariant.

As a consequence

$$\int d^4x \sqrt{-g}\,\delta_f g_{\mu\nu}(x)T^{\mu\nu}(x) = 0,$$

whenever the fields $\Phi^i(x)$ satisfy the classical equations of motion.[3] Here

$$T^{\mu\nu}(x) \equiv \frac{1}{\sqrt{-g}} \frac{\delta S}{\delta g_{\mu\nu}(x)},$$

and $\delta_f g_{\mu\nu}(x)$ is the variation of the metric under the infinitesimal coordinate transformation $x^\mu \to x^\mu + f^\mu(x)$. In the approximation where we neglect gravitational back reaction of the quantum fields on the metric, we are of course interested only in fixed background metrics. The W–T identities are only interesting when the coordinate transformations correspond to isometries of the metric, i.e. transformations that leave the form of the metric invariant. For Minkowski space this means Poincaré transformations. Another interesting possibility is conformal isometries, where

$$\delta_f g_{\mu\nu}(x) = s(x) g_{\mu\nu}(x).$$

These are of interest if the quantum field theory is invariant under Weyl transformations of the metric, $g_{\mu\nu}(x) \to \Omega^2(x) g_{\mu\nu}(x)$. The conformal isometry group of Minkowski space is isomorphic to SO(2,4). In addition to Poincaré transformations, it contains scale transformations $x^\mu \to \lambda x^\mu$, for positive λ, and special conformal transformations, whose infinitesimal form is $\delta x^\mu = b^\mu x^2 - 2b_\nu x^\mu x^\nu$. Conformal invariance implies that T^μ_μ, the trace of the stress tensor, vanishes. It's also obvious that conformal transformations cannot be symmetries of a theory with a discrete mass spectrum, unless all the particle masses are zero.

The stress tensor derived by variation with respect to the metric is symmetric. It is also conserved, as a consequence of translation invariance. This symmetric tensor (the Belinfante tensor) is equal to the Noether current only for translations for scalar fields. In general it differs from the Noether current by a divergence of an anti-symmetric tensor, the Noether ambiguity. Symmetry of the Belinfante tensor also implies that $\partial_\mu J^{\mu\nu\lambda} = 0$, where

$$J^{\mu\nu\lambda} \equiv x^\nu T^{\mu\lambda} - x^\lambda T^{\mu\nu}.$$

The corresponding conserved charges

$$J^{\nu\lambda} \equiv \int d^3x \, J^{0\nu\lambda}$$

are the boost and angular-momentum generators. Note that the boost generators contain explicit time dependence, so that the conservation law does not imply that they commute with the Hamiltonian. Indeed, it shouldn't because the commutation rules of the Poincaré algebra say that the commutator of a boost generator with the Hamiltonian is a spatial momentum component.

---

[3] In the quantum theory, using by now familiar manipulations, this translates into a Ward–Takahashi (W–T) identity in which the insertion of the indicated variation into a Green function gives us the variation of the Green function under the transformation.

It is easy to verify that the free massless theories with spin $\frac{1}{2}$ and spin 1 have traceless Belinfante tensor. This is not the case for spin zero, but we can write a new tensor

$$T_I^{\mu\nu} = T_B^{\mu\nu} - \frac{1}{6}(\partial^\mu \partial^\nu - \eta^{\mu\nu} \partial^2)\phi^2,$$

which is traceless. Note that this *improved* stress tensor is *not* invariant under transla- tion of the field $\phi \to \phi + c$, whereas the Belinfante tensor is.[4] There is an interesting connection between the lack of conformal invariance of the spin-zero Belinfante tensor and the theory of Nambu–Goldstone bosons (NGBs) that we will investigate in the next section. The field translation invariance of the massless free scalar is the simplest example of a spontaneously broken symmetry. We will learn that every independent one-parameter group of symmetries that does not leave the vacuum state invariant gives rise to a massless spin-zero particle, the NGB. The theory of NGBs contains a dimensionful constant, f defined by

$$\langle 0|J_\mu(0)|p\rangle = \frac{fp^\mu}{\sqrt{(2\pi)^3 2\omega_p}}.$$

In d space-time dimensions currents have mass dimension $d-1$ while states have mass dimension $(1-d)/2$. Thus f has dimensions $(d-2)/2$. The theory of NGBs is not conformally invariant, except perhaps for $d = 2$.[5] Correspondingly, the traceless stress tensor for the massless field *explicitly* breaks the field translation symmetry.

To summarize, the W–T identities for Poincaré symmetry are encoded in the con- servation and symmetry of the Belinfante stress tensor. Conformally invariant theories satisfy the additional property that the stress tensor is traceless. When we study renor- malization in a later section, we will learn that classical conformal invariance usually fails in the quantum theory. Conformal quantum field theories (CFTs) are few and far between, but they provide us with the proper definition of all quantum field theories in existence.

7.2 Spontaneously broken symmetries

The predictions of symmetries are powerful and have broad applications. Surprisingly, even more wonderful results appear when the symmetry is *broken spontaneously*. We use this phrase to describe what happens when, despite an exact symmetry of the Lagrangian, the Green functions of a theory do not obey the symmetry predictions we have just discussed. This is a phenomenon special to quantum field theory, resulting from the infinite volume of space. Of course, in the real world we do not know that the

---

[4] The diligent reader should take all the unproven statements in this section as additional exercises.

[5] In fact, the Coleman–Mermin–Wagner theorem tells us that spontaneous breakdown of compact symmetry groups does not occur in two dimensions. In particular, the abelian NGB theory is conformally invariant, and does not have spontaneous symmetry breakdown.

volume is infinite. Nonetheless, the predictions of spontaneously broken symmetry are valid with high precision even for finite volume as long as the dynamics is governed by local field theory, perhaps with an ultraviolet cut-off, and the volume is large in cut-off units.

To see how spontaneously broken symmetry can occur, we write the Noether formula for variable ϵ and take the limit of constant ϵ more carefully. Thus, by differentiating k times w.r.t. J, we obtain

$$\partial_\mu^x \langle J^\mu(x)\Phi^{i_1}(x_1)\dots\Phi^{x_k}(x_k)\rangle$$
$$= i\sum_n \delta^4(x-x_n)\langle\Phi^{i_1}(x_1)\dots T_j^{i_n}\Phi^j(x_n)\dots\Phi^k(x_k)\rangle.$$

This is called the Ward–Takahashi identity.

On integrating this w.r.t. x and dropping the integral of a total derivative, we obtain the invariance of the Green functions. The derivation can fail if it is illegitimate to drop the integrated total derivative. If we take all of the x_i to the same point y (in a way we will understand better when we discuss renormalization – this is called an operator product expansion), invariance will fail if and only if

$$\int d^4x\, \partial_\mu \langle J^\mu(x)\Pi^A(y)\rangle \neq 0,$$

for some (possibly composite) operator Π^A, which transforms in a non-trivial representation of the symmetry group. In Fourier space this is the statement that

$$p_\mu\Gamma^{\mu A}(p) = p^2\Gamma^A(p^2) \neq 0,$$

when $p \to 0$. Here we have used Lorentz invariance to write the Fourier transform of the Green function, $\Gamma^{\mu A}(p) = p^\mu\Gamma^A(p^2)$. We conclude that the two-point function has a pole, so the theory contains a massless particle that can be created from the vacuum both by the current and by the field Π^A. This is the celebrated Nambu–Goldstone boson [36–40]. It carries spin zero, because the right-hand side of the equation can be non-vanishing only if $\langle\Pi^A\rangle \neq 0$. If Lorentz invariance is not itself spontaneously broken (which doesn't happen in most field theories), then Π^A is a scalar field. Notice that this also assumes that the symmetry is an internal symmetry, which doesn't change the spin of fields. In space-time dimensions higher than 2, the only exception to this rule is supersymmetry. In that case the Nambu–Goldstone (NG) particle is a fermion, called the Goldstino.

We can obtain a more physical understanding of this result, and one that is useful in systems that are not Lorentz-invariant, by thinking about the implications of the non-zero vacuum expectation value (VEV) in the operator formalism. In operator language, a one-parameter group of symmetries is implemented by a one-parameter group of unitary operators $U(\omega) = e^{i\omega Q}$, where Q is Hermitian. A symmetry-violating VEV is possible only if the vacuum is not invariant under the symmetry transformation, which means that $Q|0\rangle \neq 0$. On the other hand, since Q commutes with the Hamiltonian, the states $U(\omega)|0\rangle$ all have zero energy.[6] Now consider instead the operator $Q_V = \int \omega(x) J^0(x)$, where $\omega(x)$ is a smooth function which vanishes outside of a volume V and is equal to a constant over most of the inside of this volume. We will consider taking the volume to infinity and letting the function $\omega(x)$ become more and more slowly varying. The commutator of the Hamiltonian with this operator is (we use current conservation)

$$[H, Q_V] \propto \int \nabla \omega \, J,$$

which can be made arbitrarily small as we take $V \to \infty$ and $\nabla \omega \to 0$. Thus we find, as a consequence of spontaneously broken continuous symmetry (a symmetry that does not preserve the ground state), a spectrum of localized excitations of arbitrarily low energy. These are the NG bosons. In a relativistic theory they must be spin-zero massless particles. The phenomenon of spontaneous symmetry breaking occurs in non-relativistic condensed-matter systems. In general these systems may not have any space-time symmetries, so all we can conclude is that the NG bosons have a dispersion relation satisfying $E(p \to 0) \to 0$. (Note that the energy gap in a non-relativistic system has nothing to do with particle masses.)

The picture to keep in mind for understanding what is going on in a theory with spontaneous symmetry breaking is shown in Figures 7.1 and 7.2. The first of these shows a ground state with spontaneously broken symmetries. The rotational symmetry of the arrows is broken because they all want to point in the same direction. Figure 7.2 shows a Nambu–Goldstone excitation of such a ground state: a long-wavelength wave in the directions of the arrows.

A final note: it is not necessary for an elementary field, which appears in the Lagrangian, to have a VEV. If there is *any* Green function of elementary fields that

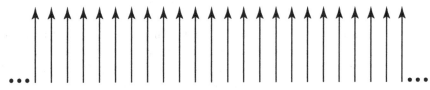

Fig. 7.1.   A state with spontaneously broken symmetry.

Fig. 7.2.   A Nambu–Goldstone excitation.

---

[6] We are ignoring mathematical rigor here. In fact the operators $U(\omega)$ are not even valid operators in the Hilbert space of our system. But the argument is valid nonetheless.

violates a symmetry, we can consider the limit where all points in this function are taken close to the same point x. As we will discuss later, in this limit the product of elementary fields has an expansion (operator product expansion or OPE) in terms of functions of coordinate differences, multiplied by local composite operators evaluated at x. For example, in free field theory

$$\phi(x)\phi(0) \sim \frac{1}{(x-y)^2} A + B : \phi^2(0) : + o(x-y).$$

If a symmetry-violating Green function is non-vanishing then some local composite operator O^A will have a non-zero VEV, and we can replay Goldstone's theorem with $S^A \to O^A$. So, like the LSZ formula, Goldstone's theorem applies to composites of Lagrangian fields, and is more general than the Lagrangian formalism.

7.3 Nambu–Goldstone bosons in the semi-classical expansion

In the semi-classical approximation, we can understand spontaneous symmetry breaking in a very direct and intuitive fashion. For simplicity, consider a simple model of a complex scalar field with a U(1) symmetry. The Lagrangian is

$$\mathcal{L} = |\partial\phi|^2 - V(\phi^*\phi).$$

The classical condition for spontaneous symmetry breaking is that the minimum of the potential (which we fix to zero by adding a constant) occurs at a non-zero value of the field. A simple example is

$$V = -\mu^2|\phi|^2 + \frac{\lambda}{2}|\phi|^4.$$

The minimum is at $|\phi|^2 = \mu^2/\lambda \equiv v^2$. There is a whole circle of minima in the ϕ plane, $\phi = ve^{i\alpha}$. When we expand the potential around any of these minima, the curvature of the potential in the circle direction vanishes, and we have a massless field.

More generally, if we introduce the vector of all real scalar fields in the theory, S^A, which transforms as $S^A \to S^A + \omega_a(T^a)^A_B S^B$, then the condition of invariance of the potential is

$$\frac{\partial V}{\partial S^B}(\omega_a T^a)^B_C S^C = 0.$$

If we differentiate this equation with respect to S^A and set $S^A = v^A$, a minimum of the potential, then we get

$$\frac{\partial^2 V}{\partial S^A \partial S^B}(\omega_a T^a)^B_C v^C = 0.$$

For every linearly independent combination of the generators T^a, that does not annihilate the VEV, we find another zero eigenvalue of the Hessian of V. But the eigenvalues of the Hessian are simply the principal curvatures of the potential, so if k generators fail to annihilate the VEV, then we have k flat directions and thus k massless particles in the expansion around the minimum.

When we expand the theory around $S^A = v^A$, the results are invariant under the original symmetry transformations, but v^A appears in all of the formulae. Thus we really retain invariance only under the subgroup H of the symmetry group G, which preserves v^A. Note that the rules for functional integration do not allow us to integrate over the constant value of $S^A - v^A$: fluctuating fields are required to fall off at infinity. The symmetry is thus spontaneously broken to H. The action of the coset G/H on the Green functions "takes us out of the Hilbert space of fluctuations of the vacuum." The variations of the fields which correspond to slowly varying transformations in G/H, which fall off at infinity, are the NG excitations.

As a final example of spontaneous symmetry breaking, let me mention QCD, with N_f massless quarks. The quark part of the Lagrangian is

$$\mathcal{L}_q = \bar{q}(i\gamma^\mu D_\mu)q.$$

For our present purposes you don't need to know anything about the QCD covariant derivative D_μ except that it is proportional to the unit matrix in Dirac spin space, and in the space of the "flavor" indices which distinguish the different quarks. It follows that the classical Lagrangian is invariant under the following $U(N_f) \times U(N_f)$ group:

$$q \rightarrow \left[U\left(\frac{1-\gamma_5}{2} \right) + V\left(\frac{1+\gamma_5}{2} \right) \right] q,$$

where U and V are arbitrary $U(N_f)$ matrices. This group is called chiral symmetry. The operator $\bar{q}q$ is invariant under the diagonal $U(N_f)$ subgroup, for which $U = V$, but transforms under elements of the coset of this subgroup. It turns out that the VEV of $\bar{q}q$ is non-zero and this part of the symmetry is spontaneously broken. It also turns out that the subgroup with $U = V^{-1} = 1$ is not a symmetry at all. The functional measure of the quark fields is not invariant under this $U_A(1)$ symmetry (this is called the chiral anomaly).

For $N_f = 2$ the Lie algebra $SU(2) \times SU(2) \sim SO(4)$. The fields $(\bar{q}q, \bar{q}\tau^a\gamma_5 q)$, where τ^a are the Pauli matrices, transform as a 4-vector under this symmetry. The VEV of $\bar{q}q$ leaves only the $SU(2)$ subgroup (isospin) invariant. There are three NG bosons, which transform as an isovector. These are the pions, and this formalism helps to explain why they are so much lighter than other hadrons. The non-zero mass of the pions is a consequence of explicit breaking of the symmetry by non-zero up and down quark masses.

7.4 Low-energy effective field theory of Nambu–Goldstone bosons

We have argued (and will argue in more detail when we discuss renormalization) that low-energy physics can always be described by an effective Lagrangian, and also that the low-energy physics of the NG bosons of a spontaneously broken symmetry is constrained by Ward identities. A shortcut to understanding what these constraints are, is to write down a Lagrangian involving the NG boson as an elementary field. If we find that, to leading order in the NG boson 4-momentum, the most general Lagrangian consistent with the symmetries contains only a few parameters, then the predictions of this Lagrangian, which follow from the symmetries, are universal.

To get an idea of how this works, let's first look at some examples. The simplest is the spontaneous breakdown of a $U(1)$ symmetry. We expect a single NG boson. Let's call its effective field G. It takes values in the manifold of vacua. The manifold of vacua is just the group $U(1)$ itself, i.e. a circle. The angle parametrizing that circle is identified with G/f. G is thus a periodic variable, with period $2\pi f$. f has dimensions of mass and is called the *decay constant* of G. The $U(1)$ symmetry is just a shift of G. The most general Lorentz-invariant, $U(1)$-invariant Lagrangian for G with the minimal number of derivatives is

$$\mathcal{L} = \frac{1}{2}(\partial_\mu G)^2. \tag{7.1}$$

The shift symmetry does not allow any non-derivative interactions and, in particular, it does not allow a mass term. Goldstone's theorem is thus incorporated automatically in this formalism. Note that the coefficient of $\frac{1}{2}$ in front of the Lagrangian is a convention. Changing it would just change the definition of f. Adding higher derivative terms would introduce new constants, but these terms are negligible at low momentum.

Now consider a fermion field ψ describing some stable massive particle in the original high-energy theory, in which symmetry breaking led to a NG boson. It can be either one of the fundamental fields of that theory, or a composite state (e.g. something cubic in the fundamental fermions). Let's suppose that, in the original theory, the interpolating field for this particle transformed chirally under the $U(1)$ symmetry: $\Psi \rightarrow e^{i\alpha\gamma_5}\Psi$. A mass term for Ψ would not be allowed by the symmetry, but in the low-energy effective theory the symmetry is broken. The symmetry transformations affect only G and no other field. Another way to say this is that we can define

$$\psi \equiv e^{-i\gamma_5 G/f}\Psi, \tag{7.2}$$

which is invariant under the symmetry. ψ can be used as the field in the low-energy effective theory, and the symmetry cannot prevent us from writing a mass term for ψ.

The most general low-energy Lagrangian coupling G to ψ is

$$\mathcal{L} = \frac{1}{2}(\partial_\mu \phi)^2 + \bar{\psi}(i\gamma^\mu\partial_\mu - m)\psi + \frac{g}{f}\partial_\mu G\bar{\psi}\gamma^\mu\gamma_5\psi. \tag{7.3}$$

I have made the assumption here that the underlying theory conserves parity symmetry, so that G is a pseudo-scalar. Note that G is always derivatively coupled: the amplitude for emission of a G particle will always be proportional to its 4-momentum. This is known as Adler's theorem [41], and is a general property of NG bosons.

A more complicated, and physically relevant, example is the breakdown of an $SU(2) \times SU(2)$ symmetry to its diagonal $SU_V(2)$ subgroup. Quantum chromodynamics (QCD), the theory of the strong interactions, contains six quark fields

q^I, $I = 1, \ldots, 6$. The quarks all have mass (with the possible and very subtle exception of the up quark), but the masses of the up and down quarks are much smaller than all the other strong-interaction energy scales. It is reasonable to imagine an approximate description in which the up and down quark masses vanish. In this limit, the classical QCD Lagrangian has a large symmetry group $U(2) \times U(2)$. If q denotes a two-dimensional vector containing the up and down quark fields, the symmetry is

$$q \to e^{i\alpha_a^L \frac{\tau_a}{4}(1-\gamma_5)} e^{i\alpha_a^R \frac{\tau_a}{4}(1+\gamma_5)} q.$$

The τ_a ($a = 0, \ldots, 3$) are the four independent two-by-two Hermitian matrices $(1, \sigma_1, \sigma_2, \sigma_3)$, and $\alpha_a^{L,R}$ are real. The transformation with $\alpha_0^L = \alpha_0^R$ (all others zero) is called baryon number. The independent axial baryon number with $\alpha_0^L = -\alpha_0^R$ is not conserved because of something called a quantum anomaly, which we will discuss later. The three-parameter subgroup of transformations with $\alpha_i^L = \alpha_i^R$, $(i = 1, \ldots, 3)$ is Wigner's isospin symmetry, from nuclear physics. We denote it by $SU_V(2)$.

There are strong theoretical arguments and a wealth of experimental data, which suggest that the VEV of $\bar{q}_i q^i$ is non-zero, spontaneously breaking the chiral symmetry group $SU(2) \times SU(2)$ to $SU_V(2)$. Let us assume that this is so. The Goldstone boson field takes values in the coset space $[SU(2) \times SU(2)]/SU_V(2)$. We can view this as the space of 2×2 unitary matrices, Σ, of determinant 1. Let $V_{L,R} = e^{i\alpha_i^{L,R} \frac{\tau_i}{2}}$. Then $\Sigma \to V_L^\dagger \Sigma V_R$ defines an action of $[SU(2) \times SU(2)]$ on the space of all Σ matrices. Furthermore, any matrix can be reached from $\Sigma = 1$ by this group action: the action is transitive. The little (stability) subgroup of a particular matrix Σ is isomorphic to that of 1, which is just the subgroup $SU_V(2)$ with $V_L = V_R$. This shows that the space of Σ matrices is just the coset $SU_L(2) \times SU_R(2)/SU_V(2)$.

The most general chirally invariant Lagrangian for Σ, with just two derivatives, is

$$\mathcal{L} = f_\pi^2 \, \text{Tr}(\partial_\mu \Sigma^\dagger \, \partial^\mu \Sigma). \tag{7.4}$$

f_π is called the pion decay constant. The pion fields are introduced by writing $\Sigma = e^{i\pi_i \frac{\tau_i}{2f_\pi}}$, where $i = 1, 2, 3$. The term quadratic in pion fields is $\frac{1}{2}(\partial_\mu \pi_i)^2$. The higher-order terms control low-energy pion–pion scattering, as well as more complicated multiparticle processes. Furthermore, when we introduce the weak interactions, the weak W bosons couple to the $SU_L(2)$ currents, which at low energy, and in the absence of heavy stable particles like the nucleon, are completely described by applying Noether's theorem to this Lagrangian. Thus f_π controls both the low-energy strong interactions of pions and the weak decays $\pi^- \to \mu + \bar{\nu}_\mu$. In addition, we can find relations between strong and weak interactions of other hadrons using only the hypothesis of spontaneously broken chiral symmetry. We will investigate some of this in the problems. The reader is urged to consult the excellent account of this subject in [42].

I now want to give a general description of the effective Lagrangian for spontaneous breakdown of a general Lie group G, with vacuum stability subgroup H. I warn the reader that this material is very general and abstract. It will be worth his/her while to work out all the details for the second example (massless QCD) described above.

The dimension of G (H) is d_G (d_H). We will let g denote a general element of G and h a general element of the subgroup. The (right) coset G/H may be described as the set of elements of G, with the equivalence relation $g \sim gh$ for any $h \in H$. It has dimension $d_G - d_H$, and this is the number of physical NGBs. We want to describe space-time fields which take values in G/H.[7] These are mappings $g(x)$ of space-time into G, with the space-time-dependent equivalence relation $g(x) \sim g(x)h(x)$. This is our first example of a *local gauge symmetry*, or gauge equivalence. We want to write Lagrangians for the field $g(x)$ that are invariant not only under the symmetry G, but also under the local transformation $g(x) \to g(x)h(x)$. The global symmetry acts by left multiplication on the group, $g(x) \to g_1 g(x)$, $g_1 \in G$.

Gauge symmetry is thus seen to be something of a misnomer. Rather than being a statement of the invariance of physics under some physical operation, it is a statement of redundancy. The coset has only $d_G - d_H$ degrees of freedom. We are going to describe it by the d_G degrees of freedom in $g(x)$ but with a gauge ambiguity, the local $h(x)$ invariance, that tells us that only $d_G - d_H$ of them are physical. The advantage of this procedure is that it allows us to avoid choosing a particular way of parametrizing the coset space. It turns out that this choice cannot be put off forever, for when we quantize the theory we have to eliminate the redundancy. However, we will see in Chapter 8 that there is an elegant method of doing this, called Becchi–Rouet–Stora–Tyutin (BRST) quantization, which is based on the classical gauge invariance we are introducing.

Now to the problem of writing down G-invariant and local H-invariant Lagrangians. There can be no terms in $\mathcal{L}$ without derivatives. Indeed, since the action of the group G on the coset space G/H is transitive, the only G-invariant function on this space is a constant. So our Lagrangian must involve derivatives of g. The derivative $\partial_\mu g$ transforms into $g_1 \partial_\mu g$ under the global transformation, but under the gauge transformation it transforms like

$$\partial_\mu g \to (\partial_\mu g + g\, \partial_\mu h\, h^{-1})h(x). \qquad (7.5)$$

This inhomogeneous transformation law is ugly and makes it hard to construct invariant Lagrangians. To fix the problem, we introduce a *gauge potential* or *H-connection* A_μ. A_μ takes values in the Lie algebra of H. It is defined to transform as

$$A_\mu(x) \to h^{-1}(x)(A_\mu(x) + \partial_\mu)h(x). \qquad (7.6)$$

We will give a geometrical explanation for this rule in Chapter 8. For the moment, it is justified by noting that the *covariant derivative*

$$D_\mu g = (\partial_\mu g - gA_\mu) \qquad (7.7)$$

transforms as

$$D_\mu g \to D_\mu gh. \qquad (7.8)$$

---

[7] Note that, in our two examples, we had an explicit representation of the distinct elements of the coset, and were able to avoid some of the machinery we are about to introduce. However, the machinery is necessary to describe the coupling of pions to nucleons in our second example.

The variables $J_\mu \equiv D_\mu g g^{-1}$ are invariant under local H transformations and transform as $J_\mu \rightarrow g_1 J_\mu g_1^{-1}$ under the global G transformation, $g \rightarrow g g_1$. We may think of them (for each space-time point and index) as elements of the Lie algebra of G, which are thus finite-dimensional matrices. There is a unique G-invariant bilinear, which is also Lorentz-invariant:

$$\mathcal{L} = f^2 \, \mathrm{Tr}(J_\mu J^\mu). \tag{7.9}$$

This defines the non-linear (σ) model on the coset space G/H. Since the multiplication law of group elements indicates that they are dimensionless, J_μ has mass dimension 1, so f must have dimensions of mass. It is called the NGB decay constant. The "σ" in *non-linear sigma model* and the "decay" in *decay constant* reflect the origins of these ideas in the history of pion physics. All relevant mass scales in NGB physics are related to f.

The Lagrangian of the non-linear model depends on two field variables, A_μ and g, but contains no derivatives of the gauge potential. Therefore, the variational equation for A_μ is purely algebraic, and we can eliminate it in terms of g. We obtain $A_\mu^m = \mathrm{Tr}[t^m g^{-1} \, \partial_\mu g]$, where t^m are the generators of H. The reader should verify that the right-hand side transforms like an H-connection under local H transformations. The Lagrangian written in terms of g is

$$\mathcal{L} = \sum_{k \in G/H} \mathrm{Tr}[t^k g^{-1} \, \partial_\mu g \, t^k g^{-1} \, \partial^\mu g].$$

It is still G-invariant and locally H-invariant. To quantize it one would have to make a choice of gauge. We will put off our discussion of quantizing gauge-invariant Lagrangians until Chapter 8.

The gauge-invariant formulation of G/H dynamics is also useful for studying the coupling of NGBs to other fields. Recalling that we are working at a scale below the scale of spontaneous breakdown of global G symmetry, we should not expect these fields to transform as representations of G, but merely of its subgroup H. If Ψ is such a field, the prescription for coupling Ψ to g is to replace all derivatives in the H-invariant Lagrangian for Ψ by gauge-covariant derivatives, using the gauge potential $A_\mu^m = \mathrm{Tr}[t^m g^{-1} \, \partial_\mu g]$. The resulting Lagrangian must be used with care. It is a good tool for calculating the emission of soft NG bosons from Ψ particles, but not, for example, for calculating the scattering of Ψ particles from each other for general kinematics. Momentum transfers must be small in order for pion exchanges to dominate nucleon scattering. In configurations for which the kinematic invariants are large compared with f_π, the low-energy effective Lagrangian does not capture all of the physics.

7.5 Problems for Chapter 7

*7.1. Use Noether's theorem to write the conserved currents associated with the O(n) or U(n) symmetry of a general invariant Lagrangrian with up to two derivatives, for scalar fields transforming in the fundamental (defining) representation of these groups. Show that the Lagrangian for massless Dirac fermions transforming in N_f copies of a representation R of any Lie group G has a $U(N_F) \times U(N_F)$ symmetry of separate unitary transformations on left- and right-handed components of the Dirac fermions. We will see that most of this symmetry remains valid when we couple the fermions to vector fields via

$$A_a^\mu \bar{\psi} \gamma_\mu t_a \psi,$$

where t_a are the representatives of infinitesimal G transformations in the R representation. A certain $U(1)$ subgroup of $U(N_F) \times U(N_F)$ is broken by a subtle quantum effect called an anomaly.

7.2. Use Noether's theorem to evaluate the conserved current corresponding to translation symmetry, for general scalar field theories coupled to the Maxwell field (couple them via the minimal substitution principle: $\partial_\mu \phi^i \to \partial_\mu - ieq_i A_\mu \phi^i$: e is the coupling constant and q_i the charge on the ith complex field). You should find the momentum is the integral of the time component of the *stress-energy tensor*:

$$P_\mu = \int d^3x \, T_{0\mu},$$

where

$$\partial^\nu T_{\nu\mu} = 0.$$

The $T_{\mu\nu}$ you find will not be symmetric. Another way to define $T_{\mu\nu}$ is to construct the action in a varying space-time metric (for the Lagrangian we are describing this is simple: simply replace the Minkowski metric $\eta_{\mu\nu}$ by $g_{\mu\nu}(x)$ everywhere, and replace $d^4x \to d^4x \sqrt{-g}$, where g is the determinant of the metric. The stress tensor is defined by

$$T_{\mu\nu}(x) = \frac{1}{\sqrt{-g}} \frac{\delta S}{\delta g^{\mu\nu}(x)},$$

and is obviously symmetric (we take $g_{\mu\nu} \to \eta_{\mu\nu}$ after taking the derivative). Show that the two definitions give the same value for P_μ as long as fields fall off sufficiently rapidly at infinity. Using the gravitational definition define

$$M_{\mu\nu\lambda} = x_\mu T_{\nu\lambda} - x_\nu T_{\mu\lambda}.$$

Show that $\partial^\lambda M_{\mu\nu\lambda} = 0$, and that $J_{\mu\nu} \equiv \int d^3x \, M_{\mu\nu 0}$ is the conserved angular-momentum generator.

*7.3. Suppose that the stress tensor is traceless:

$$\eta^{\mu\nu} T_{\mu\nu} = 0.$$

Show that there is a five-parameter set of quadratic polynomials

$$f^\mu = Cx^\mu + C^\mu_{\nu\kappa} x^\nu x^\kappa$$

(you must find the allowed form of $C^\mu_{\nu\kappa}$) such that $f^\mu T_{\mu\nu}$ is conserved. Together with the Poincaré generators, the corresponding Noether charges form the Lie algebra of the group of conformal transformations SO (2,4). Field theories invariant under these transformations are called *conformal field theories* (CFTs). In our discussion of renormalization we will see that the extreme ultraviolet and infrared behavior of any field theory is described by a pair of conformal field theories (though the IR limit may be trivial, and have only delta-function correlations). Any field theory may be constructed by perturbation theory around its UV conformal limit.

*7.4. The pion field $\pi^a(x)$ transforms as a triplet (vector) of the SU (2) isospin symmetry of strong interactions. Show that, in the limit in which we consider this symmetry to be exact, the Ward identities for Green functions, and the LSZ formula, imply that all three pions have the same mass. Now use the symmetry and the obvious invariance of the four-point function $\langle \pi^{a_1}(x_1) \dots \pi^{a_4}(x_4) \rangle$ under permutation of the pion fields to show that the 3^4 different components of this Green function are related to a much smaller number of functions of $x_1 \to x_4$. Use the LSZ formula to conclude that there are only two independent pion–pion scattering amplitudes, in the limit in which isospin is a good symmetry.

7.5. Represent the nucleon at low energy by an isospin doublet field N. This field is invariant under chiral transformations, but since it transforms under the unbroken isospin subgroup, we must write its Lagrangian with a covariant derivative:

$$\mathcal{L} = \bar{N}[i\gamma^\mu(\partial_\mu - A_\mu^a t_a) - m]N.$$

The Goldstone boson fields belong to the coset space SU(2) × SU(2)/SU$_V$(2), where the vector or diagonal subgroup is the simultaneous action of the same SU(2) transformation in both factors. In other words we have $(g_L(x), g_R(x))$, with the gauge identification $(g_L(x), g_R(x)) \sim (g_L(x)h(x), g_R(x)h(x))$. Here g_L, g_R, and h are SU(2) matrices. Show that the two fields $A_{\mu L,R} \equiv g_{L,R}^{-1}\partial_\mu g_{L,R}$ are invariant under the global symmetry and transform as gauge potentials for the gauge identification. Either one could act as the gauge potential in the covariant derivative. Use this, and the fact that strong interactions are invariant under parity (which tells you which combination of $A_{\mu L,R}$ should be substituted for A_μ^a in the nucleon covariant derivative), to write down the most general pion–nucleon Lagrangian invariant under the symmetries of QCD with two massless flavors, and containing at most one derivative of the pion field in nucleon interactions, and two in pion interactions. In addition to the nucleon mass and pion decay constant, you will have to introduce one dimensionless coupling g_A, determined by the expectation value of the axial SU(2) current in the nucleon state. This constant appears in a parity-invariant coupling of the other linear combination of $A_{\mu L,R}$ to an axial current built from nucleon fields. Use Noether's theorem to derive the SU(2)$_L$ × SU(2)$_R$ currents, recalling that in this formalism only the NGB fields $g_{L,R}$ transform under these symmetries. Now fix the gauge by carrying out a transformation with $h = g_R^{-1}$. The NGB field in this gauge is $\Sigma = g_L g_R^{-1}$ and transforms as $\Sigma \to V_L \Sigma V_R^\dagger$. Write the Lagrangian in terms of Σ (you have to be careful to transform the nucleon field by $h = g_R^{-1}$ to get this right). Show that the long-range single-pion exchange force between nuclei is determined by g_A and f_π, while pion–pion scattering depends only on f_π. Now assume that the $SU(2)_L$ current couples to W bosons. Compute the amplitudes for neutron decay and charged-pion decay in terms of the strong-interaction parameters f_π and g_A, and the Fermi constant G_F.

7.6. Write the Ward identities for the generating functional of non-abelian currents, the functional average of

$$\langle e^{i \int J_a^\mu (x) A_\mu^a (x)} \rangle \equiv e^{i W(A)}$$

for some global symmetry group G with generators T^a. Show that they are equivalent to the requirement that $W(A)$ be invariant under non-abelian gauge transformations of A. Define the non-standard Legendre transform

$$\Gamma(a_\mu^a) + M^2 (A_\mu^a - a_\mu^a)^2 = W(A_\mu^a).$$

Describe how the expansion coefficients of Γ are related to connected Green functions. Show that the Ward identities imply that Γ is a gauge-invariant functional of a_μ^a. At low energies this means that Γ is just the usual Yang–Mills action of Chapter 8. Now argue that, if G is broken to H, then we should couple the a_μ^a field to the NGBs living in the G/H coset, in a way that respects gauge invariance. Show that in the Yang–Mills approximation we get a theory of massive vector mesons interacting with massless NGBs. The vector mass matrix is not just given by M^2. There is also a contribution from the interaction with the NGBs. Compute it. Apply this formalism to the case of strong interaction chiral symmetry $G = SU(2) \times SU(2)$. It gives an approximate theory of ρ, ω, and A_1 mesons interacting with pions. The theory cannot be justified in the way that we justify the chiral Lagrangian. Even if we make the chiral symmetry exact, by setting up and down quark masses to zero, the vector mesons remain massive, with a mass not much smaller than $4\pi f_\pi$, so there is no limit in which we can exactly replace Γ by the Yang–Mills action. Nonetheless, this *vector-dominance model* of the hadronic currents is a useful approach to problems in strong-interaction physics.

8   Non-abelian gauge theory

In our discussion of the general effective field theory for NG bosons, we encountered a field theory whose target space was the coset space G/H. We realized this space in terms of fields that take values in the group manifold G, with the equivalence relation $g(x) \sim h(x)g(x)$. This is an example of a gauge equivalence, and we were forced to introduce a gauge potential $A_\mu(x)$, transforming as

$$A_\mu(x) \to h^{-1}(\partial_\mu + A_\mu)h,$$

and covariant derivatives

$$D_\mu g = (\partial_\mu - A_\mu)g,$$

in order to write down an invariant action for the NG fields.

We now want to generalize these considerations. It turns out that all non-gravitational physics at energy scales that have been explored experimentally can be described in terms of fields with a linear gauge invariance. That is, if we use the language of the field vector, $\Phi(x)$, there is a group of linear gauge equivalences, $\Phi(x) \to U(x)\Phi(x)$, where the U matrices belong to a subgroup G of the group of all unitary transformations on the field vector.

Mathematicians describe this sort of situation by saying that $\Phi(x)$ is a *section of a vector bundle over space-time* and G is the *structure group of the bundle*. For a physicist, a vector bundle is just a collection of vector spaces, one over every point in space-time, satisfying a few mathematical rules. The mathematical language is useful, because certain important non-perturbative phenomena in field theory depend in a crucial way on the mathematical theory of the topology of vector bundles.

The most important concept in the theory of vector bundles is the notion of the parallel transport matrix $U_\Gamma(x, y)$ connecting two points in space-time, along a path Γ. This is also called the Wilson line (although it was introduced into modern physics by Schwinger and Mandelstam [43–45]). One imagines that observers at different space-time points have chosen different bases in the field vector spaces, which are related by G equivalence transformations. $U_\Gamma(x, y)$ tells us how to relate the frame at x to that at y. Under a gauge transformation,

$$U_\Gamma(x, y) \to U(x)U_\Gamma(x, y)U^\dagger(y).$$

The matrix $U_P(x, x)$ for a closed path is called the *holonomy around* P *at* x. Since our matrices are always finite-dimensional, we can define the Wilson loop around P at x to be the trace of the holonomy. It is gauge-invariant.

Another way in which to make gauge-invariant objects is to construct $\Phi^\dagger(x)$ $U_\Gamma(x, y)\Phi(y)$. More generally, if the action of G on Φ is reducible, we can make a construction like this for each irreducible piece of the field vector. Similarly, we can construct Wilson loops in each irreducible representation of the gauge group G. This exhausts the collection of gauge-invariant objects in the theory, but not all of these constructions are independent. There are identities relating Wilson loops in different representations, as well as along different paths.

In field theory, the action is local, so we need to construct limits of these objects for very short paths Γ and P. This leads to the definition of covariant derivative and vector potential:

$$U_\Gamma(x, x + \mathrm{d}x)\Phi(x + \mathrm{d}x) = \Phi(x) + D_\mu\Phi(x)\mathrm{d}x^\mu \equiv \Phi(x) + (\partial_\mu - \mathrm{i}A_\mu)\Phi\,\mathrm{d}x^\mu.$$

The transformation law (7.6) of the vector potential follows from that of the Wilson line.

An infinitesimal closed loop P is characterized by the area element $\mathrm{d}x^\mu \wedge \mathrm{d}x^\nu$ that it spans. We define

$$U_P(x) = 1 + \mathrm{i}F_{\mu\nu}(x)\mathrm{d}x^\mu \wedge \mathrm{d}x^\nu.$$

Note that, according to our definitions, the 1-form A_μ and the 2-form field strength $F_{\mu\nu}$ are Hermitian matrices belonging to the Lie algebra of the group G. Under a gauge transformation, $F_{\mu\nu}(x) \to U(x)F_{\mu\nu}(x)U^\dagger(x)$.

We can construct a closed path around the area element $\mathrm{d}x^\mu \wedge \mathrm{d}x^\nu$ from four segments of open path along $\mathrm{d}x^\mu$, $\mathrm{d}x^\nu$, $-\mathrm{d}x^\mu$, and $-\mathrm{d}x^\nu$. This gives us the relation

$$[D_\mu, D_\nu]\Phi = \mathrm{i}F_{\mu\nu}\Phi,$$

which can be written

$$F_{\mu\nu} = \partial_\mu A_\nu - \partial_\nu A_\mu + \mathrm{i}[A_\mu, A_\nu].$$

The field vector, its covariant derivatives, and the field strength tensor all transform homogeneously under the gauge group. Thus it is easy to write down the most general perturbatively renormalizable action involving these fields. We break the field vector up into its scalar ϕ and spinor ψ parts[1] and write

$$\mathcal{L} = -\sum_r \left[\frac{1}{2g_r^2}\mathrm{Tr}\left(F_{\mu\nu}^{(r)}F_{(r)}^{\mu\nu}\right) + \frac{\theta_r}{32\pi^2}\mathrm{Tr}\left(F_{\mu\nu}^{(r)}F_{\alpha\beta}^{(r)}\right)\epsilon^{\mu\nu\alpha\beta}\right]$$
$$+ \psi^\dagger(\mathrm{i}\bar\sigma^\mu D_\mu)\psi + |D_\mu\phi|^2 - V(\phi) + [\psi M\psi + \text{h.c.}].$$

We have labeled the simple factors of the gauge group by the integer r; $G = G_1 \otimes \ldots$ $G_r \ldots \otimes G_K$. The field strength matrices $(F_{\mu\nu}^{(r)})^a t_a$ are in a representation of the group G_r satisfying $\mathrm{Tr}(t_a t_b) = \frac{1}{2}\delta_{ab}$. For unitary and orthogonal groups, we take it to be the fundamental representation. In other representations we have $\mathrm{Tr}_R(t_a t_b) = \frac{1}{2}D(R)\delta_{ab}$.

---

[1] The alert reader may wonder why there are no vector fields included in the field vector. The answer, which we will not have space to prove, is that renormalizability implies that the only vector fields charged under the gauge group are the Yang–Mills fields themselves.

The constant $D(R)$ is called the Dynkin index of the representation. We have chosen to write all of the fermions as left-handed Weyl fermions. When writing Feynman rules, it is best to revert to Dirac notation.

The standard model of particle physics has gauge group $G = SU(3) \times SU(2) \times U(1)$. There are three generations of left-handed Weyl fields. Each generation has the following 15 members:

- Left-handed quarks, q_L, in the $[3, 2, \frac{1}{3}]$ representation (the last number is the U(1) quantum number, called *weak hypercharge*)
- Anti-up quarks, $\bar{u}_R$, in the $[\bar{3}, 1, \frac{4}{3}]$ (the conjugates of the right-handed up quarks)
- Anti-down quarks, $\bar{d}_R$ in the $[\bar{3}, 1, \frac{2}{3}]$
- A lepton doublet, l_L in the $[1, 2, -1]$
- An anti-lepton $\bar{e}_R, \bar{\mu}_R, \bar{\tau}_R$, in the $[1, 1, 2]$ (the conjugate of the right-handed charged lepton (electron, muon, or tau) field).

There is also a single complex scalar field, the Higgs field H, in the $[1, 2, 1]$. We see that, in the standard model, all fields except neutrinos have natural right-handed partners. For the purposes of writing Feynman diagrams we introduce a Dirac neutrino field. In Dirac notation, chiral gauge couplings will have $\frac{1}{2}(1 \mp \gamma_5)$ projection factors. The right-handed component of the Dirac neutrino field will be a decoupled free field. $M(\phi)$ is the most general gauge-invariant Yukawa coupling, which might include a constant gauge-invariant fermion mass matrix, and V is the most general gauge-invariant quartic polynomial in the scalar fields. Note that we have written the scalar field Lagrangian as if all the fields were in complex representations of the gauge group. For real representations, we must multiply the scalar kinetic term by a factor of one half.

The CP-violating terms in this Lagrangian, proportional to θ_r, are all total derivatives:

$$\epsilon^{\mu\nu\alpha\beta} F^a_{\mu\nu} F^a_{\alpha\beta} = \partial_\mu K^\mu,$$

$$K^\mu = 2\epsilon^{\mu\nu\alpha\beta} \left[A^a_\nu F^a_{\alpha\beta} - \frac{1}{6} f_{abc} A^a_\nu A^b_\alpha a^c_\beta \right],$$

where

$$[t_a, t_b] = \mathrm{i} f_{abc} t_c.$$

They have no effect on perturbation theory, but do have important non-perturbative effects. Apart from these terms, CP violation could arise through the Yukawa couplings, mass terms, and scalar self-couplings of the fermions and scalars. There is an intricate interplay between gauge invariance and CP violation, because gauge invariance restricts these couplings [30]. CP is conserved when there is a field basis in which all couplings and masses are real.

8.1 The non-abelian Higgs phenomenon

In a theory with gauge group G, the scalar fields Φ will transform in a (generally reducible) representation R_S of G. In this section we will call the generators of G in this representation T^a. In later sections we will call them T^a_S. Given a point v in the (linear) space of scalars, let H be the stability subgroup of v ($hv = v$ if $h \in$ H). Its generators are t^i, which are a subset of the T^a. We can write the general scalar field as

$$\Phi(x) = \Omega(x)[v + \Delta(x)],$$

where Δ is in the subspace of field space orthogonal to all of the vectors $T^a v$. Under a gauge transformation $\Omega(x) \to V(x)\Omega(x)$, while $\Delta(x)$ is invariant.

This parametrization of field space has a gauge ambiguity $\Omega(x) \to \Omega(x)h(x)$, $\Delta(x) \to h^{-1}(x)\Delta(x)$, with $h(x) \in$ H. The new gauge group is isomorphic to H. This fact, and the close relationship of the mathematics to that of a theory with global symmetry G broken down to H by the VEV $\langle \Phi \rangle = v$, accounts for the standard terminology "spontaneously broken gauge symmetry." We say that a semi-classical expansion around the point $\Delta(x) = 0$ *spontaneously breaks the gauge group* G *to the gauge group* H. In fact what has happened is that our parametrization of the G gauge-invariant field space Δ has a *new* H gauge ambiguity.

Now define

$$B_\mu = \Omega^{-1} D_\mu \Omega \equiv a_\mu + W_\mu.$$

The Lie algebra of G is the direct sum of the Lie algebra of H and the subspace of generators orthogonal to those in H. We call these the *coset generators* for the coset G/H. The coset generators[2] form a representation of H, under the adjoint action of H on the Lie algebra of G. The second equality refers to this decomposition. a_μ involves only H generators, while W_μ is composed of coset generators. All components of B_μ are invariant under G gauge transformations. Under the new H gauge transformations a_μ transforms like a connection or gauge potential, while W_μ transforms homogeneously.

Now we can rewrite the scalar field kinetic term as

$$\frac{1}{2}(D_\mu(A)\Phi)^2 = \frac{1}{2}\Big[\Omega\Big(D_\mu(a)\Delta + W_\mu(v+\Delta)\Big)\Big]^2.$$

Using the facts that v is annihilated by all generators of H and that $\Delta^T T^a v = 0$ for all generators of G, we can rewrite this as

$$(D_\mu(A)\Phi)^2 = (D_\mu(a)\Delta)^2 + 2(W_\mu(v+\Delta))^T D_\mu(a)\Delta + (W_\mu v)^2 + (W_\mu \Delta)^2.$$

---

[2] The coset is not generally a group, but the coset generators are a linear subspace (not a subalgebra) of the Lie algebra.

On expanding around $\Delta = 0$, we find that the W_μ fields are massive vectors, with mass matrix[3]

$$(\mu_V^2)^{ab} = v^T T^a T^b v.$$

The a_μ fields do not get a mass. The fields Δ are G-gauge-invariant, and the gauge-invariant potential $V(\Phi)$, whose minima are at $\Phi = \Omega v$, will generally give mass to all of them. The Δ fields are the physical Higgs bosons.

So, given a non-abelian gauge theory with scalar fields, the space of gauge-equivalent classical vacuum states is the coset space G/H of fields $\Phi = \Omega v$. The Higgs fields Δ parametrize gauge-invariant deformations of the scalar field away from this vacuum manifold. The number of massive gauge bosons W_μ is just the dimension of G/H, and there are perturbatively massless vector potentials a_μ for the stability subgroup H of v.

8.2 BRST symmetry

It would be straightforward to derive the Feynman rules for the general gauge-invariant Lagrangian, apart from one disturbing fact. The quadratic terms in the gauge fields look like dim G copies of Maxwell's Lagrangian, and cannot be inverted to give a propagator. We have already encountered this problem in our discussion of QED, and finessed it by introducing a small photon mass. We promised a more extensive discussion of the massless limit. The time has come to fulfill that promise. The nature of this book prevents us from exploring the full extent of this subject. We will content ourselves with introducing what I will call the Becchi–Rouet–Stora–Tyutin (BRST) trick [46, 47]. The reader interested in the geometrical foundations of this trick can consult the second volume of Weinberg's monograph [42]. More detailed references may be found in [48–50]. It would be helpful at this point to do Problems 8.1–8.3. These will show you how BRST symmetry works for free-field theories.

The basic idea of the BRST trick is simple: introduce the gauge parameters in an enlarged field theory, where the gauge symmetry is an ordinary global symmetry. The tricky part is that, to make it all work, the scalar gauge parameters must be quantized as fermions. Explicitly, we introduce two scalar fermion fields $\bar{c}$ and c, and a boson field N, all transforming in the adjoint representation of the gauge symmetry. c is called the *ghost* field, $\bar{c}$ the anti-ghost, and N the Nakanishi–Lautrup Lagrange multiplier. The BRST symmetry acts as an infinitesimal gauge transformation, with gauge parameter

---

[3] Actually, we are writing these formulae for gauge potentials whose kinetic term involves a factor of the gauge coupling, so the properly normalized mass matrix includes a factor of $g^a g^b$, where g^a is the coupling of the simple factor of G to which T^a belongs.

ϵc on all of the ordinary fields in the theory (ϵ is a constant Grassmann parameter). In addition,

$$\delta_{BRST} c^a = \epsilon f_{bc}^a c^b c^c,$$

$$\delta_{BRST} \bar{c}^a = \epsilon N^a,$$

$$\delta_{BRST} N^a = 0.$$

Note that δ_{BRST}^2 applied to any of these fields vanishes, as a consequence of the Jacobi identity for the structure constants and the Fermi statistics of the c fields. It's easy to verify that the same is true for the action of δ_{BRST} on ordinary fields. The operator Q_{BRST} which implements this symmetry on the quantum Hilbert space should be nilpotent, $Q_{BRST}^2 = 0$.

We also implement a U(1) symmetry of these equations under which N^a and all ordinary fields have charge zero, while c^a and $\bar{c}^a$ have opposite charge (normalized to ± 1). The BRST charge Q_{BRST} has charge 1. This U(1) quantum number is called the *ghost number*.

The nilpotency of the BRST operator makes it very easy to add invariant ghost field terms to the Lagrangian. Simply choose a term that has the form $\delta\mathcal{L} = [Q_{BRST}, \Psi]_+$, where Ψ is a fermionic operator of ghost number -1, called the *gauge fermion*. For most purposes, the most convenient choice is $\Psi = \bar{c}^a(F^a + \frac{1}{2}\kappa N^a)$, where F^a is a function of ordinary fields. The resulting *gauge-fixing Lagrangian* has the form

$$\delta\mathcal{L} = \int d^4y\, \bar{c}^a(x) \frac{\delta F^a(x)}{\delta\omega^b(y)} c^b(y) + \frac{1}{2}\kappa(N^a)^2 + N^a F^a.$$

We can integrate out the Lagrange multiplier, and convert this to

$$\delta\mathcal{L} = \int d^4y\, \bar{c}^a(x) \frac{\delta F^a(x)}{\delta\omega^b(y)} c^b(y) - \frac{1}{2\kappa}(F^a)^2.$$

F^a should be chosen so that the ghost kinetic term is non-degenerate, if we wish to carry out a perturbation expansion. The choice $F^a = \partial_\mu A^{\mu a}$ is the most common one, because it is simple and covariant. In models in which the Higgs phenomenon occurs, a slightly more general form, involving scalar fields, is preferable.

Like any other global symmetry, the BRST symmetry will give rise to Ward identities in the quantum theory. They have the form

$$\sum_j \langle O_1(x_1) \dots [Q_{BRST}, O_j(x_j)]_\pm (-1)^{\sigma_j} \dots O_n(x_n)\rangle = 0.$$

The O_i are any operators in the theory. The peculiar features of this Ward identity come from the fact that the symmetry generator is fermionic. This accounts for the commutator/anti-commutator ambiguity, depending on the fermion parity of O_j, as well as the factor $(-1)^{\sigma_j}$ which reflects the Leibniz rule for Grassmann numbers. The most important consequence of this identity comes from the case in which O_1 is itself a BRST commutator (or anti-commutator) and the rest of the O_k are BRST-invariant. The Ward identity then implies that

$$\langle O_1(x_1) \dots O_n(x_n)\rangle = 0.$$

In particular, since the difference between any two choices of gauge-fixing Lagrangian is a BRST anti-commutator, *this identity shows that the expectation values of all BRST-invariant operators are independent of the choice of gauge-fixing Lagrangian.*

There are two obvious classes of BRST-invariant operators. The first consists of gauge-invariant functionals of the ordinary variables. The second, called BRST *trivial*, are operators of the form $[Q_{\text{BRST}}, A]_{\pm}$. It is easy to see that the two classes are disjoint. It is somewhat harder to prove [48–50] that there are no other ways of being BRST-invariant. Thus, the non-trivial BRST-invariant operators are precisely the gauge-invariant observables of the classical theory. The Ward identity shows us that their expectation values will be identical for all choices of gauge-fixing Lagrangian. This is the key result that enables us to resolve the irritating problem that, in the Higgs phase of non-abelian gauge theories, there is no choice of gauge that is simultaneously *covariant, unitary, and renormalizable.* BRST invariance guarantees the unitarity of gauge-invariant Green functions computed in a covariant gauge, because it implies that they are equal to Green functions of the same operators computed in a non-covariant, but unitary, gauge (like the axial gauge $A_3^a = 0$).

The covariant gauge Feynman rules for non-abelian gauge theory may be found in Appendix D. The problem sets will give the reader ample opportunity to practice the use of these rules. In the text we will do some perturbative calculations when we get to the discussion of the renormalization group for non-abelian gauge theory. Before entering into the intricacies of perturbative calculations, however, I want to give the reader a feeling for the physics context in which non-abelian gauge theory is important and the qualitative nature of the physical phenomena which arise from this theory.

A historical note: the ghost fields first appeared in the work of Feynman, DeWitt, and Mandelstam [51–53] and were explained in terms of a change of variables in the functional integral by Faddeev and Popov [54]. They are often called Faddeev–Popov ghosts. The BRST symmetry is the most elegant way to formulate the introduction of ghosts, and can easily be generalized to include gauge equivalences that do not form a group.

8.3 A brief history of the physics of non-abelian gauge theory

A resource for the history of the standard model is [55]. Non-abelian gauge theory is an integral part of Einstein's theory of gravitation, as was realized by Cartan [56]. Weyl [57] introduced the term gauge symmetry in an attempt to explain electromagnetism in terms of the scale ambiguity of the space-time metric (Weyl transformations). The first inkling that it had something to do with particle physics came in the Fermi theory of weak interactions.[4] Fermi postulated the existence of a charge-carrying vector

---

[4] Curiously, in 1938 Oskar Klein wrote down a Lagrangian similar to the standard electro-weak theory [58–59], but did not use his knowledge of Kaluza–Klein compactification to make it gauge-invariant.

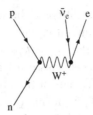

Fig. 8.1.   Fermi's theory of weak interactions.

boson W_μ^+, which could mediate the process of neutron β decay via the Feynman diagram of Figure 8.1

Fermi, the father of the idea of effective field theory, realized that, at distances long compared with the Compton wavelength of the W, this diagram would be the same as an effective four-fermion interaction:

$$L_{\text{eff}} = G_{\text{F}}(\bar{p}\gamma^\mu n \bar{\nu}\gamma_\mu e),$$

and this is what came to be known as the Fermi theory of weak interactions. Nuclear data were soon shown to be inconsistent with this simple form, and γ^μ was replaced with an arbitrary sum of Dirac matrices. It wasn't until the mid 1950s, after the ground-breaking work of Lee and Yang [60] had showed that weak interactions did not preserve parity, that Marshak and Sudarshan, Gershtein and Zel'dovitch, and Feynman and Gell-Mann [61–63] realized that, if one allowed the W boson to couple to a linear combination of vector and axial currents, then one could construct an elegant theory, which also fit the data. The stage was set for a non-abelian gauge theory of weak interactions.

In 1954, Yang and Mills [64] had introduced non-abelian gauge theory into particle physics in an attempt to account for the strongly interacting vector bosons.[5] Bludman [65] and Schwinger [66] realized that one could adapt this technology to the Fermi theory and that an SU(2) gauge theory could unify the weak and electromagnetic interactions. There were two problems with this idea. Glashow [67] showed that the weak leptonic currents did not close on an algebra with the electromagnetic current (Problem 8.14). One either had to introduce new leptons or follow the route chosen by Glashow (and the world) of introducing an SU(2) × U(1) gauge group, with the photon associated with a linear combination of one of the SU(2) generators and the U(1).

The remaining problem, that of the mass of the non-abelian bosons, was not so easily solved. Mass terms seemed to break the non-abelian gauge symmetry. Furthermore, the Euclidean propagator of a massive vector boson behaves like

$$\frac{\delta_{\mu\nu} - p_\mu p_\nu / m^2}{p^2 + m^2},$$

and the effect of the mass does not appear to vanish at high momentum.

In the meantime, developments stemming from the theory of superconductivity were leading to a solution of the problem. Following work of Schwinger on two-dimensional

---

[5] The modern approach to this idea is outlined in Problem 8.15.

quantum electrodynamics [68], Anderson [69] discussed the screening of electromagnetic interactions in condensed matter systems. Higgs showed how to incorporate the Meissner effect into a fully relativistic treatment of electrodynamics [70–72]. We will study his model in the next section. Developments followed rapidly [73–74]. Weinberg [75] and Salam [76] combined the Higgs mechanism with Glashow's SU(2) × U(1) theory to construct the modern theory of the electro-weak interaction. A few years later, 't Hooft and Veltman [77–78] proved that *spontaneously broken gauge theories*, as the new theories came to be called, were renormalizable.

I was a graduate student at the time of these developments, and it was obvious that a generalization of these ideas to the strong interactions was called for. In fact, such a generalization already existed, though it was relatively obscure, particularly in Europe. It was not the original Yang–Mills theory of vector bosons, but rather Nambu's SU(3) color gauge theory [79] invented to explain the statistics of quarks. In the quark model, baryons are bound states of three quarks, and the correct spectrum of baryons is obtained if one chooses a wave function symmetric under interchange of the three quarks. This is paradoxical because the quarks carry spin $\frac{1}{2}$. Greenberg [80] realized that the paradox could be removed if one introduced an additional three-valued label for the quarks (color) and insisted that all states be singlets under an SU(3) group transforming this label.[6] Nambu [79] invented the color gauge theory as an attempt to explain the color singlet condition dynamically. He showed that the single-gauge-boson exchange forces between quarks would lower the energy of the singlet states (just as neutral states have less electromagnetic energy than charged states). In Nambu's picture, the colored states, including quarks, would eventually be found at higher energies. It was perhaps for this reason that he also figured out a way to give the quarks integer electric charge [81], using the new degree of freedom. In the modern view of quantum chromodynamics or QCD, Nambu's calculation is just giving the short-distance part of the interquark forces. At long distances, the potential rises linearly and quarks are permanently confined. Such a potential cannot arise from particle exchange. To understand it, we will turn in the next section to the Higgs theory of relativistic superconductivity, which provides both the framework for electro-weak physics and (after a duality transformation) the explanation of quark confinement.

8.4 The Higgs model, duality, and the phases of gauge theory

Let us recall the Stueckelberg formalism for massive vector fields. We describe a massive vector B_μ as a gauge-invariant theory of a Maxwell field A_μ coupled to a scalar, with the Lagrangian

---

[6] Greenberg actually used the language of *parastatistics* rather than color. The two ideas are equivalent for classifying possible states but dynamically different. In modern language, parastatistics is the impossible limit in which the QCD scale is taken to infinity with hadron masses kept fixed.

$$\mathcal{L}_{\text{Stueck}} = -\frac{1}{4g^2}A_{\mu\nu}^2 + \frac{\mu^2}{2}(A_\mu - \partial_\mu\theta)^2.$$

Note that, unlike a conventional scalar field, θ is dimensionless.

The Higgs model is obtained by the simple observation that the Stueckelberg Lagrangian may be viewed as an approximation to a more conventional Lagrangian of a charged scalar field coupled to electromagnetism. Simply write

$$\mathcal{L}_{\text{Higgs}} = -\frac{1}{4g^2}A_{\mu\nu}^2 + \frac{1}{g^2}[|D_\mu\phi|^2 - V(\phi^*\phi)].$$

Transforming to radial coordinates for the charged field, $\phi \equiv \rho e^{i\theta}$, we obtain

$$\mathcal{L}_{\text{Higgs}} = -\frac{1}{4g^2}A_{\mu\nu}^2 + \frac{1}{g^2}[\rho^2(A_\mu - \partial_\mu\theta)^2 + (\partial_\mu\rho)^2 - V(\rho)].$$

If ρ is "frozen out" at a value $\rho^2 = \phi_0^2 = \mu^2/2$, we obtain the Stueckelberg Lagrangian. In the Higgs model, ρ fluctuates around its expectation value and quantization of the small fluctuations gives another particle, called the Higgs boson.

The renormalizable version of this Lagrangian contains two parameters, in addition to the gauge coupling, g^2. The Higgs potential is

$$V = \lambda(\phi^*\phi - \phi_0^2)^2.$$

It has a one-parameter set of minima,

$$\phi_{\text{m}} = \phi_0 e^{i\alpha},$$

which are related by gauge transformations (and thus represent a redundant parametrization of the gauge-invariant minimum $\rho = \phi_0$). Small fluctuations around a minimum, in the direction of the gauge transformation, are the degrees of freedom associated with the Stueckelberg field θ and are "eaten" by the gauge boson A_μ to form the massive field B_μ. Fluctuations perpendicular to the trough of minima are Higgs particles. The gauge-boson mass is

$$m_{\text{V}}^2 = 2\phi_0^2,$$

and the Higgs boson mass

$$m_{\text{H}}^2 = 4\lambda\phi_0^2.$$

The Stueckelberg Lagrangian is a good low-energy approximation if

$$\lambda \gg 1.$$

The quartic coupling of canonically normalized fields is $g^2\lambda/4$, so this can be true in a semi-classical regime if $g^2 \ll 1$.

In order to quantize the theory, we need to choose a gauge-fixing function F and an associated gauge fermion. The standard covariant choice $F = \partial_\mu A^\mu$ recommends itself, as does the alternative covariant choice

$$F = (\phi - \phi^*)\phi_0,$$

fixing the gauge freedom in ϕ completely.[7] This, the so-called unitary gauge,[8] is covariant and contains no unphysical degrees of freedom. The gauge potential in this gauge is actually equal to the gauge-invariant field B_μ, because the phase of ϕ is frozen. Unfortunately, in non-abelian theories, Green functions in the unitary gauge do not have a renormalizable perturbation expansion (although the scattering matrix does – see below). There is actually a convenient interpolating choice for the gauge fermion. We take

$$\Psi = \frac{1}{g^2}\bar{c}(\partial_\mu A^\mu + i\alpha\phi_0(\phi - \phi^*)).$$

This is to be used only with a weight function for the Lagrange multiplier field

$$\delta\mathcal{L} = \frac{1}{2g^2\alpha}N^2.$$

The propagators in the semi-classical expansion in this gauge contain no mixing between the scalar and gauge field for any value of α.

The Feynman rules for a general, non-abelian, Higgs model in this R_α gauge are given in Appendix D. We will not do any explicit loop computations in Higgs models in the text, but leave them for the exercises.

8.5 Confinement of monopoles in the Higgs phase

Maxwell's equations, with both electric and magnetic current sources, have the form

$$\partial^\mu F_{\mu\nu} = J_\nu,$$

$$\partial^\mu \tilde{F}_{\mu\nu} = \tilde{J}_\nu,$$

and are invariant under the duality transformation

$$F_{\mu\nu} \to \tilde{F}_{\mu\nu} = \frac{1}{2}\epsilon_{\mu\nu\lambda\kappa}F^{\lambda\kappa},$$

if we exchange the electric current, J^μ, with the magnetic one $\tilde{J}^\mu$.

To quantize the system we must make a choice. It turns out that consistency requires the coupling strengths of the two types of currents to be inversely proportional to each other. The only way we can have a controlled semi-classical expansion is to insist

---

[7] Strictly speaking $\rho(x) = (\phi + \phi^*)(x)$ is positive, but in perturbation theory it is expanded about the positive value ϕ_0, and this constraint is fulfilled automatically.

[8] Also called the unitarity gauge.

that the strongly coupled type of particle (monopoles) becomes very massive as the coupling gets strong, so that virtual loops of these particles become negligible despite the strong coupling. Indeed, naive calculations of radiative corrections to the mass of these particles suggest that it is $o(1/g^2)$, where g is the weak coupling of the electrically charged particles. In fact, as we shall see in Chapter 10, magnetic monopoles arise as classical soliton configurations, with masses of order $1/g^2$, whenever the U(1) of electromagnetism is obtained by spontaneous breakdown of a simple group. Virtual effects of these solitons are exponentially small in g, because of form factors. We will generally use the terminology "electric charges" to refer to those particles which are weakly coupled in a semi-classical expansion and "magnetic monopoles" to refer to the strongly coupled objects.

The choice we make upon quantization is to solve the magnetic Maxwell equation as

$$F_{\mu\nu}(x) = \frac{1}{2}\epsilon_{\mu\nu\lambda\kappa}\int \mathrm{d}^4 y [f^\lambda(x-y)\tilde{J}^\kappa(y)] + \partial_\nu A_\mu - \partial_\mu A_\nu,$$

where

$$\partial_\lambda f^\lambda = \delta^4(x-y).$$

We shall also insist that the support of $f^\lambda(x)$ is on a one-dimensional curve connecting the origin to infinity. f^λ is called the Dirac string. We will generally assume that the curve is in fact a straight line. Then we are constructing the magnetic monopole by viewing it as one end of an infinitely long, infinitesimally thin solenoid. Dirac argued that, classically, the solenoid is invisible, but in quantum mechanics (in anticipation of Aharonov and Bohm), charged-particle paths that encircle the solenoid would pick up a phase e^{ieg} (g is the magnetic charge), which would enable one to locate the direction of the solenoid. If that were the case, a monopole would not be a point particle. This conclusion can be avoided only if there is a universal quantization rule for the coupling strengths of electric and magnetic poles

$$eg = 2\pi N.$$

Actually, if we consider the possibility that particles carry both electric and magnetic charge (and follow Schwinger by calling such particles dyons), then the general quantization rule is

$$q_1 m_2 - q_2 m_1 = 2\pi N,$$

for some integer N. This follows because, as one particle encircles the other's solenoid, the second particle encircles the first's solenoid in the opposite direction. In this formula we have used q and m to represent the electric and magnetic charges in units of e and $1/e$, respectively, where e^2 is the semi-classical expansion parameter for Maxwell's field.

We note in passing that the modern treatment of monopole physics uses the theory of fiber bundles. Instead of introducing the singular Dirac string directly, one introduces two vector potentials, each of which describes the magnetic field of a monopole only over part of space. Each has a Dirac string singularity, but only in the part of space where it is not used. The two regions overlap, and, on the sphere at infinity, the overlap consists of a small region around some equator of the sphere. The difference between the two vector potentials is a pure gauge, but the gauge transformation has a non-trivial winding around the equator,

$$\oint dx^\mu \, \partial_\mu \omega = 2\pi m,$$

where m is the magnetic charge.

We now quantize the theory by writing the Lagrangian

$$\mathcal{L} = -\frac{1}{4e^2} F^D_{\mu\nu} F^{\mu\nu D} + A_\mu J^\mu + \frac{1}{2\alpha} (\partial_\mu A^\mu)^2.$$

We can do the path integral over A_μ, and, if we take the electric and magnetic currents to be conserved, it is independent of the choice of gauge. The result is (in Euclidean space)

$$Z[J, \tilde{J}] = e^{-\frac{e^2}{2} \int d^4x \, d^4y \, J^\mu(x) D(x-y) J_\mu(y)} e^{-\frac{2\pi^2}{e^2} \int d^4x \, d^4y \, \tilde{J}^\mu(x) D(x,y) \tilde{J}_\mu(y)}$$
$$\times e^{i \int d^4x \, d^4y \, d^4z \, \epsilon_{\mu\nu\lambda\kappa} J^\mu(x) \partial^\nu D(x-y) f^\lambda(y-z) \tilde{J}^\kappa(z)}.$$

Here $f^\lambda = n^\lambda (n^\alpha \partial_\alpha)^{-1}$, where n^α is the space-like direction of the Dirac vortex string attached to each monopole. We have imposed the minimal Dirac quantization condition and the currents are normalized to 1. When the monopole is stationary, at the origin, and we choose $n^\lambda = \hat{x}^3$, then

$$\int d^4z \, \epsilon_{\mu\nu\lambda\kappa} f^\lambda(y-z) \tilde{J}^\kappa(z) = \theta(y^3) \delta^2(\vec{y}) m \epsilon_{\mu\nu34} \delta(y^4 - \tau).$$

We have chosen the parameter along the monopole world line to be its Euclidean time coordinate. The phase is then non-vanishing for space-like charged-particle trajectories, which encircle the vortex, and is equal to e^{iqm}. This is the derivation of Dirac's quantization rule.

We conclude that a theory of light charged particles interacting with heavy magnetic monopoles makes sense, at least in the approximation that the monopoles are treated as singular classical sources. When we realize that monopoles are forced on us, as classical solitons in non-abelian theories, we will understand that this had to be the case. Those theories make it clear that a fully relativistic theory of electric charges interacting with monopoles must be consistent as well.

Now we come to an apparent conundrum. Let's consider the Higgs model, but add in a coupling to a heavy magnetic monopole. Gauss' law tells us that there must be a magnetic Coulomb field at infinity, but we have seen that there are no massless fields in the theory. How do we get a long-range field? You might have asked the same question about electric charge, but here the answer is easy. The field ϕ has a vacuum expectation value,[9] but doesn't commute with the electric-charge operator. The vacuum is therefore

---

[9] There is a subtlety here, because this field isn't gauge-invariant. Later on, when we discuss Wilson line operators, we will see how to convert this argument into one about a gauge-invariant variable.

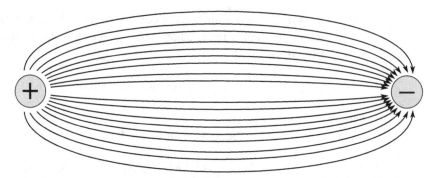

Fig. 8.2.   **A flux tube between monopoles in a superconductor.**

not an eigenstate of electric charge, but rather a superposition of different charge states. The charge fluctuations in the vacuum screen any external source of charge, so that it has no long-range field. There are no dynamical monopoles in the theory, so this cannot be the explanation of the absence of a monopole Coulomb field.

Instead, we must recognize that Gauss' law only relates the charge to the integrated magnetic flux at infinity. In ordinary empty space, the lowest-energy configuration satisfying this constraint is the rotationally invariant Coulomb field. In the Higgs vacuum, the quanta of the magnetic field are massive, and it turns out that it costs an energy per unit volume to force magnetic flux through the system. The minimal energy configuration with a single point monopole is a *flux tube* (Figure 8.2), which has energy per unit length.

To see this explicitly, let us take the monopole off to infinity, and search for the minimal energy configuration with an infinite straight flux tube and no source. The energy density for x^3 independent static configurations is

$$\mathcal{E} = \frac{1}{e^2}\left[\frac{F_{12}^2}{2} + |D_i\phi|^2 + \frac{\lambda}{2}(|\phi|^2 - \phi_0^2)^2\right]. \tag{8.1}$$

If $\lambda = 1$, we can write this as

$$\mathcal{E} = \frac{1}{e^2}\left[\frac{(F_{12} + |\phi|^2 - \phi_0^2)^2}{2} + |D_i\phi + i\epsilon_{ij}D_j\phi|^2 + \phi_0^2 F_{12} - i\epsilon_{kl}\,\partial_k(\phi^* D_l\phi)\right]. \tag{8.2}$$

Note that the last two terms are total derivatives and contribute to the total energy only through their values at infinity. We will find that $D_l\phi$ falls off exponentially, so that the contribution of these terms to the energy is just ϕ_0^2/e^2 times the total magnetic flux through the plane. We can minimize the energy by setting the two perfect squares in $\mathcal{E}$ to zero, so that this flux contribution is the total energy of the vortex. The minimum energy solutions are determined by the Bogomolnyi–Prasad–Sommerfield (BPS) [82–83] equations[10]

$$(D_i + i\epsilon_{ij}D_j)\phi = 0, \tag{8.3}$$

$$F_{12} = \phi_0^2 - \phi^2. \tag{8.4}$$

The solution with flux n is

$$\phi = e^{in\theta}f(r),$$

$$A_1 + iA_2 = -ie^{i\theta}\frac{a(r) - n}{r},$$

where

$$f' = \frac{a}{r}f,$$

$$a' = r(f^2 - \phi_0^2).$$

The boundary conditions are

$$a \to 0, \quad f \to \phi_0, \quad \text{for } r \to \infty;$$

$$a \to n + O(r^2), \quad f \to r^n(1 + O(r^2)), \quad \text{for } r \to 0.$$

It is easy to see, by linearizing the equations at infinity, that f approaches its asymptotic value exponentially. According to the second BPS equation, this means that the magnetic flux is confined to a small region around $r = 0$. ϕ_0 sets the scale for this region.

The qualitative nature of the solution is similar for other values of λ. The *Nielsen–Olesen vortex* [84] that we have constructed is the global minimum-energy configuration with fixed flux. The fact that magnetic flux penetrates the Higgs vacuum in thin tubes was first discovered, both theoretically and experimentally, in the study of superconductivity. There the expulsion of magnetic field from the superconducting ground state is called the Meissner effect, and the flux-tube solution was found by Abrikosov [85].

The significance of the Meissner effect for the physics of dynamical monopoles in the Higgs vacuum is profound. Isolated monopoles will have infinite energy. Monopole–anti-monopole pairs will experience a linear confining potential (energy per unit length for a flux tube connecting them) at large separation. They will be permanently bound into states with a characteristic size determined by the energy density of the flux tube.

At the time the Nielsen–Olesen vortex was discovered, experimental physics had revealed a puzzle about the nature of quarks. Spectroscopy had given strong indications of a quark substructure underlying hadrons. High-energy scattering experiments, both deep inelastic scattering of leptons and nucleons and electron–positron annihilation into hadrons, had given dramatic confirmation to the idea that hadrons are bound states of quarks and that quarks behave almost like free particles when they are close

---

[10] For $\lambda = 1$, the Higgs Lagrangian is the bosonic sector of a supersymmetric model, and the BPS equations are the conditions that the vortex solution preserve some of the supersymmetry.

together (asymptotic freedom). But quarks had never been seen as free particles, despite the fact that experiments were being done at energies above the characteristic scale of the strong interactions.

By 1973, the apparently free behavior of quarks in short-distance experiments had been understood in terms of quantum chromodynamics (QCD), the SU(3) color gauge theory of strong interactions. As we will learn in Chapter 9, the effective gauge coupling is scale-dependent and goes to zero at short distance. Conversely, as the separation between quarks gets larger, it grows, up until a length scale of order $(150 \text{ MeV})^{-1}$, whereupon the perturbative calculations on which this observation is based break down. Many people speculated that the coupling became infinitely strong and led to the confinement of quarks.

Actual calculations showing the confinement of quarks were first done (in four dimensions) by Wilson, in a latticized form of the gauge theory [86]. Shortly thereafter, 't Hooft [87] and Mandelstam [88] noted that one could obtain an intuitive picture of confinement by treating the Cartan subgroup of SU(3) as electromagnetism (actually two different U(1) gauge fields since the group has rank two), and postulating that the vacuum state of the theory was in a *magnetic Higgs phase*. Using electric–magnetic duality and the calculations we have done for monopoles in the Higgs phase, we see that such a state would lead to confinement of quarks by electric flux tubes.

Much work in lattice gauge theory has gone into trying to validate the 't Hooft–Mandelstam picture of confinement. The results are a trifle ambiguous. Nonetheless, this is a good qualitative picture of what is going on in QCD. The picture has been verified both in lattice models and in partially soluble supersymmetric gauge theories. We will now examine a simple lattice model, which will introduce the reader to the techniques of lattice gauge theory and exhibit the 't Hooft–Mandelstam mechanism for quark confinement in a striking manner [89–92].

In a lattice theory (we will work in Euclidean space), the functional integral is replaced by an integral over variables defined on a hypercubic (for simplicity) lattice, and derivatives are replaced by finite differences. In a *lattice gauge theory*, the fundamental variable is the parallel transporter, or Wilson line between nearest-neighbor points on a lattice. We will denote this by $U_\mu(x)$, where x is a lattice point and μ one of the eight independent directions on the lattice. It connects the points x and $x + \mu$. Under a gauge transformation,

$$U_\mu(x) \to V(x) U_\mu(x) V^\dagger(x + \mu).$$

We also insist that $[U_\mu(x)]^{-1} = U_{-\mu}(x + \mu)$. We take these variables to be matrices, which are group elements in the fundamental representation (we'll deal only with U(N) groups in this text).

Given gauge-variant fields $\phi(x)$, transforming in the representation R of the gauge group, we can write gauge-covariant finite differences

$$\phi(x) - U_\mu^{(R)}(x)\phi(x + \mu),$$

where the matrix that appears is the R representation of the group element $U_\mu(x)$. These can be combined to form gauge-invariant actions. Gauge-invariant actions constructed solely from the gauge fields are functions of traces of Wilson loops. The simplest such loop is the plaquette in the $\mu\nu$ plane,

$$U_{\mu\nu}(x) = U_\mu(x)U_\nu(x+\mu)U_{-\mu}(x+\mu+\nu)U_{-\nu}(x+\nu).$$

Since it goes around a closed curve, its trace is gauge-invariant.

All of this simplifies for abelian gauge groups, for which we can write $U_\mu(x) = e^{i\theta_\mu(x)}$ and treat θ_μ much as we would a gauge potential in the continuum. We will consider the group Z_N.

To couple a lattice gauge theory to a heavy external particle, we let $x_\mu(s)$ be the Euclidean path that the particle follows. Here s is a discrete parameter that counts the steps in the path. In order to define a gauge-invariant quantity, we insist that the path be closed. Physically this corresponds to studying a heavy charged particle by creating a particle–anti-particle pair, separating them, and bringing them back together to annihilate. The observable which computes the gauge-theory contribution to the action of this path is just the corresponding Wilson loop $W[x(s)]$. The gauge charge of the heavy particle is encoded in the choice of representation for the U_μ matrices in the Wilson loop.

To compute the potential energy of the pair a distance R apart we choose (Figure 8.3) a curve containing long straight segments along which only $x_0(s)$ changes, and the distance between the two straight segments is R.

If T is the length of the long segments and we have a linear confining potential above some distance $R_c < R$, then we expect

$$\langle W[x(s)]\rangle \to e^{-kRT},$$

as $T \to \infty$. This is a special case of a more general and more geometric formula

$$\langle W[x(s)]\rangle \to e^{-A},$$

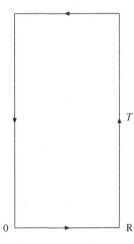

Fig. 8.3.   The Wilson loop for the static potential.

where A is the area of the minimal surface spanning the loop. Confinement is equivalent to Wilson's area law for large-area Wilson loops.

It is worth noting that, in non-abelian gauge theories, we do not expect to see confinement for heavy particles in generic representations of the gauge group. Indeed, such theories contain dynamical gauge bosons, which might be able to combine with the heavy particle to form singlet bound states. These singlets can propagate far away from each other without feeling a confining force. However, if the group is SU(N) it has a non-trivial center Z_N. The adjoint representation does not transform under the center, so it cannot screen particles in representations that do transform.[11]

Since the essence of confinement in an SU(N) theory is associated with the Z_N center, it seems worthwhile to study the pure Z_N lattice gauge theory, and that is what we will do. The variables of this theory, $\theta_\mu(x)$, are integers modulo N, one for each link on the lattice. A group element is

$$U_\mu(x) = e^{\frac{2\pi i \theta_\mu(x)}{N}}.$$

We will choose a partition function of the form

$$Z = \int \prod [d\theta_\mu(x)] \prod_{x,\mu\nu} f(\theta_{\mu\nu}(x)).$$

The "measure" $\int[d\theta]$ is just the instruction to sum each link variable over its N possible values. The gauge-invariant field strengths $\theta_{\mu\nu}$ are defined by

$$\theta_{\mu\nu}(x) = \theta_\mu(x+\nu) - \theta_\mu(x) - \theta_\nu(x+\mu) + \theta_\nu(x) \equiv \Delta_\nu\theta_\mu - \Delta_\mu\theta_\nu.$$

Like the original link variables θ_μ, these add modulo N.

We can write the most general function f of these Z_N-valued variables as a finite Fourier series

$$f(\theta) = \frac{1}{N} \sum e^{il\theta} g(l),$$

where the variable l is also Z_N-valued. If we insert the Fourier transform for each $\theta_{\mu\nu}$, and integrate by parts, we can do the θ_μ "integrals" exactly, obtaining

$$Z = \int [dl_{\mu\nu}(x)] \prod [g(l_{\mu\nu}(x))]\delta[\Delta_\nu l_{\mu\nu}(x)],$$

where we have used the Einstein summation convention for the ν index inside the delta function. The "functional delta function," $\delta[\ ]$, is the product over all links, $(x, x+\hat{\mu})$, in the lattice, of the (Z_N-valued) Kronecker delta of the variables $\Delta_\nu l_{\mu\nu}(x)$. The Wilson-loop expectation value is the ratio of two such integrals over $\theta_\mu(x)$. The argument of the functional delta function in the numerator is $\Delta_\nu l_{\mu\nu} - q w_\mu(x)$. q is the Z_N charge of the static particle, and $w_\mu(x)$ is 1 if the Wilson loop goes through the link $(x, x+\hat{\mu})$ in the positive direction, -1 if it goes through in the negative direction, and 0 if the Wilson loop does not pass through that link.

---

[11] If the theory contains dynamical particles in the fundamental representation, then any heavy source can be screened by creation of pairs of these particles. This is what happens for quarks in the real world.

To convert this to a form making its relation to electrodynamics clear, we introduce redundant variables, to write the sums over Z_N variables in terms of ordinary integer sums. Thus, we write $l_{\mu\nu} \to l_{\mu\nu} + Nq_{\mu\nu}$, where l and q now take on arbitrary integer values. Shifting l by a multiple of N can be compensated by a shift of q, so, if all functions are just functions of this combination, the double sum over integers just produces an infinite number of copies of the sum over Z_N variables. This infinity cancels out in the ratios that define expectation values. In order to preserve the Z_N character of the delta functional, we introduce an integer-valued link variable $e_\mu(x)$ and write the functional as

$$\delta[\Delta_\nu l_{\mu\nu}(x) - Ne_\mu(x)].$$

The "current" e_μ must be conserved in order to satisfy this constraint.

For convenience, we will choose the weighting function $g(l_{\mu\nu}) = e^{-g^2 l_{\mu\nu}^2}$. Other choices give similar results. When g^2 is very large, non-zero values of $l_{\mu\nu}$ are suppressed. If we introduce a Wilson loop with Z_N charge 1, this shifts the delta function in the numerator functional integral to $\delta[\Delta_\nu l_{\mu\nu} - w_\mu]$. $w_\mu = \pm q$ or zero. The delta-function constraint in the numerator now forces $l_{\mu\nu}$ to be q over some area bounded by the Wilson loop. For large g^2 we obviously want to pick the minimal area. In the denominator we simply set $l_{\mu\nu} = 0$ for large g^2. We find that the Wilson loop falls off like $e^{-g^2 A}$, where A is the minimal area it bounds.

In order to understand what happens for smaller values of g^2, we introduce the Poisson summation formula:

$$\sum_l f(l) = \sum_m \int \mathrm{d}a\, f(a) e^{2\pi ima}.$$

This formula is true because the sum over m gives rise to a sum of delta functions of a concentrated on the integers $a = l$. If $f(a) = e^{-g^2 a^2}$, we can do the Gaussian integral and obtain

$$\sum_l e^{-g^2 l^2} = \sqrt{\frac{\pi}{g^2}} \sum_m e^{\frac{-\pi^2 m^2}{g^2}}.$$

We will use this formula after solving the constraint equation for $l_{\mu\nu}$ as

$$l_{\mu\nu} = (n_\nu(n\Delta)^{-1} j_\mu - n_\mu(n\Delta)^{-1} j_\nu) + \epsilon_{\mu\nu\lambda\kappa} \Delta_\lambda l_\kappa.$$

l_μ is an integer-valued link variable, with an obvious gauge ambiguity. $j_\mu = Ne_\mu$ in the denominator functional integral and $j_\mu = Ne_\mu + qw_\mu$ in the numerator. We want to apply the Poisson summation formula to the sums over $l_\mu(x)$. In principle we must eliminate the gauge ambiguity first, and do so (for example by picking an axial gauge like $l_3 = 0$) in a way that preserves the integer nature of the variable. However, if we introduce the Poisson variables m_μ by a factor

$$e^{2\pi i \sum_x l_\mu m_\mu},$$

and insist that $\Delta_\mu m_\mu = 0$, then the answers will be independent of the choice of gauge. The result is a lattice action depending on three types of variable, e_μ, m_μ, and a_μ. The former are integer-valued conserved currents on the lattice, while the latter is a lattice version of Maxwell's field. We have

$$S = \sum_x [g^2(\Delta_\mu a_\nu - \Delta_\nu a_\mu - \epsilon_{\mu\nu\lambda\kappa}n_\lambda(n\Delta)^{-1}j_\kappa)^2 + 2\pi i m_\mu a_\mu].$$

This is a discrete analog of the action for electric and magnetic monopoles interacting via the Maxwell field. If we are interested in the small-g^2 limit, we should carry out a duality transformation and describe j_μ as an electric current and m_μ as a magnetic current. The resulting action is

$$S = \sum_x \left[\frac{1}{4g^2}(\Delta_\mu A_\nu - \Delta_\nu A_\mu - \epsilon_{\mu\nu\lambda\kappa}n_\lambda(n\Delta)^{-1}2\pi m_\kappa)^2 + i j_\mu A_\mu \right].$$

If we do the path integral over A, we get a non-local action for the currents

$$S_{NL} = \frac{g^2}{2} \sum_{x,y} [w_\mu + N e_\mu](x) D(x-y)[w_\mu + N e_\mu](y)$$

$$+ \frac{2\pi^2}{g^2} \sum_{x,y} m_\mu(x) D(x-y) m_\mu(y) + i\Phi_D,$$

where Φ_D is the Dirac phase factor we have discussed above in the continuum. $D(x-y)$ is the Green function of the four-dimensional lattice Laplacian.

We treat this expression in the following way: keep the term with $x = y$ in D as is, and rewrite the rest as a path integral over A with a modified kinetic term, whose inverse is $D(x-y) - D(0)$. The sum over e_μ and m_μ is now weighted by

$$e^{-N^2 g^2 D(0) \sum e_\mu^2(x) - \frac{4\pi^2}{g^2} D(0) \sum m_\mu^2(x)}.$$

For small $g^2 \ll 1/N^2$, the non-zero values of m_μ are highly suppressed, while the sum over e_μ is free. We do it by introducing a Lagrange multiplier to enforce the constraint $\Delta_\mu e_\mu = 0$, via a term

$$e^{i \sum \theta(\Delta_\mu e_\mu)}.$$

The sum over $e\mu$ gives an action of the form

$$S = \sum \left[F(\Delta_\mu \theta - N A_\mu) + A_\mu w_\mu + \frac{1}{4g^2}(A_{\mu\nu} - F_{\mu\nu}^D(m)) \right]^2 + (4\pi^2/g^2)m_\mu^2.$$

$A_{\mu\nu}$ is the lattice field strength and $F_{\mu\nu}^D(m)$ is the Dirac string field corresponding to the magnetic current m_μ. This looks like a Higgs model coupled to magnetic monopoles, and for small g^2 we can perform the path integral semi-classically. The result is that the magnetic-monopole loops are confined to small size and are rare. The Wilson loop just has a perimeter law.

On the other hand, for large g^2 we instead apply this treatment to the m_μ loops, getting a magnetic Higgs phase. Electric loops of e_μ are generally suppressed. However, if the Wilson loop w_μ bounds a large area, and if it has charge q (mod N) rather than charge 1, then it pays to cancel it to the smaller of q and $N - q$. We get an area law with a coefficient that is largest for $q \sim N/2$. We have already analyzed the large-g^2 regime in terms of the $l_{\mu\nu}$ variables. Now we realize that we were just describing the magnetic Higgs phenomenon. $l_{\mu\nu}$ is just the quantized electric flux which must pay an energy price per unit length to penetrate the magnetic Higgs vacuum.

Finally, when N is large, one can argue that there is a regime of g^2 where both the electric and magnetic currents, e_μ and m_μ, are suppressed, and the theory behaves like free electrodynamics. This *Coulomb phase* exists roughly when $4\pi^2 c_1 > g^2 > c_2/N^2$, where the c_i are constants of order 1. For large N, the Z_N lattice gauge model exhibits the main phases of gauge theories and the electric–magnetic duality between Higgs and confinement phases. Other possible phases exist in more complicated non-abelian theories, in which the particle whose field exhibits the Higgs mechanism at strong coupling has both electric and magnetic charge at weak coupling. These are called *oblique confinement phases*. Non-abelian theories have also been shown to exhibit conformally invariant phases that are not free-field theories. These go under the name of non-abelian Coulomb phases.

8.6 The electro-weak sector of the standard model

We now turn to the use of non-abelian gauge theory in its Higgs phase, in the theory of weak and electromagnetic interactions. Glashow's argument leads us to a model of the electromagnetic and weak interactions based on an $SU(2) \times U(1)$ gauge group, broken to $U(1)$. Salam and Weinberg introduced the simplest scalar sector which achieves this purpose, a doublet, $H(x)$ with hypercharge 1. We can write the most general field configuration as

$$H(x) = e^{i\alpha(x)} \Omega(x) \frac{1}{\sqrt{2}} \begin{pmatrix} 0 \\ v + h(x) \end{pmatrix}.$$

The four real components of H transform as a 4-vector under $SO(4) = SU(2) \times SU(2)$. The first $SU(2)$ is the gauge group, while the second is a global symmetry, broken only by the $U(1)$ gauge interactions. The Higgs potential V is a quartic polynomial

$$V = -\mu^2 H^\dagger H + \lambda (H^\dagger H)^2.$$

Note that $SU(2) \times U(1)$ gauge invariance implies that it is $SO(4)$-invariant. The minimum potential energy is at $v^2 = \mu^2/\lambda$, and the mass of the gauge-invariant excitation of h, the Higgs particle, is

$$m_H = \sqrt{\lambda/2} v.$$

If we call the gauge couplings $g_{1,2}$ then the gauge-boson mass matrix is

$$\mu_V^2 = \frac{v^2}{4} \begin{pmatrix} g_2^2 & 0 & 0 & 0 \\ 0 & g_2^2 & 0 & 0 \\ 0 & 0 & g_2^2 & -g_1 g_2 \\ 0 & 0 & -g_1 g_2 & g_1^2 \end{pmatrix}.$$

This gives

$$m_Z = \sqrt{g_1^2 + g_2^2}\, v,$$

$$m_W^\pm = g_2 v = m_Z \cos\theta_W,$$

where

$$\tan\theta_W \equiv \frac{g_1}{g_2}.$$

The latest measurements of the Weinberg angle, θ_W, from comparison of precision measurements and detailed loop calculations in the standard model, give

$$\sin^2\theta_W = 0.231\,20(15).$$

As we will learn in Chapter 9, effective Lagrangian parameters vary with energy scale and depend on the renormalization scheme chosen to define the theory. The quoted value is the value at the Z mass of the parameter defined by the *modified minimal subtraction* scheme. The measured values of the gauge-boson masses are

$$m_Z^{\text{exp}} = 91.1876(21),$$
$$m_W^{\text{exp}} = 80.403(29).$$

The tree-level relation between them is pretty well satisfied.

The origin of this relation is the SO(4) symmetry we mentioned above. One SU(2) subgroup is a gauge group. If $g_1 = 0$, then the other is a global SU(2) called custodial symmetry and would have predicted $m_W = m_Z$. It is broken by the g_1 gauge interaction. Note that the relation between m_W and m_Z depends only on the two gauge couplings. Since the couplings are small we should expect this relation to survive in the quantum theory with only small corrections. In particular, the formula for the ratio *must* approach 1 as the coupling g_1 is taken to zero.

The fourth gauge boson is the massless photon, whose coupling to particles of charge q is eq with $e = g_2 \sin\theta_W$. We will discuss the charge assignments of standard-model fermions in the section on anomalies below. Here we simply record the implied couplings to the various gauge bosons:

$$J_W^{\mu+} = \bar{u}_L^i \gamma^\mu (V_{\text{CKM}})_{ij} d_L^j + \bar{v}^i \gamma^\mu l_L^i,$$

$$J_Z^\mu = \frac{g_2}{\cos\theta_W} \bar{f}_L^M (T^3 - \sin^2\theta_W\, Q)_M \gamma^\mu f_L^M,$$

$$J_\gamma^\mu = e\bar{f}_L^M (Q)_M \gamma^\mu f_L^M.$$

We have used Dirac spinor notation, where the spinors satisfy the left-handedness condition $(1+\gamma_5)f_L^M = 0$. T^3 and Q are the diagonal weak isospin and charge matrices, and the index M runs over all the quarks and leptons. The indices i and j run from 1 to 3 and denote the three copies of the same representation of SU(1, 2, 3), which we find in nature. Note that the charged gauge-boson coupling is not diagonal in these indices. This is due to the fact that we have to diagonalize the quark mass matrix.

Fermion masses are generated by Yukawa couplings to the Higgs field. For quarks and charged leptons, the relevant terms are

$$(y_u)_{ij}\bar{u}_i H q_j + (y_d)_{ij}\bar{d}_i H^\dagger q_j + (y_l)_{ij}\bar{e}_i H^\dagger l_j + \text{h.c.}$$

The Yukawa coupling matrices are 3×3 complex matrices, because of the peculiar experimental fact that we have found three generations or families or flavors of quarks and leptons, with precisely the same couplings to the gauge group. These matrices can be brought to diagonal form, with positive elements on the diagonal, by multiplying $q, \bar{u}, \bar{d}, \bar{e}, l$ by independent unitary transformations. Note, however, that in order to diagonalize both y_u and y_d we must carry out independent unitary transformations on the two isospin components of q. Thus, the relative unitary transformation $V_u V_d^\dagger \equiv V_{\text{CKM}}$ appears, as we have shown it, in the coupling to W bosons. Note, however, the complete absence of neutral flavor-changing couplings. This is the Glashow–Iliopoulos–Maiani (GIM) mechanism [93], and it accounts for the near absence of such interactions in experiment. The mixings of neutral K, D, and B mesons are the dominant flavor-changing processes observed. At the time of writing, all such interactions in nature can be accounted for by double W-boson exchange graphs at one loop.

The matrix V_{CKM} and the quark and lepton masses are the additions to the standard-model parameter space caused by the fermionic fields. The Cabibbo–Kobayashi–Maskawa [94–95] matrix itself contains three physical mixing angles and a CP-violating phase. All experimental evidence for CP violation, with the exception of the baryon asymmetry of the Universe, can be accounted for by this phase.

In the absence of neutrino masses, there is no similar mixing matrix for leptons. Observations on neutrinos from the Sun and in the atmosphere give evidence both for neutrino masses and for mixings. Neutrino mixing goes beyond the standard model. It can be accounted for by a dimension-5 operator

$$\Delta\mathcal{L} = \frac{y_{ij}^\nu}{M_S}(Hl_i)(Hl_j),$$

where y^ν is a complex symmetric matrix. The evidence suggests that $M_S \sim 10^{14}$–10^{15} GeV, and we have some constraints on the matrix elements of y^ν.

The quark masses range between 174 GeV and 5 MeV, and charged-lepton masses run between 1.78 GeV and 0.5 MeV. The large mass ratios apparent here are among the mysteries of the standard model. Various theoretical explanations have been suggested, but none has been definitively established. The most prominent one involves the Froggatt–Nielson [96] mechanism. Various mass matrix elements are determined as powers of the expectation value of a new scalar field S divided by a new mass scale M_{FN}. One introduces a symmetry broken by the VEV of S and assigns quantum numbers to standard-model fields so that each Yukawa matrix element must be proportional to a particular power of S/M_{FN}. These ideas can also explain the peculiar *texture* of the CKM matrix: the mixing angles seem largest between generations closest in mass. Other ideas for explaining masses and mixing angles use the Kaluza–Klein picture of extra spatial dimensions.

There is a peculiar interaction between the strong CP-violating coupling $\theta_{QCD}/(32\pi^2)$ and the quark mass matrix. The argument of the determinant of $y_u y_d$ was eliminated by a U(1) rotation which has a QCD anomaly. This rotation effectively shifts the strong CP parameter. Experimentally, $\theta_{QCD} < 10^{-9}$, and its small size is one of the strangest puzzles in the standard model. If $\det(y_u y_d)$ were zero, there would be no issue, because we could use the anomaly to rotate θ_{QCD} into the Yukawa matrices and it would then vanish. But the alert reader will easily see that the non-zero pion mass shows that there is no symmetry of the standard model that could set $\det(y_u y_d) = 0$. Indeed, non-perturbative QCD physics would renormalize it away from zero even if it were zero above the scale where the QCD coupling becomes strong.[12] There are models in which extra symmetries and degrees of freedom at the TeV scale can set $\arg \det(y_u y_d)$ to zero, and in which the low-energy value of θ_{QCD} is very small [97]. However, it is unclear whether those models can explain the non-zero value of $m_u(100\,\text{MeV})$ which seems to be indicated by low-energy analysis [98].

The SU(2) × U(1) theory of electro-weak physics can account for all of the data on those interactions in a very precise way. We will leave actual calculations in this model to the exercises.

8.7 Symmetries and symmetry breaking in the strong interactions

The modern theory of the strong interactions is called quantum chromodynamics (QCD). It is an SU(3) gauge theory with N_F Dirac quark fields transforming in the triplet representation of SU(3). The quark mass matrix has the form

$$\bar{q}_R^j M_{ij} q_L^i + \text{h.c.} \tag{8.5}$$

Here $q_{L,R}^i = (1 \pm \gamma_5)q^i/2$. The mass matrix could be a general complex matrix if we do not insist on other symmetries. The kinetic terms are classically invariant under $U(N_f) \times U(N_f)$ transformations on the quarks. Using these we can make the mass matrix diagonal, with positive entries. However, as we will discuss, the $U_A(1)$ transformation

$$q^i \rightarrow e^{i\alpha\gamma_5} q^i$$

does not leave the functional measure of the quarks invariant. It changes the Euclidean action by (all traces in this section are taken in the fundamental representation of SU(3) unless otherwise noted)

$$\delta S = (2N_f)i\frac{\alpha}{32\pi^2} \int \text{tr}(F_{\mu\nu}\tilde{F}^{\mu\nu}),$$

---

[12] Refer to Chapter 9 for the notion that couplings change with energy scale.

where $\tilde{F}^{\mu\nu} = \frac{1}{2}\epsilon^{\mu\nu\lambda\kappa}F_{\lambda\kappa}$. We will see in the final chapter of this book that finite action configurations of the gauge field have $[1/(32\pi^2)] \int \mathrm{tr}(F_{\mu\nu}\tilde{F}^{\mu\nu})$, which is quantized in integer units, so unless $2N_f\alpha = 2\pi n$ for the transformation which makes the determinant of the quark mass matrix real, we should expect a term

$$\frac{\theta_{\mathrm{QCD}}}{32\pi^2} \int \mathrm{tr}(F_{\mu\nu}\tilde{F}^{\mu\nu})$$

in the QCD action. θ could be any number between 0 and 2π, but a dimensional-analysis estimate of the neutron electric dipole moment suggests that $\theta < 10^{-9}$. There have been several proposed explanations of why θ_{QCD} should be so small, none of which is completely satisfactory. This is called *the strong CP problem*.

Apart from the mysterious parameter θ_{QCD}, QCD appears to have a dimensionless coupling, which appears in

$$\frac{1}{4g^2} \mathrm{tr}(F_{\mu\nu}F^{\mu\nu}).$$

However, in the course of renormalization of the theory, this will be replaced by a scale Λ_{QCD}.[13] Experiments on high-energy scattering suggest that this scale is of order 150 MeV. We have already alluded to the fact that the properties of pions, the lightest hadron, can be best understood if we assume that two of the quark masses, m_u and m_d, are much smaller than Λ_{QCD}. Data on K and η mesons are also compatible with the idea that the strange-quark mass can be treated as small. In the limit in which we set all of these masses to zero, QCD has an SU(3) × SU(3) chiral symmetry. Much of low-energy hadron physics can be understood if we assume that this symmetry is broken spontaneously to the diagonal vector subgroup SU(3), with the explicit breaking of the latter symmetry by quark masses treated in first-order perturbation theory.[14] The light pseudo-scalar mesons are interpreted as the pseudo-Nambu–Goldstone bosons of this broken symmetry.

Our purpose in this section is to outline some of the theoretical arguments which support this conclusion. We will assume that quark confinement occurs in QCD. The arguments of the last section gave us a plausible picture of why this is so. More compelling is the experimental fact that, although we can see evidence for the quark structure of hadrons in experiments at very high energy and momentum transfer, and in the structure of the hadron spectrum itself, no experiment has ever produced an isolated quark.

The first part of our argument is that the vector SU(3) currents do not suffer spontaneous breakdown. To see this, note that if we gave the up, down, and strange quarks equal mass, m, SU(3) would still be a good symmetry. We can then ask whether this

[13] This mysterious-sounding process is called *dimensional transmutation*. It is a consequence of the fact that the dimensions of operators in quantum field theory receive quantum corrections, as we will see in Chapter 9.

[14] One also has to take into account breaking of the isospin subgroup, SU(2), by electromagnetism.

symmetry is spontaneously broken, that is, whether there could be a NG boson pole in the two-point function

$$\int d^4 x \, e^{ipx} \langle J_A^\mu(x) J_B^\nu(0) \rangle.$$

In the functional integral formalism this is calculated as the average of

$$\mathrm{Tr}(\gamma^\mu T_A S(x, 0) \gamma^\nu T_B S(0, x)),$$

over gauge field configurations. $S(x, 0)$ is the quark propagator in an external gluon field. The measure of integration over gluons is

$$e^{-\frac{1}{4g^2} \int \mathrm{tr} \, F^2} \det(i\gamma^\mu D_\mu + im),$$

which is positive (Problem 8.12). Note that we have assumed $\theta_{\mathrm{QCD}} = 0$. The result we will obtain is true for at least a small range of θ_{QCD} around 0.

The heart of the argument, which is due to Vafa and Witten [99], is that $S(x, 0)$ falls off exponentially at large x, with a bound on the exponent that is independent of the gauge configuration. This is obvious in perturbation theory around vanishing gauge field and for gauge fields that fall off rapidly at infinity. Vafa and Witten proved that it is true in general. It follows from general properties of probability measures that the averaged two-point function also falls exponentially, so the two-point function cannot have a pole at zero momentum. Thus, the vector SU(3) is not spontaneously broken for any finite m. But this means that it cannot be broken for $m = 0$, because turning on m does not break the symmetry, so NG bosons present at $m = 0$ would persist for finite m. To prove that the axial symmetries *are* spontaneously broken for $m = 0$, we will have to make a short detour through the subject of *anomalies*.

8.8 Anomalies

There are a variety of examples in quantum field theory where classical symmetries or gauge equivalences do not survive quantization [100–106]. In four-dimensional space-time, this happens only for symmetries acting on chiral fermions.[15] Therefore, consider M Weyl fermions, coupled to a U(M) gauge potential

$$\mathcal{L} = \psi^\dagger i \bar{\sigma}^\mu D_\mu \psi \equiv \psi^\dagger i \bar{\sigma}^\mu (\partial_\mu - A_\mu) \psi.$$

Here $A_\mu = i A_\mu^a \lambda_a$, where the λ_a are the Gell-Mann basis for all Hermitian $M \times M$ matrices. At this point we will not specify which of the A_μ^a are dynamical (that is, are

---

[15] And to conformal symmetry. The breaking of classical conformal invariance is part of the theory of renormalization. In the language we will learn in Chapter 9, most marginal operators in field theory are marginal only to leading order. Conformally invariant quantum field theories typically occur only for isolated values of the couplings, called fixed points of the renormalization group.

to be integrated over in the functional integral) and which are just external sources, which can be used to generate Green functions containing global symmetry currents.

The Euclidean Weyl matrices $\sigma^\mu = (\mathrm{i}, \sigma)$ map left-handed into right-handed fermions. Note that in Euclidean space the rotation group is $SU(2) \times SU(2)$ (rather than the Lorentzian $SL(2, C)$), so the $(1, 2)$ and $(2, 1)$ representations are not related by complex conjugation. $\bar{\psi}$ is an independent right-handed Euclidean Weyl field. The Weyl matrices satisfy

$$
\begin{aligned}
\sigma_\mu \sigma_\nu^\dagger &= \delta_{\mu\nu} + \mathrm{i}[\delta_{\mu 0}\delta_{\nu i} - \delta_{\nu 0}\delta_{\mu i} + \epsilon_{\mu\nu i}]\sigma^i \\
&\equiv \delta_{\mu\nu} + \mathrm{i}\eta^{\mathrm{R}}_{\mu\nu i}\sigma^i, \\
\sigma_\mu^\dagger \sigma_\nu &= \delta_{\mu\nu} + \mathrm{i}[-\delta_{\mu 0}\delta_{\nu i} + \delta_{\nu 0}\delta_{\mu i} - \epsilon_{\mu\nu i}]\sigma^i \\
&\equiv \delta_{\mu\nu} + \mathrm{i}\eta^{\mathrm{L}}_{\mu\nu i}\sigma^i.
\end{aligned}
$$

The three-dimensional Levi-Civita symbol $\epsilon_{\mu\nu i}$ is defined to be zero if any of its indices is 0. The 't Hooft symbols $\eta^{\mathrm{L,R}}$ form a basis for the space of (anti) self-dual tensors. The operators $(\sigma D)(\sigma^\dagger D) = \Delta_{\mathrm{R}}$ and $(\sigma^\dagger D)(\sigma D) = \Delta_{\mathrm{L}}$ map the spaces of right- and left-handed fermions into themselves. Their determinants are defined in a finite and gauge-invariant manner by the formula

$$
\ln \mathrm{Det}\left(\frac{\Delta_{\mathrm{L,R}}}{\Delta^0_{\mathrm{L,R}}}\right) = \int_{s_0}^\infty \frac{\mathrm{d}s}{s}\, \mathrm{Tr}(\mathrm{e}^{s\Delta_{\mathrm{L,R}}} - \mathrm{e}^{s\Delta^0_{\mathrm{L,R}}}).
$$

$\Delta^0_{\mathrm{L,R}}$ are the values of the operators at vanishing A_μ. The trace includes a sum over spin indices, internal symmetry indices, and an integral over Euclidean space-time. These formulae are valid if neither of the operators has a normalizable zero mode. In that case the two operators are negative definite and the integrals converge at their upper ends. When zero modes exist, the formula is valid for the restrictions of $\Delta_{\mathrm{L,R}}$ to their non-zero eigenspaces. The determinants are given by the formal limit $s_0 \to 0$. There are divergences in this limit coming from the small-s singularities of

$$
\langle x|\mathrm{e}^{s\Delta_{\mathrm{L,R}}}|x\rangle.
$$

These are all concentrated in an infinitesimal neighborhood of the point x and can be absorbed into divergent coefficients times local functions of A_μ and its derivatives (this contributes to one-loop renormalization). By construction, everything is gauge-invariant. This tells us that any violation of gauge invariance in the determinant is due to the unitary operator which maps between left- and right-handed Hilbert spaces and is therefore a pure phase. If we have a parity-invariant gauge theory, that is we gauge only a subgroup of $U(M)$ under which the M representation is real, then this phase cancels out, and the determinant is positive and gauge-invariant.

The possible violations of local gauge invariance are restricted to an infinitesimal neighborhood of a point x, since we can choose gauge transformations that differ from the identity only in the neighborhood of x. Thus

$$
\delta \ln \mathrm{Det}[\mathrm{i}\sigma^\mu D_\mu] = \int \mathrm{d}^4 x\, \delta A_\mu^a(x) \langle J^{\mu a}(x)\rangle = D_\mu^{ab}\langle J^{\mu b}(x)\rangle.
$$

$J^{\mu a} \equiv i\bar{\psi}\sigma^\mu(\lambda^a/2)\psi$. Gauge invariance is equivalent to covariant conservation of the current induced in the vacuum by the background gauge potential. We must expect that any answer we get is independent of any cut-off parameters like s_0, which we introduce to make the calculation of the determinant finite, in the limit that $s_0 \to 0$. In this conclusion we anticipate the result of renormalization theory: *any divergence in renormalizable local field theory can be eliminated by changing a finite number of parameters in the Lagrangian and redefining operators by a finite linear redefinition.* Dimensional analysis, and the requirement that the answer be odd under parity, then restricts us to

$$D_\mu^{ab}\langle J^{\mu b}(x)\rangle = \epsilon^{\mu\nu\alpha\beta}\mathcal{A}(M)(d^{abc}\ \partial_\mu A_\nu^b\ \partial_\alpha A_\beta^c + g^{abcd}\ \partial_\mu A_\nu^b A_\alpha^c A_\beta^d$$
$$+ h^{abcde}A_\mu^b A_\nu^c A_\alpha^d A_\beta^e).$$

The coefficients must be group-covariant numerical tensors. In fact, d^{abc} must be proportional to the unique invariant in the symmetric product of two adjoint representations of $U(M)$. This is defined by

$$\lambda^a\lambda^b = (if^{abc} + d^{abc})\lambda^c,$$

where f is the totally anti-symmetric structure constant and d is totally symmetric. The equality follows from the fact that the λ^a are a basis for all Hermitian matrices and the symmetry properties from cyclicity of the trace. We have defined $\mathcal{A}(M)$ so that d^{abc} are precisely these coefficients.

If we require the non-conservation equation to be covariant under gauge transformations, we get the unique answer

$$A^a(x) \equiv D_\mu^{ab}\langle J^{\mu b}(x)\rangle = \epsilon^{\mu\nu\alpha\beta}\mathcal{A}(M)d^{abc}G_{\mu\nu}^b G_{\alpha\beta}^c.$$

However, in the non-abelian case, this covariance argument is inconsistent with an even more basic requirement. The functional differential operators

$$G^a(x) = \frac{1}{i}D_\mu^{ab}\frac{\delta}{\delta A_\mu^b(x)}$$

satisfy the algebra

$$[G^a(x), G^b(y)] = if^{abc}G^c(x)\delta^4(x-y).$$

This leads to the Wess–Zumino (WZ) consistency condition [106]

$$G^a(x)A^b(y) - G^b(y)A^a(x) = if^{abc}A^c(x)\delta^4(x-y).$$

The covariant formula doesn't satisfy this condition. However, it can be shown that the difference between the covariant formula and the consistent one we will derive below is the variation of a local functional. Thus, the ambiguities of renormalization can turn one into the other. By contrast, we will see that no such local redefinition of the action can set $A^a = 0$.

The simplest derivation of the anomaly equation starts by regulating the formula for $D_\mu^{ab}\langle J^{\mu b}\rangle$. We use fermion functional integration to write

$$\int d^4x \, D_\mu^{ab} \langle J^{\mu b}(x) \rangle = i \int d^4x \, \mathrm{tr}\left(\frac{\lambda_a}{2} \langle x| (\sigma^\mu D_\mu)^{-1} \sigma_\mu [D^\mu, \omega]|x \rangle \right).$$

Here ω is the space-time-dependent, Hermitian-matrix-valued, infinitesimal gauge parameter. The commutator is taken in the sense of differential operators and functions, as well as in the sense of matrices. The lower-case tr means trace only over spin and gauge indices.

We now use two formulae for $(\sigma^\mu D_\mu)^{-1}$:

$$(\sigma^\mu D_\mu)^{-1} = \Delta_R^{-1} \sigma_\mu^\dagger D^\mu = -\int_{s_0}^\infty ds \, e^{s\Delta_R} \sigma_\mu^\dagger D^\mu,$$

$$(\sigma^\mu D_\mu)^{-1} = \sigma_\mu^\dagger D^\mu \Delta_L^{-1} = -\sigma_\mu^\dagger D^\mu \int_{s_0}^\infty ds \, e^{s\Delta_L}.$$

We use these two forms in the two different terms of the commutator. With s_0 finite, we are perfectly justified in using cyclicity of the trace to write

$$\int d^4x \, D_\mu^{ab} \langle J^{\mu b}(x) \rangle = -i \int d^4x \, \mathrm{tr}\left(\frac{\lambda_a}{2} \int_{s_0}^\infty \langle x|e^{s\Delta_R} \Delta_R \omega - \Delta_L e^{s\Delta_L} \omega|x \rangle \right).$$

We note that $\Delta_L e^{s\Delta_L} = (d/ds)e^{s\Delta_L}$ and $e^{s\Delta_R}\Delta_R = (d/ds)e^{s\Delta_R}$, which enables us to do the s integral:

$$\int d^4x \, D_\mu^{ab} \langle J^{\mu b}(x) \rangle = i \int d^4x \, \mathrm{tr}\left(\frac{\lambda_a}{2} \langle x|e^{s_0\Delta_R} - e^{s_0\Delta_L}|x \rangle \right).$$

Now we recall the Baker–Campbell–Hausdorff formula

$$e^{s(A+B)} = e^{sA} e^{sB} e^{\frac{s^2}{2}[A,B]} \prod_{n\geq 3} e^{s^n C_n},$$

which is valid for any two operators A and B. The operators C_n are sums of n-fold multiple commutators of A and B. In our present computation, we take $A = D^2$ and $B_{L,R} = \eta_{L,R}^{\mu\nu a}(\sigma_a/2)F_{\mu\nu}$. Note that the spin traces of $B_{L,R}$ are both zero, as are the traces of the commutators $[A, B_{L,R}]$. In our computation we are taking the small-s_0 limit of the difference between the exponentials of Δ_L and Δ_R. Taking the trace properties into account, we find that the leading term is

$$\int d^4x \, D_\mu^{ab} \langle J^{\mu b}(x) \rangle = \frac{is_0^2}{2} \int d^4x \, \mathrm{tr}\left(\frac{\lambda_a}{2} \langle x|e^{s_0 D^2}|x \rangle (B_R^2 - B_L^2)(x)\omega(x) \right).$$

Any surviving contribution must come from the singularity of $\langle x|e^{s_0 D^2}|x \rangle$, as $s_0 \to 0$. By using the BCH formula to expand around the operator with zero gauge potential, we can see that the leading singularity comes from the substitution $D^2 \to \partial^2$, whence

$$\langle x | e^{s\, \partial^2} | x \rangle = \int \frac{\mathrm{d}^4 p}{(2\pi)^4} e^{-s_0 p^2} = \frac{1}{16\pi^2 s_0^2}.$$

The limit $s_0 \to 0$ is thus finite and gives

$$\int \mathrm{d}^4 x \, D_\mu^{ab} \langle J^{\mu b}(x) \rangle = \frac{\mathrm{i}}{32\pi^2} \int \mathrm{d}^4 x \, \mathrm{tr}\left(\frac{\lambda_a}{2} (B_R^2 - B_L^2)(x)\omega(x) \right).$$

This has the gauge-covariant form we anticipated, and is odd under parity. On doing all the traces we find

$$\int \mathrm{d}^4 x \, D_\mu^{ab} \langle J^{\mu b}(x) \rangle = -\frac{\mathrm{i}}{32\pi^2} d^{abc} \int \mathrm{d}^4 x \, \epsilon^{\mu\nu\alpha\beta} G_{\mu\nu}^b G_{\alpha\beta}^c.$$

Here we have normalized all of the λ^a matrices by $\mathrm{tr}(\lambda^a \lambda^b) = 2\delta^{ab}$, and defined $\lambda^a \lambda^b = d^{abc}\lambda^c + \mathrm{i}f^{abc}\lambda^c$. In particular, $d^{ab0} = 1$, and the generator of U(1), $\lambda^0/2 = I/M$. The individual fermions are often chosen to have U(1) charge 1, in which case the U(1) SU$(M)^2$ anomaly equation is multiplied by a factor of M.

Another formula for d^{abc} is $d^{abc} = 2\,\mathrm{tr}(T^a[T^b, T^c]_+)$. This formula is important because we often gauge a subgroup G of U(M), under which the M-dimensional representation breaks up into a direct sum of complicated representations of G. The anomaly for G will always be proportional to this trace. However, the normalizations of generators need not be the same as those we use here. A fairly standard convention in physics is to always normalize the generators in the lowest-dimensional representation of G by $\mathrm{tr}(T^a T^b) = \frac{1}{2}\delta^{ab}$. This defines the normalization of the structure constants. In any other representation $\mathrm{tr}_R(T^a T^b) = D(R)\delta^{ab}$, where $D(R)$ is called the *Dynkin Index* of the representation. Similarly, the symmetrized cubic trace which appears in any representation is proportional to the same invariant tensor (the analog of d^{abc} for U(M)) with a coefficient called $A(R)$, the *anomaly of the representation*. Since $d^{aaa} \neq 0$ for all groups with anomalies and some values of a, and the structure of d^{abc} is representation-independent, anomaly coefficients can be computed for any representation by finding a single generator such that $\mathrm{tr}\, T^3 \neq 0$ and being careful about normalization conventions.

The Hermitian generators T_a of a Lie algebra in a representation R are generally complex matrices, and the matrices $(-T_a^*)$ give another unitary representation called the conjugate representation $\bar{R}$. It may happen that these two representations are equivalent, i.e. $(-T_a^*) = UT_aU^\dagger$, for a unitary matrix U. The product of a representation and its conjugate always contains the trivial representation. If R is unitarily equivalent to $\bar{R}$ this means that there is an invariant product on R itself, $F_i^* F_i = F_i (U^\dagger)^{ij} F_j$. If this product is symmetric, the representation is called real. There is a basis in which all of the T_a are imaginary and anti-symmetric. If it is anti-symmetric (symplectic) then the representation is called *pseudo-real*. In either case it is easy to see that the anomaly vanishes. The anomaly coefficients are real, because the generators are Hermitian. Thus $\mathcal{A}(R) = -\mathcal{A}(\bar{R})$. However, the coefficients are invariant under unitary equivalence, so $\mathcal{A}(R) = \mathcal{A}(\bar{R})$ for real or pseudo-real representations.

A simple example of a pseudo-real representation is the doublet of SU(2). We have $\sigma^* = -\sigma_2 \sigma \sigma_2$. Since there is exactly one SU(2) representation of each integer dimension, we see that all representations of SU(2) are either real or pseudo-real. There are no anomalies in pure SU(2) gauge theories.[16] More generally, the anomalies vanish in any representation of O(n) with $n \neq 6$. The argument is simple. The anomaly coefficient

$$\mathrm{tr}(T^{mn}[T^{kl}, T^{ij}]_+)$$

must be a numerical tensor that is anti-symmetric in each pair of indices and symmetric under interchange of the pairs. All numerical tensors in SO(n) can be built from products of Kronecker and Levi-Civita symbols. Only when $n = 6$ is there a tensor, ϵ^{mnklij}, with the appropriate symmetries. This is one of the famous "accidents" of the Cartan classification so(6) = su(4), and the fundamental representation of SU(4) is the chiral spinor of SO(6). All other SO(n) gauge theories are anomaly-free. The same is true for the unitary symplectic groups Sp($2n$) and all of the exceptional groups.

8.8.1 The consistent anomaly equation

For a U(1) gauge group, the anomaly equation we have derived *does* satisfy the WZ consistency condition. For the abelian case, our equation becomes

$$\partial_\mu \frac{\delta}{\mathrm{i}\,\delta A_\mu(x)} \ln \mathrm{Det}[\mathrm{i}\sigma^\mu D_\mu] = \frac{\mathrm{i}}{2\pi^2} \int \mathrm{d}^4x\, \omega(x) F_{\mu\nu}\tilde{F}^{\mu\nu}.$$

We can find an action that gives rise to this formula by the following peculiar trick: introduce a five-dimensional manifold X_5 whose boundary is ordinary four-dimensional Euclidean space-time. Introduce *any* smooth five-dimensional gauge potential whose boundary value is $A_\mu(x)$, with $A_5 = 0$ on the boundary. Define the *Chern–Simons* (CS) action by

$$S_{\mathrm{CS}} = -\frac{\mathrm{i}}{2\pi^2} \int \mathrm{d}^5x\, \epsilon^{MNKLR} A_M F_{NK} F_{LR}.$$

Note that this does not depend on the five-dimensional metric on X_5. Under a five-dimensional gauge transformation, this goes into

$$\delta S_{\mathrm{CS}} = -\frac{\mathrm{i}}{2\pi^2} \int \mathrm{d}^5x\, \epsilon^{MNKLR}\, \partial_M \omega F_{NK} F_{LR}.$$

On integrating by parts we see that the gauge variation comes only from the boundary, and exactly reproduces our anomaly equation.

We can understand the reason for the existence of the CS action by looking at a *six*-dimensional U(1) gauge theory. The gauge-invariant operator

$$\epsilon^{M_1 M_2 M_3 M_4 M_5 M_6} F_{M_1 M_2} F_{M_3 M_4} F_{M_5 M_6}$$

---

[16] There is something called a global anomaly [107] in certain models. An example is an SU(2) gauge model with Weyl fermions in an odd number of doublet representations. The fermion determinant in such models is an odd function on the space of gauge potentials modulo gauge transformations, and the partition function vanishes. These models are inconsistent.

can be written

$$\epsilon^{M_1 M_2 M_3 M_4 M_5 M_6} F_{M_1 M_2} F_{M_3 M_4} F_{M_5 M_6}$$
$$= \epsilon^{M_1 M_2 M_3 M_4 M_5 M_6} \partial_{M_1}(A_{M_2} F_{M_3 M_4} F_{M_5 M_6}).$$

The fact that the five-dimensional gauge variation of $\mathcal{L}_{CS}$ is a total divergence follows from the fact that the gauge-invariant six-dimensional operator is the total divergence of the lift of the CS Lagrangian density to a current density in six dimensions. This fact shows us how to generalize the CS trick to non-abelian theories. Simply define the non-abelian CS invariant by

$$\text{tr}[\epsilon^{M_1 M_2 M_3 M_4 M_5 M_6} F_{M_1 M_2} F_{M_3 M_4} F_{M_5 M_6}]$$
$$\equiv \epsilon^{M_1 M_2 M_3 M_4 M_5 M_6} \partial_{M_1} C_{M_2 M_3 M_4 M_5 M_6},$$

and use the five-dimensional restriction of $C_{M_1 \ldots M_5}$ to define the non-abelian CS action. By construction, its gauge variation on a 5-manifold with boundary will depend only on the boundary values of the fields. Its variation is parity odd, dimension 4 (contains no dimensionful parameters), and satisfies the WZ condition. It will also reproduce our direct calculation for abelian subgroups.

We will show below that there can be no local four-dimensional action whose gauge variation reproduces the anomaly equation. It is a little harder to show that there is no other WZ-consistent form of the anomaly equation that cannot be written as the variation of a four-dimensional action. Furthermore, the difference between our covariant form of the anomaly equation and the consistent form *can* be written as such a variation. Thus, the anomaly cannot be removed by some other method of regulation, and its form is completely determined by its value for abelian gauge fields (but one must examine all U(1) subgroups). Some of the proofs of the statements of this paragraph can be found in Chapter 22 of [42] and references therein.

8.8.2 The use of anomalies: consistency of gauge theories

The most striking consequence of the existence of anomalies is that some classical gauge theories are inconsistent at the quantum level. The lack of gauge invariance translates into either a violation of Lorentz invariance, in manifestly unitary gauges like the axial gauge, or a violation of unitarity, in covariant gauges. The first thing we should do is to check whether our favorite gauge theory of the real world, the standard $SU(1, 2, 3)$ model, is anomaly-free. We know that the $SU(2)^3$ and $SU(3)^3$ anomalies cancel out, because the representations of these groups are real or pseudo-real. We are left with the $U(1)^3$, $U(1) SU(2)^2$, and $U(1) SU(3)^2$ anomalies. These give the following constraints on weak hypercharges:

$$6Y_q^3 + 3Y_{\bar{u}}^3 + 3Y_{\bar{d}}^3 + 2Y_l^3 + Y_{\bar{e}}^3 = 0,$$
$$3Y_q + Y_l = 0,$$
$$2Y_q + Y_{\bar{u}} + Y_{\bar{d}} = 0.$$

There is one more constraint, which has to do with the consistency of the coupling of the theory to gravity. In order to couple spinors to gravity, we have to introduce an SO(1, 3) gauge potential, ω_μ^{ab}, to relate the spinor basis at one point of the space-time manifold to those at any other. The ab indices refer to the adjoint representation of SO(1, 3). In Einstein's theory of gravitation this is accomplished by introducing an orthonormal frame or *Vierbein*, $e_a^\mu(x)$, in the tangent space to space-time at space-time coordinate x. The coordinate-frame space-time metric is $g_{\mu\nu} = e_\mu^a e_\nu^b \eta_{ab}$, where η is the Minkowski metric in the tangent space. e_ν^a is the inverse matrix to e_a^ν. This defines the connection via the equation

$$\partial_\mu e_\nu^a - \partial_\nu e_\mu^a = \omega_\mu^{ab} e_\nu^b - \omega_\nu^{ab} e_\mu^b.$$

One can solve this equation algebraically for ω_μ^{ab}. If $e_\mu^a \to \Lambda_b^a(x) e_\mu^b$ under local Lorentz transformations, it is easy to see that ω_μ^{ab} transforms as a Lorentz connection. The Lagrangian for Weyl spinors coupled to gauge fields and gravity is

$$\bar{\psi} i e_a^\mu \sigma^a (\partial_\mu - A_\mu - i\omega_\mu^{ab}\sigma_{ab})\psi.$$

σ_{ab} are the matrices representing infinitesimal Lorentz transformations. A derivation quite analogous to what we have done above shows that there will be U(1) gravity[2] anomalies unless the traces of all U(1) generators vanish. Thus, in the standard model we must have

$$6Y_q + 3Y_{\bar{u}} + 3Y_{\bar{d}} + 2Y_l + Y_{\bar{e}} = 0.$$

We can use the three linear equations to eliminate Y_l, $Y_{\bar{e}}$, and $Y_{\bar{d}}$. Finally, we use the cubic equation to eliminate $Y_{\bar{u}}$ in terms of Y_q. Remarkably, it turns into a quadratic equation, whose solutions are

$$Y_{\bar{u}} = -Y_q \pm 3|Y_q|.$$

It is then obvious that all of the hypercharges must be quantized in units of Y_q. Recall that the normalization of all hypercharges is determined only once we have fixed the gauge coupling g_1.

There are two amazing things about this result. First, we have derived charge quantization from consistency conditions. Second, the consistency conditions require a non-trivial cancelation between quark and lepton contributions. Finally, the real-world hypercharges actually satisfy the conditions. All of this could be simply explained if the standard model emerged from the Higgs mechanism in a simple group at a high energy scale, $M_U \gg m_W$. This possibility was first investigated by Georgi and Glashow. The idea is to search for the simplest possibility, a simple group of rank 4 (the rank of the standard model) that contains SU(1, 2, 3). The list of simple groups of rank 4 consists of SU(5) SO(8), SO(9), Sp(8), and F_4. The only one with an SU(1, 2, 3) subgroup is SU(5). SU(5) has no irreducible 15-dimensional representation, but the $R = \bar{5} \oplus 10$ representation[17] and its conjugate are 15-dimensional reducible representations. The

---

[17] The 10 is the anti-symmetric product of two 5 representations.

$\bar{5} + 10$ has the right quantum-number assignments to fit the standard model. The $\bar{d}$ and lepton multiplets fit in the $\bar{5}$ and the rest of the particles are in the 10.

8.8.3 Violation of global symmetries

If we have a simple, non-abelian gauge group G, then the only possible anomalies in global symmetry currents have the form

$$\partial_\mu (J^\mu)^i = \frac{a^i}{32\pi^2} G^A_{\mu\nu} \tilde{G}^{\mu\nu A}.$$

The anomaly coefficients have the form

$$a_i = \sum t^i_R d_R,$$

where the sum is over irreducible representations of G and we are working in a basis where the generator T^i of the global symmetry is diagonal. From this formula it is clear that if T^i is the (traceless) generator of a simple non-abelian global symmetry then the anomaly vanishes. Further, while a generic model may have several U(1) global symmetries, their anomalies are all proportional to each other so only one linear combination of the U(1)s is anomalous. If our non-abelian gauge group has k simple factors, then, barring fortuitous cancelations, there will be k anomalous U(1) symmetries.

There are two important applications of anomalous violation of global symmetries in the standard model. The first is the anomaly in the $U_A(1)$ symmetry of QCD with massless quarks. In this case the gauge coupling g_3^2 is not really a free dimensionless parameter. Rather, as we will see in our chapter on renormalization, g_3 varies with the momentum scale. It is small at very high momenta, but eventually becomes strong. The real parameter of massless QCD is a scale Λ_{QCD} at which perturbation theory in g_3 breaks down. There is thus no sense, except in processes at energy $\gg \Lambda_{QCD}$, in which the violation of the $U_A(1)$ symmetry is small. Rather, the RHS of the anomaly equation is just estimated from dimensional analysis.

This is a good thing, because our hypothesis of non-zero $\langle \bar{q} q \rangle$, which is necessary to the successful predictions of chiral symmetry for pion and kaon physics, implies that, if $U_A(1)$ is a symmetry, it is spontaneously broken. This would predict a ninth pseudo-Goldstone boson in the hadron spectrum, which could be created by the singlet axial current. The lightest particle with the right quantum numbers is the η'. It is no lighter than the proton. Chiral perturbation theory predicts that its mass is less than twice the pion mass. The anomaly saves us from this disaster, by telling us that $U_A(1)$ is not a symmetry of QCD.

The God of the standard model must be Jewish, for there is a price to pay for this pleasant victory. When we rotated away the overall phase of the quark mass matrix, we used a $U_A(1)$ transformation. The anomaly equation tells us that this simply shifts the CP-violating parameter θ_{QCD} in the gauge-field Lagrangian. Since experimental evidence shows that CP is violated in the real world,[18] we are left with no explanation of why this parameter is small. We might have tried to argue, on the basis of perturbation theory, that the operator $G^a_{\mu\nu}\tilde{G}^{\mu\nu a}$ was a total divergence, with no effect on physics. The anomalous resolution of the problem of the ninth Goldstone boson requires us to abandon such a conjecture. Furthermore, as we will see when we study instantons, there is strong mathematical evidence that the total divergence does have an effect on non-perturbative physics. The strong CP problem is one of the deepest puzzles of particle physics. Both naive and more sophisticated estimates of the neutron electric dipole moment indicate that $\theta < 10^{-9}$. Theoretical explanations of this small number have been proposed, but none has yet been experimentally verified.

The second anomalous violation of a global symmetry in the standard model is the violation of baryon number by $\mathrm{BSU}(2)^2$ anomalies. The only particles in the standard model which have both baryon number and non-singlet $\mathrm{SU}(2)_L$ are the left-handed quark doublets. The reader should, at this point, be able to see that there is a baryon-number anomaly, as well as a lepton-number anomaly, but that $B - L$ is conserved. Because the electro-weak couplings are small at the scale of the W-boson mass, electro-weak perturbation theory is a good approximation. B violation is not seen in any order in the perturbation expansion. Instanton methods, which we will learn about in the last chapter, show that the violation of B in few-particle reactions is of order

$$\Delta B/(VT) \sim m_W^4 e^{-8\pi^2/g_2^2} \sim m_W^4 e^{-180}.$$

This is of course invisible in laboratory experiments. It can be shown that, if the temperature of the early Universe reaches something of order 100 GeV, then electro-weak B-violating processes become much more probable. They may be part of the story of how the Universe developed an asymmetry between baryons and anti-baryons.

8.8.4 Anomaly matching and massless spectrum

Our final example of the use of anomaly equations uses non-anomalous global symmetries in a theory with a strongly interacting gauge invariance. We will work with the particular example of massless QCD with $\mathrm{SU}(N)$ gauge group and N_F species of massless quarks. In particular, we will concentrate on the case (approximately) relevant to the real world: $N = N_F = 3$. Taking into account the $U_A(1)$ anomaly, this theory has a $G_F = \mathrm{U}(1) \times \mathrm{SU}(N_F) \times \mathrm{SU}(N_F)$ global symmetry, called the flavor group of the standard model (in the approximation of N_F zero quark masses). Imagine trying to gauge this symmetry, but take all the gauge couplings very small.[19] Such a gauge theory is inconsistent, because of the anomalies. Note that the anomaly equations for G_F are independent of the $\mathrm{SU}(N)$ gauge fields, and so remain valid after one has

---

[18] And the phase in the CKM matrix gives a nice explanation of all observations to date.

[19] Once renormalization is taken into account, gauge couplings vary with energy scale. The statement that all the couplings are small means that they remain small throughout the energy range relevant to the following discussion.

done the functional integral over these fields.[20] Let us add a collection of massless left-handed fermions, called spectators, in representations of G_F, which cancel out all of the anomalies.

There is strong evidence that for N_F/N sufficiently small (the critical value is of order $11/2$) a small $SU(N)$ gauge coupling in the extreme ultraviolet grows until an energy scale Λ_{QCD}. Below that scale a perturbative description in terms of quarks and gluons is invalid. In fact, it is believed that the theory exhibits confinement: neither quarks nor gluons appear in the spectrum of asymptotic particle states. Instead, the particle states are created from the vacuum by color-singlet composite operators composed of N quarks, N anti-quarks, or a quark–anti-quark pair. The latter are called mesons, the former, baryons. Collectively they are known as hadrons.

Far below the scale Λ_{QCD} there must be a consistent effective G_F gauge theory containing the spectators and massless hadrons (if any). We immediately conclude that there must be massless hadrons, because otherwise there could be nothing to cancel out the anomaly of the spectator fields. One possibility is massless baryons in anomalous representations of G_F. We will look into this possibility in a moment. The other possibility is illustrated for $U(1)$ subgroups of G_F by the following Lagrangian:

$$\mathcal{L} = \frac{f^2}{2}(\partial_\mu \phi)^2 + K\phi F_{\mu\nu}\tilde{F}^{\mu\nu}.$$

ϕ is an angle variable and represents the NGB of spontaneous $U(1)$ breakdown. The $U(1)$ symmetry is realized as a shift of ϕ. For an appropriate choice of K, this Lagrangian can reproduce an anomalous symmetry transformation, which cancels out that of the spectators. Note that, if G_F is explicitly broken by quark masses, then the massive pseudo-NGB will decay into "photons" of the $U(1)$ subgroup, with an amplitude completely determined by the anomaly K and the decay constant f. Such a computation in fact provides a quantitatively correct description of the decay of the π^0 into two Maxwell photons. The non-abelian generalization of this anomalous NGB Lagrangian was discovered by Wess and Zumino in the same paper [106] as that in which they derived their consistency condition for anomalies. We will investigate it in the exercises.

So we appear to have a dichotomous choice of massless hadron spectrum: NGBs or massless baryons constrained by anomaly cancelation. Let's consider the second possibility. For $N = 3$, the baryon fields have the form

$$B^{abc}_{\alpha\beta\gamma} = \epsilon_{ijk}q^{ai}_\alpha q^{bj}_\beta q^{ck}_\gamma$$

and

$$\bar{B}^{\alpha\beta\gamma}_{abc} = \epsilon^{ijk}\bar{q}^\alpha_{ai}\bar{q}^\beta_{bj}\bar{q}^\gamma_{ck}.$$

Here we have used the notation that lower or upper Greek indices represent left-handed Weyl spinors. Lower early Latin indices are $SU(N_f)_L$ fundamental representations,

---

[20] There is a much more rigorous version of this quick and dirty argument, called the Adler–Bardeen theorem [105].

whereas lower mid-alphabet Latin indices are the fundamental of SU(3). Upper Latin indices are anti-fundamental $SU(N_f)_R$ or $SU(N)$ representations. The spin indices must be combined to spin 1/2. The other possibility, massless spin-$\frac{3}{2}$ fields with gauge interaction, can be realized only in supergravity in anti-de Sitter space. The gauge group representation in that case is non-chiral and does not have anomalies. Furthermore, the spin-$\frac{3}{2}$ particles are supersymmetric partners of the graviton; so, if they were composites, the graviton too would have to be composite. Weinberg and Witten [108] showed that a composite graviton cannot arise in a Lorentz-invariant local field theory.

We can immediately see that in the case $N = N_f = 3$ the baryons cannot cancel out the spectator anomaly. For $N = 3$ the baryons which transform under $SU(3)_L$ are of the form $\epsilon_{ijk} q_L^i q_L^j q_L^k$. The only complex representation of $SU(3)_L$ which appears in the product of three three-dimensional representations is the totally symmetric 10. If we consider the generator

$$T = \text{diagonal}(1, 1, -2)$$

in the fundamental representation, then

$$T_{10} = \text{diagonal}(3, 3, -6, 3, 3, 0, 0, 0, -3, -3).$$

If we have L baryons in the 10 representation, this gives an $SU(3)^3$ anomaly proportional to $-6(27)L$. The quark anomaly is instead -18 (including the three from color) in the same units. The anomalies match only if $L = \frac{1}{9}$, which is absurd. We conclude that $U(1) \times SU_L(3) \times SU_R(3)$ *must break down spontaneously to its anomaly-free* $U(1) \times SU_V(3)$ *subgroup.* This is in fact the symmetry-breaking pattern observed for QCD in the real world.

Another class of cases in which baryons cannot do the job is even N, with arbitrary $N_F < N_F^{(c)}$, where baryons are bosons. In fact, using one more general property of parity-symmetric gauge theories coupled to massless fermions, one can prove that anomaly matching is not possible for any N and N_F. Let us give a mass to a single quark. I claim that all baryons containing that quark field are massive. The proof is due to Vafa and Witten, and follows the same line of argument as that we gave above to prove that vector-like symmetries were not spontaneously broken. Do the functional integral in steps, first doing the integral over fermion fields. The two-point function of baryons in an external field is given by

$$\epsilon^{i_1 \ldots i_N} \epsilon_{j_1 \ldots j_N} G_{i_1}^{j_1}(x, y) \ldots G_{i_N}^{j_N}(x, y),$$

where we have suppressed the spin and flavor indices on the quark Green functions. For the baryons in question, at least one of the Green functions is that of the massive quark. To all orders in perturbation theory in the external field, this Green function falls off exponentially at infinity. Vafa and Witten [99] proved that this was true rigorously and that the exponential falloff was independent of the gauge field configuration. In a parity-symmetric gauge theory the fermion determinant $\text{Det}[i\gamma^\mu D_\mu] = \text{Det}[-\Delta_{L,R}]$ for all configurations that do not have normalizable zero modes and is zero for those which do. Thus the full baryon Green function is an integral of the external field Green function with a positive weight function. The field-independent bound on the exponential falloff shows that the baryon Green function falls off exponentially in Euclidean space. This implies that the baryon operators containing the massive quark cannot create massless particles from the vacuum.

As a consequence, given a solution to the anomaly-matching condition for some value of (N, N_F), which is postulated to give a massless baryon spectrum in that version of QCD, we should get another solution for $(N, N_F - 1)$, with the same baryon multiplicities (the analog of the integer L for $N = N_F = 3$). It is straightforward but tedious to show that these conditions lead to the conclusion that there are no physically consistent solutions of the anomaly-matching conditions, even though there are mathematical solutions for particular values of N and N_F. Thus, all versions of QCD spontaneously break their global symmetry to its anomaly-free vector $U(N_F)$ subgroup. If, as occurs in the real world, part of the anomalous symmetry is weakly gauged, with the anomaly canceled out by spectators (i.e. leptons in the real world), then we also get formulae (from the Wess–Zumino Lagrangian alluded to above) for the decay amplitudes of NGBs to weakly coupled gauge bosons.

Let us finally note that our current use of the anomaly equation shows that the anomaly cannot be removed by adding a local term to the four-dimensional Lagrangian. Indeed, we have seen that the anomaly equation seems in some cases to imply a massless NGB pole in the Green functions of currents. In fact what one can show [109–110] is that the anomaly equation implies that Green functions involving three currents, one of which is anomalous, have a momentum-space discontinuity disc $A \propto \delta(q^2)$, where q is the momentum carried by the anomalous current. The NGB pole is one way to achieve such a discontinuity; the other is a simple triangle diagram of massless fermions. In this diagram there is non-zero phase space for on-shell massless fermions to travel exactly parallel to each other, mocking up the discontinuity caused by a massless scalar pole. No other diagram has finite measure for such configurations, and this constitutes another proof of the Adler–Bardeen theorem that the anomaly is given exactly by the expressions we derived by naive manipulations, without any higher-order corrections. The fact that the anomaly equation implies massless singularities in Green functions shows us that one cannot find a local Lagrangian counterterm that removes it.

8.9 Quantization of gauge theories in the Higgs phase

We now leave the subject of anomalies and turn to the quantization of gauge theories in the Higgs phase. We imagine a gauge group G, which, abusing language, is broken down to a subgroup H by the VEV, v_j, of a multiplet of scalar fields ϕ_j. Define

$$f_i^a = (T_{R_S}^a)_i^j v_j$$

and

$$\phi_i = v_i + \Delta_i.$$

We use canonically normalized fields, whose covariant derivative is

$$D_\mu \phi_i = \partial_\mu \phi_i - g_a (T_{RS}^a)_i^j \phi_j.$$

The couplings g_a are the same for all generators belonging to the same simple factor of the gauge group.

If we expand the classical Lagrangian around v_i, there will be kinetic terms mixing gauge bosons and scalars. It is convenient to choose a gauge-fixing term to eliminate this mixing. This defines the R_κ gauges, to which we have alluded in our discussion of the abelian Higgs model. After integrating out the Lagrange multiplier fields N^a we have a gauge-fixing Lagrangian

$$\mathcal{L}_{GF} = \frac{1}{2\kappa}(\partial^\mu A_\mu^a - \kappa g_a f_i^a \Delta_i)^2 + \bar{c}^a[\partial_\mu D_{ab}^\mu - \kappa g_a g_b f_i^a (T^b)_i^j (v_j + \Delta_j)]c^b.$$

In the last term we modify the summation convention and consider the term $g_a g_b f_i^a (T^b)_i^j (v_j + \Delta_j)$ as a matrix with two adjoint indices.

In the R_κ gauges, all components of the vector fields propagate. The physical components have mass matrix

$$(m_A^2)^{ab} = g_a g_b f_i^a f_i^b,$$

while the time-like components, as well as the ghosts, have mass matrix κm_A^2. The propagator is

$$\frac{-i}{k^2 - m_A^2}\left[\eta^{\mu\nu} - \frac{k^\mu k^\nu}{k^2 - \kappa m_A^2}(1 - \kappa)\right].$$

For finite values of κ this falls off at high momenta, but if we take the $\kappa \to \infty$ limit it looks like a massive vector propagator. On inspecting the gauge-fixing term, we see that in this limit we appear to get the unitary gauge. The unphysical components of the Δ_i fields are projected out. Indeed, the would-be NGBs have mass matrix

$$(m_{NGB}^2)_{ij} = \kappa f_i^a f_j^a g_a^2,$$

because the full mass matrix for the fields Δ_i is

$$m_{NGB}^2 + M^2,$$

where M^2 is κ-independent and operates in the subspace of fields orthogonal to the would-be NGBs. M is the mass matrix for physical Higgs-boson excitations.

In Problem 8.3 you will show that the physical linearized excitations of a gauge theory in R_κ gauge are BRST-invariant states. As a consequence, the fields of the model create these states from the vacuum to all orders in perturbation theory. Furthermore, the linearized theory tells us precisely how to project the fields A_μ^a, c^a, and Δ_i onto physical states. We simply take the three physical polarization states of massive gauge bosons, the transversely polarized states of massless gauge bosons, and components of Δ_i perpendicular to the subspace of would-be Goldstone bosons (i.e. satisfying $f_i^a \Delta_i = 0$). No ghosts, time-like gauge fields, or would-be NGBs are allowed.

In scattering theory, fields are evaluated along the trajectories of asymptotic particles at infinity. Thus, if we make these projections on the asymptotic fields, we are defining BRST-invariant operators. The LSZ formula converts such asymptotic expressions into residues of poles in momentum-space Green functions. We conclude that, if we evaluate the *non-gauge-invariant* Green functions of the fields A_μ^a and Δ_i, but search for poles in perturbation theory that are near the zeroth-order masses of physical particle states, and project onto the proper subspaces of on-shell fields, then these residues are gauge-invariant and independent of κ. In other words, the S-matrix for physical states is a BRST-invariant, and therefore κ-independent, quantity. As $\kappa \to \infty$, unphysical states go off to infinite mass and the theory becomes manifestly unitary. Thus, the scattering matrix satisfies perturbative unitarity.

Another way to understand Problem 8.3, and the remarks of the preceding paragraph, is to introduce gauge-invariant fields via

$$\phi = \Omega(v + \Delta),$$
$$B_\mu = \Omega^\dagger(\partial_\mu - gA_\mu)\Omega,$$
$$\Psi = \Omega_F^\dagger \psi.$$

Δ has fewer components than ϕ. It parametrizes the space of field configurations orthogonal to the space of gauge transforms of the classical vacuum. ψ is the multiplet of fermion fields in the theory, and Ω_F is the action of the gauge group on that multiplet. It is easy to see that, in R_κ gauge, with $\kappa \to \infty$, the massive physical fields are equal to these gauge-invariant fields to leading order in the semi-classical expansion. For the massless vectors in the unbroken subgroup H, we must still make a transverse projection. We may think of defining the S-matrix directly in terms of these gauge-invariant operators.

The problem is that, as we have noted, the $\kappa \to \infty$ and large-momentum limits do not commute for general Green functions. The perturbation expansion describing Green functions of the fields is not renormalizable in the $\kappa \to \infty$ limit. We will discuss this further in Chapter 9.

8.10 Problems for Chapter 8

*8.1. Show that the action of the BRST symmetry on ordinary fields is nilpotent.

*8.2. Quantize free Maxwell electrodynamics using the BRST method, using the gauge-fixing Lagrangian $\mathcal{L}_{GF} = [Q_{BRST}, c(\partial_\mu A^\mu + \kappa N/2)]$. Use Noether's theorem to construct Q_{BRST} in terms of the fields, and then from that expression get the expression in terms of creation and annihilation operators for four photon polarizations and the ghost fields (the ghosts are quantized like a complex scalar with $b = c^\dagger$). Show that the BRST-invariant states, which are not of the BRST trivial form $|s\rangle = Q_{BRST}|t\rangle$, are precisely the physical, transverse, photon polarizations.

*8.3. Repeat the previous exercise for a general gauge theory in the Higgs phase, using R_κ gauge. Show that only physical polarization states of massive vectors, transverse polarizations of massless vectors, and scalar excitations orthogonal to the subspace of would-be Goldstone excitations are BRST-invariant.

*8.4. Using BRST quantization for pure Yang–Mills theory, in a general covariant gauge with parameter κ, compute the leading-order contribution to the expectation value of a Euclidean Wilson loop in gauge group representation R, of rectangular shape $T \times L$, with

$$T \gg L.$$

You should find an answer independent of κ, of the form

$$\mathrm{Tr}\, e^{-T \frac{cg^2}{L} t_R^a t_{\bar{R}}^a}\, e^{-T\infty}.$$

The physical interpretation of this computation is that the logarithm is T times the energy of a pair of static sources for the gauge field, separated by a distance L, one in the representation R and the other in the complex-conjugate representation $\bar{R}$. The trace is taken over the tensor product of these two representations. Expand $(t_R^a + t_{\bar{R}}^a)^2$ to write $t_R^a t_{\bar{R}}^a$ in terms of Casimir operators for the two representations and for all representations that appear in the tensor product. The trace tells you that the Wilson loop represents a bunch of energy eigenstates where the heavy-particle–anti-particle color indices are combined into different irreducible representations of the gauge group. Each such state has a Coulomb attraction or repulsion with a coefficient that depends on the irreducible representation. There is also an infinite self-energy term, which is L-independent. Show that the most attractive potential occurs when the indices are combined to give a singlet. This was Nambu's argument that QCD explained why color-singlet combinations of quarks and anti-quarks were the lowest-lying states. For baryons one first combines two quarks into an anti-quark representation, via the SU(3) Clebsch–Gordan formula $3 \times 3 = \bar{3} + 6$, and then makes the singlet combination with the third quark. This leads to a picture of the structure of a baryon in which two quarks are closer together than they are to the third. The picture is valid for baryons with large orbital angular momentum.

There are two things to be careful of in this problem. Take care with the path ordering of the exponential defining the Wilson loop, but argue that in leading order in g it can be neglected, except for the change of representation from R to $\bar{R}$ for the particle at the origin. Also the Wilson loop is originally defined in terms of a single trace in the R representation. Show that taking the limit $T \gg L$ effectively replaces this by a trace over the tensor-product. One way to see this is to take $T = \infty$ from the start. Then the object we are computing is defined as the trace over the tensor product Hilbert space of the product of two infinite Wilson lines, one going forward in time and the other backward in time. Show that the path ordering accounts for the change of representation.

*8.5. Show that $\mathrm{tr}_R (T^a [T^b, T^c]_+)$ vanishes for any real or pseudo-real representation of an arbitrary compact Lie group.

*8.6. Given $2M$ chiral fermions, consider the U(1) subgroup of U($2M$) whose generator is diag$(1 \ldots 1, -1, \ldots -1)$, where the diagonal blocks are $M \times M$. The commutant of U(1) in U($2M$) is $G_F \equiv U(1) \times SU(M) \times SU(M)$ (commutant means maximal subgroup commuting with U(1)). Show that the subgroup of G_F U(1) $\times$ SU$_D(M)$ is anomaly-free. Now consider the five-dimensional Chern–Simons action for the gauge group G_F. The variation of this action under a finite gauge transformation is

$$S_{CS} \to S_{CS} + \int d^4x \, \mathcal{L}_{WZ}(A, \Omega),$$

where $\mathcal{L}_{WZ}$ is a local function of the four-dimensional boundary values of A and Ω and their first derivatives. Compute $\mathcal{L}_{WZ}$. Now use the above remark about anomaly freedom to conclude that $\mathcal{L}_{WZ}$ depends only on the coset element corresponding to $\Omega(x)$ in $G_F/[U(1) \times SU_D(N)]$. That is, it is invariant under transformations of the form $\Omega(x) \to \Omega(x)h(x)$. Up until now we have considered $\Omega(x)$ to be the boundary value of the five-dimensional gauge transformation. Now switch gears and consider $\mathcal{L}_{WZ}$ to be a Lagrangian for gauge fields in G_F, coupled to NGBs in corresponding to the spontaneous breakdown of G_f to $U(1) \times SU_D(N)$. If we now do G_F gauge transformations $g(x)$ on the gauge fields, accompanied by $\Omega(x) \to g(x)\Omega(x)$, show that $\mathcal{L}_{WZ}$ reproduces the anomaly of the fermions.

*8.7. Show that the fermion content of the $\bar{5} + 10$ representation of SU(5) reproduces a single generation of the standard model, if we regard the latter as the SU(1, 2, 3) subgroup of SU(5).

*8.8. Consider an SU(N) gauge theory with a Weyl fermion ψ_{ij} in the second-rank anti-symmetric tensor representation. Show that there is an integer N_F such that the anomaly of SU(N) is canceled out if we add N_F Weyl fermions in the fundamental N representation (N_F negative means fermions in the $\bar{N}$ representation). Find all global classical symmetries of the resulting gauge-theory action and describe which ones are anomalous. Compute the anomaly in the three-point function of three non-anomalous global symmetries.

*8.9. If we embed the standard-model (with a single generation of quarks and leptons) gauge theory into SU(5), there are two global classical U(1) symmetries. Show that one linear combination of the U(1) generators has no U(1)SU(5)^2 anomalies. Show that one can add an extra SU(5) singlet Weyl fermion, transforming under this U(1), in such a way that the U(1)^3 anomaly also cancels out. The group SO(10) is of rank 5. Show that it has an SU(5) $\times$ U(1) subgroup, by considering the spinor representation, where by the SO(10) generators are constructed as commutators of ten Euclidean Dirac–Clifford matrices satisfying

$$[\Gamma_\mu, \Gamma_\nu]_+ = 2\delta_{\mu\nu}.$$

Define

$$\sqrt{2}a_i = \Gamma_{2i-1} + i\Gamma_{2i}, \quad i = 1, \ldots, 5,$$

and show that

$$[a_i, a_j]_+ = 0, \quad [a_i, a_j^\dagger]_+ = \delta_{ij}.$$

This is the algebra of five independent fermion-creation and -annihilation operators. The 25 matrices $a_i^\dagger a_j$ form the Lie algebra su(5)⊕u(1). Use this construction to show that the representation of the Dirac matrices is 32-dimensional. Show that the product of all the Dirac matrices anti-commutes with each Dirac matrix and thus commutes with all of the SO(10) generators. The 32-dimensional SO(10) spinor representation breaks into two irreducible 16-dimensional representations. Show that these are the subspaces where either an odd or an even number of creation operators acts on the state satisfying

$$a_i|s\rangle = 0.$$

Imagine that a Higgs field breaks SO(10) down to U(1) × SU(5). Show that one of these 16-dimensional representations consists of the standard model plus the singlet we introduced above to cancel out anomalies. The anomaly-free U(1) generator we introduced above is just the extra U(1) in SO(10). If you wish, you can also find a Higgs field and an effective potential for it, which performs the breaking of SO(10) to SU(5) × U(1).

*8.10. Show that baryon and lepton number have an SU(2) anomaly in the standard model, but that $B - L$ is anomaly-free. Referring to the previous exercise, show that $B - L$ is the extra U(1) that commutes with SU(5) in SO(10).

*8.11. Compute the amplitudes, at tree level in the SU(2) × U(1) theory of electro-weak interactions, for an electron–positron pair to annihilate into a pair of charged W bosons. You may make the approximation that the electron is massless, because this interaction can occur only at energies much larger than the electron mass. It is convenient to do the computation separately for left-handed and right-handed electrons, because these helicity states couple differently to the weak-interaction gauge bosons. Do the computations in a general R_κ gauge and show explicitly that the result is κ-independent. Comment on the energy dependence of the production amplitude for longitudinally polarized W bosons.

*8.12. Show that the determinant of the Euclidean Dirac operator i$\not{D}$ + im is formally positive, by using the Weyl representation.

*8.13. Carry out the leading-order computation of the potential between static sources in a non-abelian gauge theory in a general covariant gauge, and verify that the answer is independent of κ.

*8.14. The charge-changing weak currents involved in muon decay are

$$J_+^{mu} = \bar{e}\gamma^\mu(1 - \gamma_5)\nu_e + \bar{\mu}\gamma^\mu(1 - \gamma_5)\nu_\mu$$

and its Hermitian conjugate. Compute the commutator $[J_+^0, J_-^0]$, using canonical commutation relations, and show that the result is not the electromagnetic current. This was Glashow's argument that the electro-weak gauge group had to be SU(2) × U(1).

*8.15. Write the Ward identities for the generating functional of non-abelian currents, the functional average of

$$\langle e^{i\int J_a^\mu(x)A_\mu^a(x)}\rangle \equiv e^{iW(A)}$$

for some global symmetry group G with generators T^a. Show that they are equivalent to the requirement that $W(A)$ be invariant under non-abelian gauge transformations of A. Define the non-standard Legendre transform

$$\Gamma(a_\mu^a) + M^2(A_\mu^a - a_\mu^a)^2 = W(A_\mu^a).$$

Describe how the expansion coefficients of Γ are related to connected Green functions. Show that the Ward identities imply that Γ is a gauge-invariant functional of a_μ^a. At low energies this means that Γ is just the usual Yang–Mills action of Chapter 7. Now argue that, if G is broken to H, then we should couple the a_μ^a field to the NGBs living in the G/H coset, in a way which respects gauge invariance. Show that in the Yang–Mills approximation we get a theory of massive vector mesons interacting with massless NGBs. The vector mass matrix is not just given by M^2. There is also a contribution from the interaction with the NGBs. Compute it. Apply this formalism to the case of strong interaction chiral symmetry $G = SU(2) \times SU(2)$. It gives an approximate theory of ρ, ω, and A_1 mesons interacting with pions. The theory cannot be justified in the way that we justify the chiral Lagrangian. Even if we make the chiral symmetry exact, by setting up- and down-quark masses to zero, the vector mesons remain massive, with a mass not much smaller than $4\pi F_\pi$, so there is no limit in which we can exactly replace Γ by the Yang–Mills action. Nonetheless, this *vector-dominance model* of the hadronic currents is a useful approach to problems in strong-interaction physics.

9 Renormalization and effective field theory

We are finally ready to face up to the fact that most of the expressions we write in the Feynman-diagram solution of quantum field theory are nonsensical, which is to say infinite. The purpose of this introductory section is to demonstrate the simple and general counting rules which prove that this is so, and to give the reader a conceptual orientation to the general theory of renormalization, which we will discuss in the rest of this chapter.

We will begin with the conceptual. The formalism of quantum field theory makes statements about arbitrarily short distances in space-time and arbitrarily high energies. At a given moment in history experiment will never probe arbitrarily short distances or high energies. We should expect that the formalism will have to change to fit the data as we uncover more empirical facts about short distances. This has happened before. Hydrodynamics is a field theory, but we know that at short enough distance scales it is not a good description of water. A better one is given (in principle) by the Schrödinger equation for hydrogen and oxygen nuclei interacting with electrons, and that in turn is superseded by the standard model of particle physics as we go to even shorter distances and higher energies. Yet only a fool would imagine that one should try to understand the properties of waves in the ocean in terms of Feynman-diagram calculations in the standard model, even if the latter understanding is possible "in principle."

This simple parable illustrates the idea of an *effective field theory* (EFT). EFT is a description of phenomena at wavelengths longer than some effective cut-off scale l_c and energies below some energy cut-off E_c. In a theory that is (even approximately) relativistically invariant, the cut-offs are related by $l_c \sim E_c^{-1}$. In principle, any EFT is a quantum theory, but it may be that, as in the case of hydrodynamics, the classical approximation is valid throughout the range of validity of the EFT itself. The basic idea of an effective field theory is that physics in a certain energy regime, and a certain resolution for length scales, should be described most simply by a set of effective degrees of freedom appropriate to that scale and depend on any underlying, more "fundamental" description only through a small set of parameters governing the dynamics of these degrees of freedom.

The basic idea of renormalization is to parametrize the *effect* of all possible modifications of the theory at higher energy and shorter distance in terms of all possible interactions between the effective degrees of freedom. The non-trivial part of renormalization theory is the mathematical demonstration of universality of long-distance, low-energy physics. That is, one shows that, starting from any set of interactions at the cut-off scale, the low-energy, long-distance physics depends on only a few *relevant* parameters. The mechanics of this demonstration involves a procedure for reducing the cut-off scale and finding new Lagrangians that reproduce all of the same low-energy physics. Continuous changes in the cut-off lead to a flow on the space of all possible interactions

$$\frac{\mathrm{d}g^i}{\mathrm{d}t} = \beta^i(\{g^j(t)\}),$$

where t parametrizes an infinitesimal rescaling of the cut-off. This is called the *renormalization group (RG) flow*, despite the fact that it is irreversible, and therefore only a semigroup. For a large class of models, the RG can be shown to be a gradient flow, so that the general asymptotic behavior is a *fixed point*. The relevant parameters describe the manifold of unstable directions of the fixed point. They are the parameters which have to be tuned in order to obtain low-energy physics that is independent of the details of the physics at the cut-off scale, and all dependence on the cut-off is encoded in the value one chooses for these parameters.

All possible cut-off-independent infrared behaviors of a certain set of degrees of freedom thus fall into universality classes corresponding to the possible fixed points of their RG flow. The fixed-point Lagrangians obviously have to be invariant under space-time scaling transformations. For most interesting models they are actually invariant under the larger group of space-time conformal transformations, (SO$(2, d)$ for a d-dimensional Minkowski space-time). Such conformal field theories (CFTs) can often be constructed without recourse to Lagrangians and Feynman diagrams. Indeed, Lagrangians and Feynman diagrams are concepts appropriate only to the *Gaussian CFTs*, a fancy name for free massless field theories. This leads to a possible mathematical definition of all possible QFTs as perturbations of CFTs by turning on non-zero values of relevant parameters. But we are getting ahead of ourselves, and will return to these issues when we understand a little more of the technicalities of the RG.

The basic idea behind the derivation of the RG flow equations is the concept of *integrating out degrees of freedom*. We can illustrate this in a simple example from classical physics by imagining two harmonic oscillators with frequencies Ω and $\omega \ll \Omega$, coupled by an anharmonic interaction of the form gXx^2. We can try to solve the equations of motion for this system by first solving for X and writing equations of motion for x that take into account the evolution of X. We've cleverly chosen the interaction so that the first step can be done in closed form. We find the equations

$$\ddot{x}(t) + \omega^2 x(t) = 2gx(t) \int \mathrm{d}s[G(t - s)x^2(s)],$$

where the boundary conditions on $X(t)$ are encoded in the choice of Green function G:

$$(\partial_t^2 + \Omega^2)G(t - s) = \delta(t - s).$$

In QFT we will always assume that the high-frequency degrees of freedom are in their ground state, so the Green function is the one defined by Feynman, which is the analytic continuation of the Euclidean propagator. Integrating out degrees of freedom is always done in the Euclidean path integral.

The effective equations of motion for $x(t)$ are non-local, but if we are interested only in times scales much longer than Ω^{-1} we can make the expansion

$$G \approx \frac{1}{\Omega^2}\delta(t-s) - \frac{1}{\Omega^4}\,\partial_t^2\delta(t-s) + \cdots.$$

We get an infinite series of higher-derivative terms in the approximately local equations for x. When the solution for x contains only frequencies $\omega_i \ll \Omega$, the higher-order terms in the series are negligible, and we refer to them as *irrelevant*.

9.1 Divergences in Feynman graphs

To begin our consideration of ultraviolet divergences, we study the ϕ^n scalar field theory. A 1PI Feynman graph with E external legs and V vertices has $I = (nV - E)/2$ internal lines. The number of loop integrals is

$$L = I - V + 1 = 1 + ((n-2)V - E)/2.$$

We evaluate the graphs in momentum space and look at the region of integration where all loop momenta are large. Then we can drop all masses and external momenta in the internal propagators. In this regime, the integral looks like

$$\int \frac{\mathrm{d}^{4L}p}{p^{2I}},$$

and it diverges if $4L \geq 2I$. In terms of V and E, this inequality is

$$4 + 2((n-2)V - E) \geq nV - E,$$

or

$$4 \geq (4-n)V + E.$$

We see a striking fact: for $n = 3$, only a finite number of orders of perturbation theory will have such a divergence. For $n = 4$, the degree of divergence is independent of the order of perturbation theory, whereas for $n > 4$ it increases with the order. We also note that, for a given order of perturbation theory, the degree of divergence decreases as we increase the number of external legs. Furthermore, if we differentiate a graph with respect to external momenta, the degree of divergence decreases by one for each derivative we take.

Indeed, all of these facts are a simple consequence of dimensional analysis, and the fact that masses on internal lines are negligible in the high-momentum regime.[1] The coupling λ_n of a ϕ^n interaction has mass dimension $4 - n$. A connected E-point function of scalar fields has mass dimension E, and its Fourier transform has dimension $4 - 3E$ if we leave off the momentum-conservation delta function. An E-point 1PI function thus has dimension $4 - E$. The overall number of powers of momentum in the Feynman graph integrals is thus completely determined by dimensional analysis. An interaction with positive (negative) mass dimension will have fewer and fewer (more and more) divergent integrals as we go to higher and higher order. We call interactions of positive mass dimension *super-renormalizable*, those of negative dimension, *non-renormalizable*, and those of dimension 0, simply *renormalizable*. For reasons that will become apparent later, these terms are often replaced by *relevant*, *irrelevant*, and *marginal*.

In purely mechanical terms, the basic idea of renormalization is that, at any order in perturbation theory, the divergences all reside in a finite number of E-point functions, and are polynomial functions of momenta. The latter remark follows from the fact that, if we take a finite number of derivatives of the Feynman graph, then the superficial divergences go away, because we have introduced more internal propagators. As a consequence, we can find local interactions of the form $c_n(\lambda_n)^V \partial^{a_1}\phi \ldots \partial^{a_E}\phi$, which, when added to the theory, can subtract away all of the divergences. In order to do this, we must choose the coefficients c_n to be infinite. A more sensible mathematical procedure would be to define a cut-off theory, which effectively gets rid of the divergences above some high scale Λ, and allow the c_n to depend on Λ in such a way that the limit $\Lambda \to \infty$ exists.

It is not clear that this simple subtraction procedure actually works, for we have dealt only with the region of integration where all momenta are large. A 1PI Feynman graph contains many 1PI sub-graphs. Even if the overall degree of divergence of the graph is negative, we might have a sub-graph whose integrals diverge when all other momenta in the graph are held fixed. The second key idea of perturbative renormalization theory is that "sub-divergences have already been taken care of by subtractions at lower orders in perturbation theory." This is far from clear in purely graphical terms. To understand why it is true (still in a purely mechanical sense) we will return to the coordinate-space definition of Feynman graphs.

Divergent loop integrals fall into two distinct classes. The first is associated with loops that consist of a single propagator attached to the same vertex (ear diagrams), Figure 9.1(a). The second consists of loops that include several vertices like Figure 9.1(b). Both can be understood in terms of the singularities in operator products in free-field theory. Indeed, interactions like ϕ^n are not well-defined operators in the free-field theory. Wick's

---

[1] This is not true for the masses of massive gauge bosons in unitary gauges. As a consequence of this fact, the unitary gauge Lagrangians are in fact non-renormalizable as quantum field theories of Green functions. However, one can show that the S-matrix in these theories is identical to that of the physical particles of the R_K-gauge Lagrangian, which is renormalizable. From the point of view of the R_K gauge the non-renormalizable divergences of Green functions of unitary gauge fields arise because these are really gauge-invariant non-polynomial functions of the R_K gauge fields.

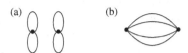

Fig. 9.1.   Examples of divergent diagrams.

theorem shows us that products of operators at different points diverge as the points are taken to coincide. This leads to the idea of an *operator product expansion* (OPE):

$$\phi(x)\phi(y) = \sum C_N(x - y)O_N(y).$$

In free, massless field theory, dimensional analysis shows us that, if the operator O_N has mass dimension d_N, then $C_N(x) \sim x^{-2+d_N}$. These operators are called $:\phi\, \partial^{\mu_1} \ldots \partial^{\mu_{N-2}}\phi:$, *the normal ordered product of ϕ with its partial derivatives.* The function C_{N-2} has the tensor transformation properties to make the OPE Lorentz covariant. Notice that O_0, the identity operator, also appears in the list. It is the only one with a divergent coefficient. However, we can now go on to consider higher-order powers of ϕ and its derivatives. The result is an infinite collection of finite operators, which we will again label O_N. There are now many ways to get an operator with a given mass dimension. The generalized OPE has the form

$$O_N(x)O_M(y) = \sum C_{NM}^K(x - y)O_K(y),$$

where we have again suppressed the tensor properties of operators and their coefficient functions. The *c*-number functions in the OPE are called Wilson coefficients. C_{NM}^K scales like $x^{-d_N-d_M+d_K}$. Wick's theorem guarantees that these are in fact operator relations. That is, the behavior of an n-point Green function when k of the points are taken together is universal, and can be written as singular coefficients depending on the coinciding points, multiplied by a function of the coincident point and the other $n - k$ points in the function.

In general, we see that, even after we have defined finite operators O_N, their products when points are taken together will often diverge. Now consider the application of all of this to the perturbation expansion. The series corresponds to integrals over interaction vertices, defined as naive powers of the field at the same point. We can expect divergences corresponding to "improper normal ordering" (the correct definition of local composite operators) as well as to singular products of normal ordered operators (since we must integrate over coinciding interaction points). The former correspond to "ear" diagrams. They can be understood by writing out ϕ^n in terms of normal ordered operators and subtracting the divergent terms.

The OPE allows us to understand and deal with the coinciding point divergences in a similar way. We have to evaluate

$$\int \langle\, :\phi^n(x_1): \ldots : \phi^n(x_k): \,\rangle.$$

Near coinciding points, we can use the OPE to write this in terms of Wilson coefficients and finite operators. This tells us that we can always subtract away the divergences by adding local operators to the Lagrangian. Furthermore, the sub-divergences that occur at order k, when $p < k$ points coincide, will have precisely the form of the Lagrangian we subtracted at order p when this particular divergence first showed up.

So we have understood, at least in a hand-waving and mechanical way, how it is that divergences in the field-theoretic perturbation expansion can be absorbed into (formally infinite) redefinitions of the Lagrangian. We can also see that the structure of these subtractions will be quite different for irrelevant perturbations, versus marginal or relevant ones. For irrelevant perturbations, we will generate every interaction consistent with the symmetries of the problem, if we go to high enough order of perturbation theory. For relevant or marginal interactions, the subtraction process stops with a finite set of operators. This means that irrelevant interactions have no (cut-off-independent) meaning. When we add an infinite counterterm to subtract a divergence, we have no rule to tell us how much of a finite coefficient one should add to the infinite one. Thus, theories with irrelevant interactions seem to make no predictions at all. Later we will see that this judgement is a bit too harsh, but that will have to wait until we understand the meaning of renormalization, not just its mechanics.

I want to end this section by noting that the definition of normal ordered operators is ambiguous in *massive* free-field theory. The Wilson coefficients are no longer required to be pure power laws. Indeed, the massless Wilson coefficients are multiplied by functions of $m|x - y|$, which are non-singular as $m|x - y|$ goes to zero. As a consequence, it is impossible to identify a normal ordered operator as the coefficient of a certain power of $|x - y|$ in the short-distance expansion. There is an ambiguity because we can replace an operator of a given mass dimension by a polynomial in m, with higher powers of m multiplying operators that would have lower mass dimension in the massless theory. A standard definition of normal ordered products is to reorder the creation and annihilation operators so that all creation operators stand to the left of all annihilation operators. This is in fact the origin of the name normal ordered operator. We should remember that this is just an arbitrary convention. It is an example of what we will call a *renormalization-scheme ambiguity* in the general theory of renormalization.

9.2 Cut-offs

The cavalier manipulations of infinite quantities, which entered into our heuristic mechanical description of renormalization, are clearly unacceptable.[2] A mathematical way of dealing with the problem is to introduce some kind of *cut-off*.

The simplest way to do this is to replace space, or Euclidean space-time, by a lattice with spacing Λ^{-1}. In fact, one way in which field theories arise in physics is in the long-distance approximation to second-order phase transitions in condensed-matter

[2] They led to one of the more memorable quips in the mathematical physics literature, due to R. Jost, "In the 1950s, under the influence of renormalization theory, the mathematical sophistication required of physicists was reduced to a rudimentary knowledge of the Latin and Greek alphabets."

systems. In that context, a crystal lattice is a physical part of a more microscopic description of the system. Another, more brutal, cut-off scheme is to simply refuse to continue the momentum-space integration in Feynman diagrams beyond some scale Λ. This procedure, while often convenient, does extreme violence to locality in space, just as a sharp cut-off in space gives rise to violent effects at high momentum. A gentler momentum cut-off is achieved by including all momenta but choosing an inverse propagator $K(p^2)$,[3] which, while retaining the form $p^2 + m^2$ at moderate momenta, grows extremely rapidly (faster than any power) above $p^2 = \Lambda^2$. We can even choose a propagator that is a smooth function, vanishing identically for $p^2 > \Lambda^2$. By looking at Feynman graphs we can see that any such choice of propagator will render all graphs with local interactions UV-finite, because the vertices are polynomials in momenta. From the functional-integral point of view, we are choosing the Gaussian part of the action to suppress the contribution of field configurations with large momenta.

The standard model of particle physics, and most proposed extensions of it, are gauge theories. While it is not strictly necessary, it is extremely convenient to have a cut-off procedure that preserves gauge invariance. Non-perturbatively, only lattice regulators have this property, but for perturbative calculations there are two useful Lorentz-invariant formulations. The first is Pauli–Villars regulation, with the related proper-time cut-off of Schwinger, which we will discuss later. The second is dimensional regulation. Dimensional regulation proceeds by noting that most expressions in Feynman diagrams appear to be analytic in the dimension of space-time.[4] Only the integration measure $d^4p/(2\pi)^4$ identifies which dimension we are in.

Dimensional regularization (DR) can be defined formally as follows: any Feynman graph in d dimensions involves multiple integrals over d-dimensional momentum space of expressions that can always be interpreted as functions of invariants made by dotting external and/or internal momenta into each other. We define the dimensional continuation of those integrals as a linear functional obeying

$$\int d^d q\, f(q+p) = \int d^d q\, f(q),$$

$$\int d^d q\, f(Rq) = \int d^d q\, f(q),$$

for any proper rotation matrix R (interpreted as meaning that the answer involves only invariant tensors) and

$$\int d^d q\, f(aq) = a^{-d} \int d^d q\, f(q).$$

Similarly, the Kronecker delta $\delta_{\mu\nu}$ which appears in these formulae is interpreted as a symbol whose trace is d, and whose contraction with two vectors is their dot product.

---

[3] Propagator cut-offs are usually introduced in Euclidean momentum space, though in principle we could distort the propagator only in spatial momentum directions.

[4] The exceptions are the Levi-Civita symbol and the Dirac γ_5 matrix.

This defines d-dimensional integration up to a multiplicative constant, which is fixed by computing a simple Gaussian integral

$$\int \mathrm{d}^d q \, \mathrm{e}^{-aq^2} = \left(\frac{\pi}{a}\right)^{\frac{d}{2}}.$$

These manipulations define analytic continuations of the scalar invariants of any Feynman graph not involving $\epsilon_{\mu_1 \ldots \mu_d}$, to complex values of d. We will see that the values of these graphs are finite, analytic functions, with multiple poles at integer values of d. One other apparent non-analyticity in d comes from closed spinor loops, since the dimension of the spinor representation is $d_{\mathrm{spin}} = 2^{[\frac{d}{2}]}$. This can be dealt with by simply defining d_{spin} to be any analytic function of d that takes on the correct value in $d = 4$ (or whatever other dimension we happen to be interested in). A way to understand this is to note that if we introduce N_{F} copies of every fermion, in a $\mathrm{U}(N_{\mathrm{F}})$-symmetric way, then every closed fermion loop will introduce a power of $N_{\mathrm{F}} d_{\mathrm{spin}}$. As we continue away from $d = 4$ we can modify N_{F} so that this product is anything we want, subject to the condition that $N_{\mathrm{F}} = 1$, and $d_{\mathrm{spin}} = 4$ when $d = 4$.

It should be emphasized that DR is a formal, *perturbative* regularization scheme. One should not imagine that it represents e.g. the result of quantum field theory on some space of fractal dimension d, analytically continued to complex d. The result $\int \mathrm{d}^d q/q^2 = 0$, which we will encounter below, shows us that DR does not correspond to a positive measure. That said, DR is by far the most convenient method of computing anything in quantum field theory, in the perturbative regime. In DR, UV divergences show up as poles at discrete values of the space-time dimension. We will use DR extensively below and provide many examples of such poles.

The physically minded reader will be asking what the physical meaning of all of these varied cut-off schemes is. The correct answer is that most of them have no physics behind them at all. They are useful because the answers to correctly defined quantum field theory are universal: they do not depend on the cut-off scheme. More generally, we will argue that, if one restricts attention to an expansion in powers of momentum over the cut-off, then all effects of one kind of cut-off can be encoded in a choice of "irrelevant" operators in a different cut-off scheme. This is the notion of *effective field theory*.

The fact that we get a length, the Planck length (10^{-33} cm), by combining quantum mechanics and gravity, suggests that the physical cut-off on field theories of particle physics comes from the quantization of gravity. There are strong indications that theories of quantum gravity do not have precisely defined local observables and we do not yet understand either the correspondence principle by which these theories admit an approximate description in terms of local field theory or the circumstances in which the approximation is valid. The wonderful thing about the RG approach to field theory is that it allows us to discuss physics below the Planck energy scale without getting involved in the conceptual intricacies of quantum gravity. The RG shows us that, once we are in a situation in which it is valid to consider space-time geometry as a fixed background, then the analysis of local experiments in space-time depends on quantum gravitational effects only through a few parameters in a low-energy effective field theory.

9.3 Renormalization and critical phenomena

The deep understanding of renormalization and the renormalization group came from the work of Fisher, Kadanoff, Wilson, and other condensed-matter physicists, working on the problem of continuous phase transitions, or *critical phenomena*. We all know that many forms of ordinary matter undergo phase transitions as we vary control parameters like pressure, temperature, and external magnetic fields. Water turns into steam, solids melt, ferromagnets magnetize, etc. etc. The thermodynamics of a system is characterized by a free energy per unit volume, $F(T, P, H, \ldots)$, which depends on the control parameters. Typically F depends analytically on these parameters, and, for systems with a finite number of degrees of freedom, we can prove that this is so. Mathematically, phase transitions occur only for infinite volume (unless we have an infinite number of fields in our system at each point), and show up as discontinuities in derivatives of F w.r.t. e.g. the temperature. Ehrenfest defined the order of a transition as the order of the derivative which became singular or discontinuous at the phase transition.

Landau, who invented the first general theory of phase transitions (generalizing van der Waals' treatment of the liquid–gas system), realized that the main difference was between first-order transitions and all the others, which are called continuous transitions. Landau's theory is extremely simple. He postulated that the different phases of the system were characterized by an *order parameter* Φ distinguishing the phases. For example, in a ferromagnet Φ is the magnetization. In the liquid–gas transition, it is related to the difference between the density of the phases and the critical density at which the transition occurs.

The order parameter is defined so that its value is small near the transition. It is a macroscopic property of the system, so it does not have thermal fluctuations. Its value is therefore determined by minimizing a free-energy function $F(\Phi, T, \ldots)$ w.r.t. Φ, at fixed values of the control parameters. Near the transition, we can expand

$$F = a(T)\Phi^2 + b(T)\Phi^3 + c(T)\Phi^4 + \cdots.$$

The coefficients in the expansion depend analytically on the control parameters, of which we have indicated only the temperature. Keeping only the indicated terms, and assuming that $c > 0$, we have a quartic polynomial, with two local minima. As we vary T, it might happen that the global minimum jumps from one to the other. This is a first-order phase transition: F is continuous, but its first derivative is not, because we have jumped from one branch of the function $\Phi(T)$ to another when the free energies of the two branches cross each other.

Landau then realized that one could explain second-order transitions by assuming that b was zero (either by fine tuning some other control parameter, or because the system had a built in $\Phi \to -\Phi$ symmetry) and that, at the critical temperature, $a(T_c) = 0$. At this point two minima coalesce into one. The magnetization and free energies have square-root branch points in $T - T_c$.

Landau's theory, called *mean field theory*, has two remarkable properties: it is completely universal, depending only on the properties of polynomials, not on the detailed nature of the order parameter or the system it characterizes; and it predicts fractional power-law exponents for physical properties near continuous phase transitions. Experimentalists rushed to test these properties. The experiments were difficult, because one had to get very near the critical point to measure the exponents.

The result of this experimental work was exciting and puzzling. Continuous phase transitions did exhibit remarkable universality. For example, the critical point in the liquid–vapor phase diagram in the P, T plane has the same critical exponents as the so-called Ising ferromagnets (ferromagnets with a single preferred axis of up–down polarization). However, there was not complete universality. For example, XY ferromagnets with a whole preferred plane of polarization, or those in which there is no preferred axis, had different critical exponents. Furthermore, while the exponents were not wildly different from the Landau square roots, they were definitely different. Clearly something was right about Landau's ideas, but something much more intricate and interesting was going on.

This was confirmed by another set of phenomena,[5] which were observed by scattering light and neutrons off of substances undergoing second-order transitions. These phenomena could be explained by assuming that correlation functions of local quantities, like the local magnetization, obeyed scaling laws of the form

$$\langle M(x)M(y)\rangle \sim |x-y|^{-2\beta},$$

at very large distances and temperatures at the critical point. By contrast, away from critical points, three-dimensional condensed-matter systems generally exhibit what is called a *correlation length*

$$\langle M(x)M(y)\rangle \sim \frac{e^{-\frac{|x-y|}{l_c}}}{|x-y|}.$$

The three new phenomena of *universality classes*, rather than a completely universal *mean field theory*, *non-mean field critical exponents*, and *infinite correlation length* at the critical point, received a common explanation in terms of the renormalization group.

Kadanoff's first crude form of the renormalization group is called the *block spin transformation*. Kadanoff first reasoned that, if critical phenomena had universal aspects, then interesting results for real systems could be extracted from simple models like the Ising model. This model consists of a classical spin, which can take only the values ± 1, situated on the sites of a d-dimensional hypercubic lattice (where $d = 1, 2$ or 3 for systems realizable in condensed-matter laboratories). The Hamiltonian is

$$H_0 = -J \sum \sigma(x)\sigma(x+\mu),$$

where the sum runs over all nearest-neighbor points in the lattice.

Kadanoff's great insight was to realize that, if critical phenomena really had something to do with an infinite correlation length, then the microscopic dynamics was

---

[5] The most spectacular of which is *critical opalescence*.

irrelevant and there should be some sort of long-wavelength effective theory that captured the phenomena. Landau's theory could be viewed as a sort of guess at what this long-wavelength theory was (a so-called Gaussian model), but the experiments showed that this guess was not correct. How should one systematically calculate the effective long-wavelength Hamiltonian?

The slogan that works, as it does for so many other things, is *little steps for little feet*. If you're working on a lattice with spacing a, and you want to investigate phenomena on a length scale much bigger than a, first go to $2a$, then from there to $4a$, and so on. Kadanoff did this by defining blocks on the lattice, and defining Φ_i to be the average of the spins on the ith block. He imagined doing the partition sum over all spins, with the Φ_i held fixed. One then gets a new system, whose variables are the Φ_i, defined on a lattice of spacing $2a$. The Hamiltonian takes the form

$$H_1 = \sum K_n^1(x_1, \ldots, x_n)\Phi(x_1)\ldots\Phi(x_n),$$

where the sums now go over all lattice points and all n.

It isn't easy to calculate the functions K_n^1, but it is easy to argue that they all have short-range correlations, because the original Hamiltonian did, and, by keeping the Φ_i fixed, we have not allowed any long-wavelength fluctuations. That is, the sum over Φ_i, *which we have not yet done*, contains both configurations where the Φ_i randomly fluctuate from point to point on the lattice with spacing $2a$ and Φ_i configurations of longer wavelength. It is only by summing over the latter that we can produce long-range correlations in the system. We can now iterate the procedure to get new effective theories (with the same partition function) for lattices of larger and larger spacing. After k steps we have

$$H_k = \sum K_n^k(x_1, \ldots, x_n)\Phi(x_1)\ldots\Phi(x_n),$$

where the functions K_n^k have short-range correlations on a lattice of spacing $2^k a$. Note that we have used the same name for the variables at all steps but the first. Actually these variables all have different discrete ranges, but as k gets large it is clear that we can let the variables be continuous and replace sums over Φ by integrals. Furthermore, although all the Φ variables are bounded between ± 1 we can abandon this constraint and replace it by a potential that favors Φs in this compact range. Our system now resembles a lattice field theory, with the statistical-mechanical Hamiltonian playing the role of the Euclidean action.

As we take k to infinity, one of two things happens. If we are at a value of the temperature for which the correlation length is finite, then the functions K_n^k become of shorter and shorter range in lattice units, once we take the spacing larger than the correlation length. Eventually, they are all Kronecker deltas, and there are no interactions between the lattice points. We call this limit the *trivial fixed point of the renormalization group*. But suppose that we are at the critical temperature, for which the correlation length is infinite. In this case, Kadanoff realized that the only way to get a sensible limit was to assume that the Hamiltonian approached a fixed point, in terms of appropriately rescaled variables $\hat{\Phi}$. That is, the fixed-point Hamiltonian had to be invariant under scale transformations of space combined with a rescaling of Φ. This then predicts scale-invariant correlation functions, as observed experimentally. It further predicts that the scaling exponents are characteristic of the fixed-point Hamiltonian and independent of the microphysics. This explains universality.

However, unlike Landau's theory, Kadanoff's block spin idea can accommodate different universality classes. For example, there is no reason for the RG transformation for a system with Ising symmetry to be the same as that for planar or fully rotation-invariant spin symmetry. Detailed calculation shows that it is not. Thus, block spin or renormalization theory explains the qualitative nature of critical phenomena. It remained for Fisher, Wilson, and others to appreciate the connection to Euclidean field theory, and to use field-theoretic tools to calculate detailed numerical agreement of the theory with a wide variety of experiments. Rather than following them on this fascinating journey, we will now return to the application of these ideas to relativistic quantum field theory itself.

9.4 The renormalization (semi-)group in field theory

Consider a scalar field theory with the following action:

$$
S = \frac{1}{2} \int \frac{d^4 p}{(2\pi)^4} \phi(p)\phi(-p)K(p^2/\Lambda^2)
$$

$$
+ \sum_{n=1}^{\infty} \int d^4 p_1 \ldots d^4 p_n \, g_n(p_1, \ldots, p_n; \Lambda) \delta^4\left(\sum p_i\right) \phi(p_1) \ldots \phi(p_n)
$$

$$
+ \int d^4 p \, J(p)\phi(p),
$$

where $K(p^2, \Lambda^2)$ is smooth, behaves like p^2 for small p, and blows up exponentially at $|p| > \Lambda$. The g_n are smooth functions bounded by polynomials in the momenta. The requirement of smoothness guarantees (by virtue of the Riemann–Lebesgue lemma) that interactions are of short range in position space. That is, the functions $g_n(x_1, \ldots, x_n)$, while not necessarily local, fall off exponentially with distance, on a scale of order Λ^{-1}. We want to study physics below some scale $\Lambda_R \ll \Lambda$. We therefore take the source function to vanish for $p^2 > \Lambda_R^2$. We call the first quadratic term S_0 and everything else but the last linear term S_I.

It is easy to see that this action has a finite perturbation series for any choice of the functions in the action. The exponentially falling propagators make every loop integral converge in the UV, because the measures of integration and the functions g_n grow at most like polynomials. Because we have chosen to perturb around a massless theory, we might find IR divergences, but we know that these can be eliminated by shifting a mass term from the function g_2 in S_I to S_0. They have nothing to do with the UV problem.

The basic idea of the RG is to rescale the cut-off to a lower value, without changing the physics below Λ_R. In this way we can eventually get an action that contains only momentum scales of order Λ_R or smaller, which has the same physics as the original action. We approach this by writing a differential equation for the required change in the action for an infinitesimal scale change. By integrating this equation we obtain the required elimination of degrees of freedom.

It should be noted that in using the Euclidean formalism we are implicitly assuming that the integrated degrees of freedom are in their ground state. That is, they are excited only by their interactions with the low-energy degrees of freedom. We have not sent in any incoming high-energy waves. This follows from the fact that the free Euclidean Green function is the analytic continuation of the *vacuum expectation value* of fields. Any other Green function for the Lorentzian Klein–Gordon operator would have wave packets propagating at infinity.

We now want to find an equation for S_I that allows us to vary Λ without changing the physics. Let $\partial_t \equiv \Lambda \, \partial/\partial\Lambda$. I claim that the equation

$$\partial_t[\mathrm{e}^{-S_I}] = -\frac{1}{2}\int \frac{\mathrm{d}^4p}{(2\pi^4)}\, \partial_t K^{-1}(p)\left[\frac{\delta^2 \mathrm{e}^{-S_I}}{\delta\phi(p)\delta\phi(-p)}\right]$$

does the trick. Indeed, if we take the scaling derivative of the numerator of the generating functional Z, and use this formula, we get

$$\partial_t I[J] = \int [\mathrm{d}\phi]\mathrm{e}^{-(S_0+\int \phi(p)J(-p))}$$

$$\times\left[-\frac{1}{2}\int \frac{\mathrm{d}^4p}{(2\pi^4)}\, \partial_t K(p)\phi(p)\phi(-p)\mathrm{e}^{-S_I}\right.$$

$$\left.-\frac{1}{2}\int \frac{\mathrm{d}^4p}{(2\pi^4)}\, \partial_t K^{-1}(p)\left[\frac{\delta^2 \mathrm{e}^{-S_I}}{\delta\phi(p)\delta\phi(-p)}\right]\right].$$

We can use functional integration by parts to throw the functional derivatives acting on e^{-S_I} onto $\mathrm{e}^{-S_0+\int \phi J}$. The terms where the derivatives act on the source give us integrals of the form $\int J(p)\partial_t K^{-1}(p)$, which vanish, because these two functions have disjoint support. The terms where they act on S_0 give

$$\mathrm{e}^{-S_0-S_I-\int \phi J}\left[-\frac{1}{2}\int \frac{\mathrm{d}^4p}{(2\pi^4)}(-K(p)\partial_t K^{-1}(p)\delta^4(0) + \phi(p)K(p)\partial_t K^{-1}(p)K(-p)\phi(-p))\right].$$

The infinite momentum-space delta function in the first term is a factor of the space-time volume. It corresponds to the renormalization of the vacuum energy density from the modes that are dropped when we lower the cut-off. It will cancel out against a similar contribution from the denominator when we compute

$$\partial_t Z[J] = \frac{\partial_t I[J]}{I[0]} - Z[J]\frac{\partial_t I[0]}{I[0]}.$$

Using the identity $\partial_t K^{-1} = -K^{-2}\,\partial_t K$, we see that the second term cancels out against the explicit scale derivative of S_0, so that $Z[J]$ is unchanged.

We can rewrite the exact RG equation for S_I as

$$\partial_t S_I = \frac{1}{2} \int \frac{\mathrm{d}^4 p}{(2\pi^4)} \, \partial_t K^{-1} \left[\frac{\delta S_I}{\delta\phi(p)} \frac{\delta S_I}{\delta\phi(-p)} - \frac{\delta^2 S_I}{\delta\phi(p)\delta\phi(-p)} \right].$$

In terms of the coefficient functions in the expansion of S_I, this has the form

$$\frac{\mathrm{d}g_n(P_n)}{\mathrm{d}t} = \int \frac{\mathrm{d}^4 p}{(2\pi)^4} \sum g_{k+1}(P_k, p)\partial_t K^{-1}(p) g_{n-k+1}(P_{n-k}, -p)$$

$$+ \int \frac{\mathrm{d}^4 p}{(2\pi)^4} \, \partial_t K^{-1}(p) g_{n+2}(p, -p, P_n).$$

The sum in the first term is over all ways of partitioning the momenta (which we have labeled by a single letter P_n) into two groups P_k and P_{n-k}, each containing the indicated number of single-particle momenta.

The two contributions to the variation of S_I have an interpretation in terms of Feynman diagrams (Figure 9.2), one a loop diagram and one a tree. Note that the internal lines in these diagrams are proportional to $\partial_t K$, which is non-zero only for momenta near the cut-off. We interpret this as integrating out degrees of freedom in an infinitesimal shell near the cut-off. Because the phase space is so small, higher loop diagrams are negligible.

The RG equation for S_I is of the form

$$\partial_t X = -\frac{1}{2} \nabla^2 X,$$

where X is the functional e^{-S_I} and

$$\nabla^2 = \int \frac{\mathrm{d}^4 p}{(2\pi)4} \, \partial_t K^{-1}(p) \, \frac{\delta^2}{\delta\phi(p)\delta\phi(-p)}.$$

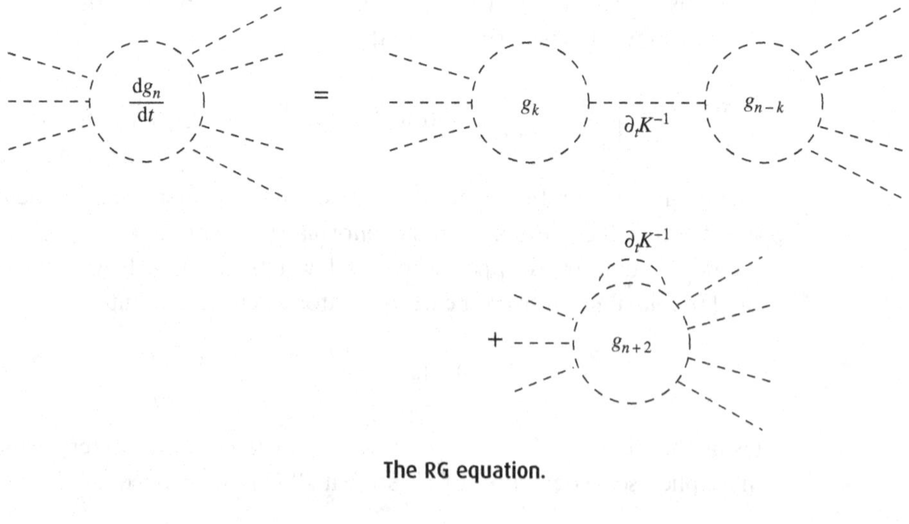

Fig. 9.2.   The RG equation.

This looks like a heat equation on the space of all functionals X, and, like all heat equations, it is a gradient flow. Not all functionals can be written as the exponential of an S_I, which has an expansion in terms of polynomially bounded smooth functions g_n. This requirement defines a curved sub-manifold on the infinite-dimensional linear space of functionals X on which the heat flow takes place. The explicit form of the equations for g_n in terms of Feynman diagrams tells us that the sub-manifold of quasilocal cut-off actions it preserved by the heat flow.

As noted above, the RG equations for the functions g_n have the form

$$\partial_t g_n(p_1, \ldots, p_n) = \sum g_k(p_{s_1} \ldots p_{s_{k-1}}, p) \partial_t K^{-1}(p) g_{n-k+1}(p, p_{s_k} \ldots p_{s_n})$$
$$- \int \frac{\mathrm{d}^4 p}{(2\pi)^4} \, \partial_t K^{-1}(p) g_{n+2}(p_1 \ldots p_n, p, -p).$$

The sums in the first term are over all choices of $k < n$ and over all permutations s_i. We employ the convention that the sum of all momenta appearing inside a vertex function g_n is zero. By expanding the functions $g_n(p_1, \ldots, p_n)$ in a power series around zero, we can convert these equations into an infinite number of equations for the expansion coefficients. Label the latter by g^I. The equations have the form

$$\partial_t g^I = \beta_J^I g^J + \beta_{JK}^I g^J g^K = \beta^I(g).$$

The fact that the flow is also gradient means that there is a positive-definite metric $G_{IJ}(g)$ such that $P_I \equiv G_{IJ} \, \partial_t g^I$ satisfies

$$\partial_I P_J = \partial_J P_I,$$

so that $P_I = \partial_I V$. Gradient flow can have only two types of asymptotic behavior, because the potential V increases along the flow. It can have a runaway to infinite values, or it can hit a fixed point. Renormalized quantum field theories correspond to fixed-point behavior of the RG flow.

Since the flow is gradient, its asymptotic behavior is governed by a fixed point where $\beta^I(g_*) = 0$. Let $D^I = g^I - g_*^I$. Then

$$\dot{D}^I = \lambda_J^I D^J + \beta_{JK}^I D^J D^K,$$

where

$$\lambda_I^J = \beta_I^J + \beta_{JK}^I g_*^K.$$

If we diagonalize λ, calling its eigenvalues $4 - d^I$, we see that the initial values of the coefficients of irrelevant operators, with $d^I > 4$, disappear exponentially from the flow as $t \to \infty$. The marginal and relevant operators, with $4 - d^I \le 0$, asymptotically satisfy a closed set of highly non-linear RG equations obtained by eliminating the irrelevant operators in terms of the others. The coefficients of irrlevant operators in the effective Lagrangian at $\Lambda_R \ll \Lambda$ are completely determined by the relevant and marginal ones.

Actually some of the marginal operators may actually be *marginally irrelevant*. That is, when non-linear corrections to the flow are taken into account, they may also be irrelevant. However, since their initial conditions are forgotten at a rate that is a power law, rather than an exponential, in *t* (and thus logarithmic in scale), the proper treatment of marginally irrelevant operators depends on the context. In a mathematical treatment, in which we try to take the cut-off to infinity, we must treat them like irrelevant operators. However, in the real world, the ultimate cut-off scale is unlikely to be larger than $M_P \sim 10^{19}$ GeV, the Planck energy scale defined by combining Newton's constant with Planck's. In theories with large or warped extra dimensions, it might be much lower. Even without such exotic theoretical input, it is easy to imagine that the elementary fields of our current effective field theories are composites of a deeper theory hiding just around the corner of the next collider experiment. Thus, we should be careful not to dismiss irrelevant, and especially marginally irrelevant, operators in too cavalier a fashion. Discovering that the effective field theories we need to explain current experiments contain irrelevant operators simply tells us that there is a scale of new physics not too far away. Marginally irrelevant operators may be pointing to the need for changes in our models only at energies that are exponentially far away in scale.

In fact, if we are considering the possibility of an energy scale just above our current experimental resolution, we should always expect to find irrelevant operators with coefficients proportional to inverse powers of that nearby scale in our effective description of nature. The *non-renormalizability* of these effective theories is telling us that there are missing pieces of our current theory and that we can expect to encounter new phenomena at that nearby scale. This is for example the case for the non-linear sigma model describing low-energy pion physics. The extra physics is the entire rich structure of QCD. Similarly, the non-renormalizability of Fermi's theory of weak interactions told us to expect new physics by about 250 GeV. In fact, as a consequence of the small dimensionless parameters of the electro-weak gauge theory, the new physics actually appeared below 100 GeV, at the W-boson mass.

This section contains the key idea of renormalization theory, which is summarized in the following picture (Figure 9.3). The RG flow stops at fixed points, whose critical surface, or basin of attraction, has low co-dimension in the space of all couplings.

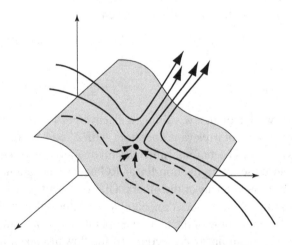

Fig. 9.3.   Renormalization-group trajectories near a critical surface.

The infinite-cut-off limit of the theory at the fixed point is conformally invariant. Initial coupling values on the critical surface all flow to the fixed point. Near the fixed point, infinitesimal directions on the critical surface are controlled by *irrelevant or non-renormalizable* perturbations of the Lagrangian. Flows beginning off the critical surface do not reach the fixed point, but if the initial couplings are close to the critical surface the flow comes close to the fixed point before moving away. We can obtain a non-scale-invariant infinite-cut-off limit by tuning the initial couplings to the critical surface as we take the cut-off to infinity. Near the fixed point, perturbations off the critical surface are called relevant or super-renormalizable operators. The general definition of a quantum field theory is to take the cut-off to infinity, tuning the coefficients of relevant operators to zero in order to obtain finite Green functions.

First-order analysis of the RG equations near the fixed point shows that there are generally marginal operators, which don't flow at all in first order. Higher-order analysis divides these into truly marginal, marginally relevant, and marginally irrelevant operators. Truly marginal operators indicate the existence of a line or higher-dimensional surface of fixed points. If we really insist on taking the cut-off to infinity, marginally relevant and irrelevant operators are just like relevant and irrelevant ones. On the other hand, in the real world, we expect the cut-off above which all quantum field theories fail to be no bigger than 10^{19} GeV. Thus, marginally irrelevant couplings, if they are sufficiently small, would be expected to appear in our theories.

9.4.1 Changing the cut-off

We now want to argue that IR physics is in fact independent of the way we choose to cut-off the theory. We begin by investigating changes in the function $K(p^2/\Lambda^2)$. By definition, these changes all take place at scales above the original cut-off. Therefore, once we have flowed to a momentum much lower than the cut-off they don't affect the RG equations. Thus, a change in K can be absorbed into a change in the initial conditions for the flow, and affects the low-energy physics only through the values of the relevant and marginal parameters.

A more drastic change in cut-off scheme is a Euclidean lattice. Here momentum space is compactified on a torus, with cycles of size $2\pi/a$. However, we can work instead with a non-compact momentum space, if we insist that the propagator vanish identically above $|P| = 2\pi/a$.[6] We do this by multiplying the lattice propagator by a smooth function. Now we can repeat the argument of the previous paragraph.

The upshot of these arguments and their generalization is that we can impose the cut-off in any way we like, once we have identified the fixed point of the RG that is relevant to our theory. In perturbation theory, we are always working with specific

---

[6] For Dirac fermions, depending on the way we implement the lattice Lagrangian, we have to proceed with care. Some versions of the lattice Dirac equation actually describe multiple copies of a continuum fermion. The extra copies come from momenta of order π/a. We must take care to incorporate all of these copies in the low-energy theory if we really want to describe the continuum limit of the lattice theory.

Gaussian fixed points, so any way of cutting off Feynman diagrams is acceptable, and different cut-off schemes will simply lead to different parametrizations of the space of relevant and marginal interactions. In words we will use later, they just lead to different *renormalization schemes*. The exact RG equations we have discussed are conceptually important, but perturbative calculations generally use the DR scheme, which leads to simpler computations.

9.5 Mathematical (Lorentz-invariant, unitary) quantum field theory

The RG enables us to understand what we mean by a mathematically well-defined, Lorentz-invariant local quantum field theory. The classification of operators according to their relevance fixes the dimension of the so-called critical sub-manifold associated with a fixed point. A general RG flow will miss the fixed point because, nearby the fixed point, some relevant couplings will be non-zero. The number of relevant (including marginally relevant) couplings is the co-dimension of the critical manifold in the space of all couplings.

One way to obtain a cut-off-independent quantum field theory is to choose initial conditions on the critical manifold, and take the cut-off to infinity. By construction, the IR limit is cut-off-independent. The theory we obtain in this way is *scale-invariant*, because dimensional analysis tells us that, if we rescale the metric of space-time, we can always compensate by rescaling the cut-off. If we consider field theory in a general background metric (which we can always do for local Lagrangians), rescaling is the operation

$$\int \mathrm{d}^4x \, g^{\mu\nu}(x) \, \frac{\delta Z}{\delta g^{\mu\nu}(x)}.$$

The definition of the stress-energy tensor of a quantum field theory is that this rescaling is just equal to

$$\int \mathrm{d}^4x \, g^{\mu\nu}(x) \langle T_{\mu\nu}(x) \rangle.$$

That is, the stress tensor is just the source for the gravitational field. In most cases, one can show that this variation can vanish only if it vanishes locally, that is, if the theory is Weyl-invariant (invariant under local rescalings of the metric). The group of local rescalings which can be compensated by coordinate transformations so that they leave the Minkowski metric $\eta_{\mu\nu}$ invariant is called the conformal group. Theories with $\eta^{\mu\nu} T_{\mu\nu} = 0$ are called conformal field theories (CFTs). For particle-physics applications we deal primarily with unitary CFTs, though certain non-unitary CFTs are important in statistical mechanics and string theory.

There is a more general way to get a cut-off-independent result from a fixed point with relevant directions. Consider flows that start near, but not on, the critical surface. Now, take the cut-off to infinity, but, as we do this, let the initial conditions approach the critical surface in such a way that we get a finite limiting theory. To put it another way: for fixed values of the cut-off and of the near critical initial condition $g^I(0)$ for the relevant parameters, the flow passes near the fixed point at some energy scale $\mu \ll \Lambda$ and then flows away. μ is a function of Λ and $g^I(0)$. By dimensional analysis it has the form $\mu = f(g^I(0))\Lambda$, with f vanishing on the critical surface. We tune the initial condition as $\Lambda \to \infty$ so that μ remains finite in the limit. This is one condition on all of the relevant parameters. With N as the co-dimension of the critical surface, we obtain a cut-off-independent theory that depends on the energy scale μ and $N - 1$ dimensionless parameters. The finite parameters which control how we have taken the infinite-cut-off limit are called the renormalized couplings of the theory. There is an infinite number of ways to parametrize the renormalized theory, and we shall discuss a few of them.

This picture of a general quantum field theory as a RG trajectory that passes infinitesimally close to the fixed point CFT suggests a new way of constructing the whole set of $\Lambda \to \infty$ theories associated with a given RG fixed point from the fixed-point CFT. We can read off the dimensions of operators in the CFT by looking at their response to scale transformations. This enables us to construct the list of relevant operators without knowledge of the RG flow for finite cut-off. We now do standard perturbation theory

$$\langle 0|TO_1(x_1)\dots O_n(x_n)|0\rangle_G = \langle 0|TO_1(x_1)\dots O_n(x_n)e^{i\int d^4x\, G^I O_I(x)}|0\rangle_0.$$

In this formula the O_i in the Green function can be any operator in the theory, while the G^I are zero for all except relevant operators.

This is essentially what we are doing when we do renormalization in ordinary perturbation theory, except that in that case we first resum the results of perturbation in quadratic relevant operators (mass terms) because it is easy to do. Special attention must be paid to marginally irrelevant operators. In principle, they should not be included in the list of non-zero G^I. However, since the signature of their initial conditions vanishes only logarithmically in the infrared, it is hard to see the difference between marginally relevant and irrelevant operators with small coefficients until energy scales exponentially larger than the renormalization scale μ.

To see how renormalization works from this point of view, note that all UV divergences in this perturbation series come from places where points coincide. Away from such places the Green functions are all finite. Infrared divergences could be cut off by putting the whole system in a large Euclidean box. There are two kinds of UV divergence. The first have to do with singularities that occur when one or more of the integrated points from the expansion of the interaction collide with one of the operator insertions $O_i(x_i)$. Recall that in free massless CFT we had to define composite operators by a careful limiting procedure (normal ordering). The divergences associated with the coincidence of perturbative vertices and operator insertions can be removed by a redefinition of the operators O_i, since they are localized near the points where these operators sit. We will not deal explicitly with these divergences, though the methods for dealing with them follow the same line of reasoning as that which we will see below.

The other sort of UV divergence comes from the collision of two integrated vertices. We deal with it by doing an operator product expansion. If x and y are the space-time positions of the two colliding vertices we write

$$\int d^4x \, d^4y \sum |x - y|^{\Delta_K - \Delta_I - \Delta_J} G^I G^J C_{IJ}^K O_K(x).$$

We have again used a shorthand in which a power of $|x - y|$ stands for any tensor structure with the same scaling dimension. Divergences in the integral over $(x - y)$ will occur whenever $\Delta_K - \Delta_I - \Delta_J \leq -4$, as long as the angular integral does not vanish. The latter condition tells us that only scalar operators O_K can contribute to the divergence.[7]

9.6 Renormalization of ϕ^4 field theory

We're now ready to do an explicit perturbative example of renormalization. The CFT of a single free massless scalar has four relevant/marginal perturbations: $: \phi^n :$ $n = 2, 3, 4$ and $: (\partial_\mu \phi)^2 :$.[8] The last of these is equivalent, after partial integration, to $: \phi \partial^2 \phi :$, which vanishes. It does not really change the theory we obtain in the limit, but only the normalization of the field. However, it is necessary to retain this operator in the action in order to get cut-off-independent Green functions.

The term ϕ^3 is the only one of these operators which is not invariant under the $\phi \to -\phi$ symmetry of the fixed-point theory. We will have a lot more to say about renormalization and symmetry later. For now, we can notice that the exact RG equations do not generate interactions odd in ϕ from even ones. Thus, we can think of a restrictive class of theories that retain this symmetry, and drop the cubic operator from our list of relevant terms.

There are two ways of speaking about perturbative renormalization. In the first, which I prefer, and will use, one writes a Lagrangian[9]

$$\mathcal{L} = \frac{1}{2} : [(\partial_\mu \phi_0)^2 - m_0^2 \phi_0^2 - 2\frac{\lambda_0}{4!}\phi_0^4] :.$$

One then defines a rescaled field $\phi = Z^{-\frac{1}{2}}\phi_0$, and allows the coefficients Z, m_0^2, λ_0 to be cut-off-dependent in order that the field ϕ has finite Green functions. The alternative method is to split the Lagrangian into $\frac{1}{2} : [(\partial_\mu \phi)^2 - m^2\phi^2 - (\lambda/4!)\phi^4] : + \delta\mathcal{L}$, where m is the physical mass of the particle and λ a physical on-shell scattering amplitude at the symmetric point in momentum space. The interaction then contains "counterterms,"

---

[7] Strictly speaking we should insert a regulator. As long as this is done in a rotation-invariant manner, the conclusion that only scalars contribute is valid.

[8] There is also ϕ itself, but we can use the symmetry $\phi \to \phi + a$ of the free theory to eliminate this in terms of the other three powers.

[9] We could actually omit the normal ordering here. The normal ordering of the quadratic operators just shifts the vacuum energy. That of the quartic term also shifts the mass renormalization condition.

Fig. 9.4.   s-, t- and u-channel diagrams for the one-loop four-point function.

which are used to cancel out loop corrections to the physical mass and coupling and the normalization of the single-particle matrix element of the field. One can define an analogous counterterm method for any choice of definition of the renormalized parameters.

Now let us consider loop corrections to the quantum action. Since we are using a normal ordered interaction, there is no one-loop correction to the 1PI two-point function. For $n \geq 5$ the one-loop correction to the n-point function is cut-off-independent (the loop integrals are finite). The one-loop, 1PI Euclidean four-point function is given by a sum of three diagrams (Figure 9.4).

They all have the form

$$\frac{1}{2}\lambda_0^2 \int \frac{d^4 p}{(2\pi)^4} \frac{1}{(p^2 + m_0^2)[(p - q)^2 + m_0^2]},$$

where $q = p_1 + p_2, p_1 - p_3, p_1 - p_4$, in the three different diagrams. The combinatoric factor $\frac{1}{2}$ comes from the symmetry of the two internal lines in the diagram.

To this order, the 1PI two-point function of ϕ is $Z^{-1}(p^2 + m^2)$. Since it is independent of λ_0 it must reproduce the physics of the free theory. We identify $Z = 1$ and $m_0^2 = m^2$, the physical mass of the free particle. Thus, any divergence in the four-point function is related to the cut-off dependence of λ_0. In order to get explicit expressions, we have to decide how we are going to cut off the integral. The general theory of the RG assures us that we can do this any way we like. The functional form of the bare parameters in terms of renormalized parameters and the cut-off will depend on the cut-off procedure, as well as on the definition of the renormalized parameters (usually called the *renormalization scheme*).

We will use the dimensional regularization scheme. Rewrite the propagators using the Schwinger proper-time trick,

$$\frac{1}{P} = \int ds\, e^{-sP},$$

and then do the Gaussian momentum integrals, using the formula

$$\int \frac{d^d p}{(2\pi)^d} e^{-sp^2} = \left(\frac{\pi}{4s}\right)^{\frac{d}{2}}.$$

The result is

$$\frac{1}{2} \int_0^\infty s\, ds \int_0^1 dx \left(\frac{\pi}{4s}\right)^{\frac{d}{2}} e^{-s[x(1-x)q^2 + m^2]}.$$

We have arrived at this formula by making the change of variables $s_1 = (1 - x)s, s_2 = xs$, where s_1 is the Schwinger parameter for $(p^2 + m^2)^{-1}$ and s_2 the parameter for the propagator that depends on external momenta. Note that, at this point, the space-time dimension d is just a parameter, and can take on arbitrary complex values. We will see that the integrals actually define analytic functions of d, with poles at $d = 4$. Assuming that this is so, we can simplify things by rescaling the s variable to make the argument of the exponent simply s. We obtain

$$\frac{1}{2} \int_0^\infty s \, ds \int_0^1 dx \left(\frac{\pi}{4s}\right)^{\frac{d}{2}} [x(1-x)q^2 + m^2]^{\frac{d}{2}-2} e^{-s}.$$

The s integral now gives $\Gamma(2 - d/2)$, which, as we claimed, has a pole at $d = 4$. Note that, if we take derivatives w.r.t. q^2 or m^2, we bring down a power of $(d - 4)$, which cancels out the pole. Thus, the divergence resides only in the constant term.

The tree-level 1PI four-point function is just λ_0, so, if we write $\lambda_0 = \lambda + \lambda^2 f(d)$ where λ is finite, and f has a pole which cancels out the pole in the one-loop computation, then we will get a finite answer as a function of λ.

We continue to discuss the renormalization program by turning to the two-loop, two-point function.[10] There is one graph, Figure 9.5, with a symmetry factor $1/3!$.

It can be evaluated by introducing three Schwinger parameters,

$$\Sigma_2(q) = \frac{\lambda_0^2}{(2\pi)^{2d} 3!} \int ds_1 \, ds_2 \, ds_3$$
$$\times \int d^d p_1 \, d^d p_2 [e^{-s_1(p_1^2+m_0^2)} e^{-s_2(p_2^2+m_0^2)} e^{-s_3((q-p_1-p_2)^2+m_0^2)}].$$

We do the momentum integrals using the Gaussian integral formula

$$\int d^d p \, e^{-ap^2+bp} = \left(\frac{\pi}{a}\right)^{\frac{d}{2}} e^{\frac{b^2}{4a}}.$$

To display the result, I will write the integrand as it appears after each Gaussian integral. We will change variables to $s_i = sx_i$, $\sum x_i = 1$, with $ds_1 \, ds_2 \, ds_3 = s^2 \, ds \, dx_1 \, dx_2 \, dx_3 \, \delta(1 - \sum x_i)$. I will also omit the prefactor $\lambda_0^2/[(2\pi)^{2d} 3!]$. After the p_1 integration we obtain

$$s^2 \left(\frac{\pi}{s(x_1+x_3)}\right)^{\frac{d}{2}} e^{-sm_0^2} e^{-sp_2^2 \frac{x_1 x_2 + x_2 x_3 + x_1 x_3}{x_1+x_3}} e^{2sp_2 q \frac{x_1 x_3}{x_1+x_3}} e^{-sq^2 \frac{x_1 x_3}{x_1+x_3}}.$$

Two-loop mass and wave-function renormalization.

[10] We will not calculate the two-loop 1PI four-point function, which should be done to complete the renormalization program. The interested reader should try it, and will see that the calculation is quite simple.

After the p_2 integral, this simplifies to

$$s^2 \left(\frac{\pi}{s}\right)^d \left(\frac{1}{x_1 x_2 + x_2 x_3 + x_1 x_3}\right)^{\frac{d}{2}} e^{-s\left(m_0^2 + q^2 \frac{x_1 x_2 x_3}{x_1 x_2 + x_2 x_3 + x_1 x_3}\right)}.$$

Next, we do the s integral, which produces a factor of

$$\Gamma(3 - d)\left(q^2 \frac{x_1 x_2 x_3}{x_1 x_2 + x_2 x_3 + x_1 x_3} + m_0^2\right)^{d-3}.$$

The only divergence we can get in the x_i integrals comes at points where two of the variables approach 0 while the other is forced to be 1 by the delta function. Near such limits, the integrand behaves like $\mathrm{d}x\, \mathrm{d}y\, W/(x + y)^{\frac{d}{2}}$, where x, y are the two variables which go to zero. The coefficient W is independent of q^2 because the coefficient which multiplies q^2 vanishes at the dangerous points. Thus, the answer contains single and double poles at $d = 4$, which are momentum-independent, and a single pole proportional to q^2.

The divergent part of the effective action coming from this diagram is given by

$$\Gamma_{\mathrm{div}} = \frac{\lambda_0^2}{3!(4\pi)^d}\Gamma(3 - d) \int \mathrm{d}^4 x[(\partial_\mu \phi)^2 I_1 + m_0^2 \phi^2 I_2].$$

$I_{1,2}$ are given by the following parametric integrals:

$$I_1 = \int_0^1 \mathrm{d}x \int_x^1 \mathrm{d}y\, \frac{x(y - x)(1 - y)}{\left(x(y - x) + y(1 - y)\right)^3},$$

$$I_2 = \int_0^1 \mathrm{d}x \int_x^1 \mathrm{d}y\, \frac{1}{\left(x(y - x) + y(1 - y)\right)^{\frac{d}{2}}}.$$

I_1 is finite, while I_2 has a single pole at $d = 4$. The minimal subtraction (MS) renormalization scheme is defined by keeping only the pole terms in the relation between renormalized and bare fields and parameters. In defining the divergent part of the above integrals in the MS scheme, we keep both the pole and the finite correction in I_2, and throw away terms of order $d - 4$. The residues of all poles in Γ_{div} are given by integrals we can evaluate in terms of elementary functions, but the calculation is a bit tedious.

We have demonstrated that all the divergent terms in the effective action correspond to one of the three operators $: (\partial_\mu \phi)^2 :$, $: \phi^2 :$, $: \phi^4 :$, and therefore expect that cut-off dependence of the coefficients of these terms in the bare Lagrangian will render the Green functions independent of the cut-off in the limit that it goes to infinity. That is, a Lagrangian of the form

$$\mathcal{L} = \frac{1}{2} : [(\partial_\mu \phi_0)^2 - m_0^2 \phi_0^2 - \frac{\lambda_0}{2 \cdot 3!}\phi_0^4] :,$$

with appropriate cut-off dependence in m_0 and λ_0, will give finite Green functions for the field $\phi = Z^{-\frac{1}{2}}\phi_0$, with cut-off-dependent Z. The renormalization procedure consists of tuning the parameters in such a way that the rescaled field ϕ has cut-off-independent Green functions.

Our calculations have shown that

$$\Gamma_{\text{div}}[\phi] = \frac{a}{2}\lambda_0^2(\partial_\mu\phi_0)^2 + \frac{b}{2}\mu^2\lambda_0^2\phi_0^2 + \frac{c}{3!}\lambda_0^2\phi_0^4,$$

where

$$a = \frac{\lambda_0^2}{3!(4\pi)^d}\Gamma(3-d)I_1, \qquad b = \frac{\lambda_0^2}{3!(4\pi)^d}\Gamma(3-d)\frac{2m_0^2}{\mu^2}I_2, \qquad c = \frac{3}{16\pi^2(d-4)}.$$

We note that, in d dimensions, the coupling λ_0 has dimensions (mass)$^{4-d}$. To eliminate the cut-off dependence in the quantum action, we define $\lambda_0 = \mu^{4-d}(\lambda + A\lambda^2)$, $m_0^2 = m^2 + B\mu^2$, and $\phi_b = \sqrt{Z}\phi$, with $Z = 1 + D\lambda^2$. The parameters λ (which is taken to be dimensionless) and m^2 are held fixed as we take away the cut-off $d \to 4$, and the ϕ field should have a finite quantum action. In these formulae, we have introduced a parameter μ^2, called the *renormalization scale in dimensional regularization*. It is necessary to rewrite functions of the dimensional constant λ_0 in terms of the dimensionless parameter λ.

The relevant and marginal terms in the quantum action are

$$\frac{1}{2}(1 + a\lambda_0^2)(\partial_\mu\phi_0)^2 + \frac{1}{2}(m_0^2 + bm_0^2\lambda_0^2)\phi_0^2 + \frac{\lambda_0}{3!}(1 + c\lambda_0)\phi_0^4.$$

We write this in terms of renormalized fields and parameters, dropping terms of higher order than λ^2. The coefficient of the kinetic term is $(1 + D\lambda^2)(1 + a\lambda^2) \approx 1 + (D+a)\lambda^2$. That of the mass term is

$$(1 + D\lambda^2)(m^2 + B\mu^2\lambda^2 + b\mu^2\lambda^2) \approx m^2 + \lambda^2(Dm^2 + (B+b)\mu^2),$$

while the quartic term has coefficient

$$\frac{\lambda}{3!}(1 + (A + 2D + c)\lambda).$$

We first choose $D + a$ to have a cut-off-independent limit. Then B can be chosen so that $Dm^2 + (B + b)\mu^2$ has a cut-off-independent limit. Finally, A is chosen so that the renormalized quartic coupling is finite.

Note that this procedure does not introduce cut-off dependence into the $o(\lambda^2)$ finite part of the 1PI action. Γ_{finite} is initially written in terms of the bare parameters and bare fields, and is of order $\lambda_0 \sim \lambda$. When written in terms of renormalized fields and parameters, it has divergent pieces, but they are of order λ^3 and higher. The basic claim of renormalization theory is that these terms cancel out exactly against sub-divergences in higher-order Feynman graphs. The combinatorics of this appears daunting, but a proof that it works was carried out by a number of authors. Our Wilsonian approach through the exact RGE assures us that it *must* work.

It is worthwhile pausing here to remark on the precise connection between our discussion of the exact Wilsonian RGE and dimensional regularization. In the context of the RGE we have remarked that we can easily replace one cut-off by another, changing the kinetic function $K(p)$ or going over to a lattice cut-off, without changing the qualitative nature of the RG flow, its fixed points, or the dimensions of operators at those points. Dimensional regularization is just another way of cutting off the divergences, albeit one that is defined only perturbatively. We should expect to be able to perform renormalized computations near the Gaussian fixed point using DR, with couplings that are related to those of the Wilsonian RGE by a finite redefinition

$$\lambda_{DR} = \lambda_W + \sum k_n \lambda_W^n.$$

Returning to our computation, we note that requiring the quantum action to be finite does not determine the parameters A, B, D uniquely. Each choice of this finite ambiguity determines another way of parametrizing the renormalized field theory, called a *renormalization scheme*. One conceptually simple scheme called *on-shell renormalization* chooses the ambiguity so that m^2 is exactly the position of the zero in the 1PI two-point function, and the vacuum to one-particle matrix element of the renormalized field is 1; $-i\lambda$ is then chosen to be the value of the invariant $2 \rightarrow 2$ scattering amplitude at the symmetric point in momentum space, $s = 4m^2, t = u = 0$.

This is not always the most convenient scheme. It requires us to be able to find the masses of particles in perturbation theory. In QCD and other asymptotically free gauge theories this is impossible. It also ties our renormalization scheme to quantities in Minkowski space. For many purposes, a Euclidean renormalization scheme, which is not directly tied to physical masses, is better. For perturbative calculations, the most efficient scheme is the so-called $\overline{MS}$ or modified minimal subtraction scheme. This exploits the way in which dimensional regularization isolates cut-off dependence in terms of poles in analytic functions. The minimal subtraction scheme is defined by letting A, B, D, and analogous higher-order coefficients have precisely the poles in $d - 4$ that they need in order to give finite Green functions. No finite parts are admitted in the definition of these coefficients. It turns out that this prescription leads to a proliferation of transcendental numbers (related to the derivative of the Euler gamma function at $t = 1$) in formulae for finite amplitudes in terms of the MS parameters. The $\overline{MS}$ scheme allows specific finite parts in A, B, D, etc., which get rid of these transcendentals. The reader is urged to consult the book by Peskin and Schroeder [33] for many examples of computations in the $\overline{MS}$ scheme.

9.7 Renormalization-group equations in dimensional regularization

We have seen that one of the peculiarities of dimensional regularization is that it introduces an extra dimensionful parameter μ apart from the coefficient of the quadratic term in the action. We know that this is a good thing, and that it is appropriate to introduce such a parameter, *the renormalization scale* in any regularization scheme. The renormalization scale serves as a surrogate for the cut-off in the renormalized theory. If we recall our discussion of the exact RGE, general QFTs are defined by unstable flows near a fixed point, and a renormalization scale was introduced to mark the momentum scale at which the behavior of a given QFT deviates from its scale-invariant progenitor. We remarked that it was always possible to exchange the renormalization scale μ for (a power of) the coefficient of one of the relevant operators in the theory. It is possible, but not necessary. We can, instead, over-parametrize the theory in terms of μ and all of the G^I. Then we expect an equation telling us how the G^I must change with μ in order to describe the same physical situation. This is the remnant of the RG flow near the fixed point, and is also called the RGE *of the renormalized theory*.

To derive the RGE using dimensional regularization, we note that a bare 1PI Green function (we suppress the momentum dependence of the Green function for the moment) satisfies

$$\mu \, \partial_\mu |_{\lambda_0} \Gamma^0_n(m^2_0, \lambda_0, \mu) = 0$$

and

$$\Gamma^0_n(m^2_0, \lambda_0, \mu) = Z^{-\frac{n}{2}} \Gamma_n(m^2, \lambda, \mu).$$

The first of these equations expresses the fact that μ does not appear in the bare Green functions, written in terms of λ_0. We now apply the first equation to the second, and obtain

$$\left(\mu \, \partial_\mu + \beta \, \partial_\lambda + \beta_{m^2} \partial_{m^2} - \frac{n}{2} \gamma \right) \Gamma_n = 0.$$

In this equation, the μ derivative is taken at fixed λ and m^2, and we have defined

$$\gamma \equiv \mu \, \partial_\mu (\ln Z), \quad \beta \equiv (\mu \, \partial_\mu)\lambda, \quad \beta_{m^2} \equiv (\mu \, \partial_\mu)m^2,$$

where the derivatives are taken at fixed λ_0, m_0. These coefficient functions are finite functions of the renormalized parameters in any renormalization scheme. Finiteness of Γ_n implies that the μ-dependent piece of $\ln Z, \lambda$, and m^2 is finite. Since μ always appears raised to the power $k(d-4)$ in kth order in the bare perturbation expansion, the single-pole parts of the expressions for λ, m^2, and Z in terms of bare parameters are not μ-dependent. Furthermore, the multiple poles must all cancel out in the scaling derivatives, which define β_λ, β_{m^2}, and γ, since these are all finite. In the MS and $\overline{\text{MS}}$ schemes, the renormalization constants have a μ-independent finite part. Thus, in this scheme, β, β_{m^2}, and γ are μ-independent and come only from the pole part of their definition. By dimensional analysis, we must have $\beta = \beta(\lambda)$, $\gamma = \gamma(\lambda)$, and $\beta_{m^2} = m^2 \gamma_{m^2}(\lambda)$.

We can solve these equations in the following way. In words, the equation says that if we rescale μ we get a term proportional to Γ_n, as if Γ_n were just rescaling, plus other terms which can be compensated for by varying the couplings. This corresponds to the idea that the renormalization scale is playing the role that the cut-off played in the exact RGE. Near the fixed point, a change in energy scale is compensated for by a change of scale of the field and a change in the marginal and relevant couplings. So we make an *Ansatz*:

$$\Gamma_n(e^{-s}\mu, \lambda, m^2) = e^{F(s)} \Gamma_n(\mu, \lambda(s), m^2(s)).$$

On plugging this in, and comparing with the RGE, we find that any function satisfying this relation will be a solution of the RGE if

$$\frac{dF}{ds} = \frac{n}{2}\gamma(\lambda(s)),$$

$$\frac{d\lambda}{ds} = \beta(\lambda(s)),$$

and

$$\frac{dm^2}{ds} = m^2\gamma_{m^2}(\lambda(s)).$$

The boundary conditions on these equations are

$$F(0) = 0, \qquad \lambda(0) = \lambda, \qquad m^2(0) = m^2.$$

So the solution is

$$\Gamma_n(e^{-s}\mu, \lambda, m^2) = e^{\frac{n}{2}\int_0^s du\,\gamma(\lambda(u))}\Gamma_n\Big(\mu, \lambda(s), m^2 e^{\int_0^s du\,\gamma_{m^2}(\lambda(u))du}\Big).$$

Now consider the momentum dependence of Γ_n. In momentum space (and including the delta function of momentum conservation) Γ_n has engineering dimension $-n$ in mass units. Thus it is of the form $|p|^{-n}$ times a function that depends on ratios of the dimensionful quantities,

$$\Gamma_n(\mu, \lambda, m^2, e^t p_i) = e^{-nt}\Gamma_n(e^{-t}\mu, \lambda, e^{-2t}m^2, p_i).$$

We now use the solution of the RGE to rewrite this as

$$\Gamma_n(\mu, \lambda, m^2, e^t p_i) = e^{-nt}e^{\frac{n}{2}\int_0^s du\,\gamma(\lambda(u))}\Gamma_n(\mu, \lambda(t), e^{-2t}e^{\int_0^s du\,\gamma_{m^2}(\lambda(u))du}m^2, p_i).$$

We can interpret this equation as saying that, as we change the momentum scale, the naively dimensionless coupling λ "runs," while the renormalized field and the mass acquire $\lambda(t)$-dependent corrections to their dimensions. You should understand that engineering dimensions never change, and that engineering-dimensional analysis is always valid. However, the behavior of the theory under rescaling of momentum depends on dimensionless ratios of momenta to the renormalization scale, and can therefore be changed by quantum corrections.

The fact that $\beta(\lambda) \neq 0$ means that our classical assessment of λ as a marginal parameter is wrong. It does flow along RG trajectories, and its behavior near the free-field fixed point is determined by the leading term $\beta = b_0\lambda^2 + o(\lambda^3)$. In this approximation, the RGE is a linear equation for $1/\lambda$,

$$\frac{d\frac{1}{\lambda}}{dt} = -b_0,$$

whose solution is

$$\lambda(t) = \frac{\lambda}{1 - b_0\lambda t}.$$

We see that, if b_0 is positive (recall that $b_0 = 3/(16\pi^2)$), then, as momentum gets larger, $\lambda(t)$ increases, whereas as it gets smaller the coupling decreases. Such a theory is called *infrared-free* because it is well approximated by a perturbation expansion in $\lambda(t)$ at low momentum, even if λ is not small.

But there is something confusing and sick here. Remember, we defined a QFT by tuning relevant parameters to zero as we took the cut-off to infinity. The RG trajectory approaches very close to the scale-invariant fixed-point theory (which in our case is massless free-field theory) and then runs away from it at the renormalization scale μ, which is tuned to remain finite as the cut-off goes to infinity. If we look back along the RG trajectory, from the scale μ to the UV, we can see only the fixed point. So a renormalized theory defined by perturbing the massless free fixed point must behave like the free-field theory at very high momenta. We have found the renormalized coupling getting strong in this limit, which is a contradiction.

The clue to this behavior is infrared freedom. $\lambda(t)$ is getting weak in the IR. In other words, it is behaving like an irrelevant (marginally irrelevant) coupling. Its value in the renormalized theory should be fixed by the other couplings. But the only other coupling is the mass. We can solve the massive free-field theory exactly, and we find no connected four-point function. Therefore the renormalized ϕ^4 coupling is zero (this is often referred to as *triviality* of ϕ^4 theory).

Why then were we able to find a finite perturbation expansion to all orders in the renormalized coupling λ? The clue lies in noting that the place where the coupling gets strong in the UV (it blows up there but we can no longer trust the perturbation expansion at the singularity) is at t of order $1/\lambda$. If λ is small this is a momentum scale exponentially higher than the mass of the particles in the theory $o(m^2)$. So we can put a cut-off in the theory at an exponentially large scale, and get perfectly predictive physics by ignoring the cut-off, up to terms of order $e^{-c/\lambda}$. In other words, theories with marginally irrelevant parameters are almost as good as mathematically well-defined QFTs, as long as the renormalized values of the marginally irrelevant couplings are small enough to be in the perturbative regime. The point where the perturbative coupling blows up is called the Landau pole. As long as one uses a renormalized perturbation expansion at energies way below the Landau pole, one is on safe ground.

9.8 Renormalization of QED at one loop

As a further example, we will treat the renormalization of quantum electrodynamics at one loop. Power counting at the Gaussian fixed point for electrodynamics in any covariant gauge suggests that the interaction $A_\mu J^\mu$ is marginal when J^μ is taken to be the current of either spin-zero or spin-$\frac{1}{2}$ particles. The additional $A_\mu^2 \phi^* \phi$ interaction necessary to the gauge-invariant Lagrangian for charged spin-zero particles is also marginal. The scalar- and fermion-mass terms are relevant, and the only other gauge-invariant marginal operator is a quartic coupling $(\phi^* \phi)^2$. If we have multiple scalars and fermions the only additional marginal couplings for generic values of the charges of different fields would be of the form $\lambda_{ij}(\phi_i^* \phi_i)(\phi_j^* \phi_j)$. For special values of the charges we could also have Yukawa couplings, cubic couplings of the scalars, or both. However, even when the charges allow these couplings, we can invoke a $\phi_i \rightarrow -\phi_i$ symmetry to forbid them.

Wilson has emphasized that general renormalization theory does not require us to use a cut-off that preserves symmetries or gauge invariances (redundancies) of the classical Lagrangian, at least as long as the Gaussian fixed-point theory has only a finite number of relevant and marginal perturbations.[11] We simply introduce any cut-off we like and then tune the relevant and marginal parameters to make the quantum theory satisfy the quantum Ward identities. This could fail only if there were violations of the Ward identities, which cannot be removed by adding a local counterterm to the action. Schwinger–Adler–Bell–Jackiw anomalies, which we studied in Chapter 8, are the unique example of the latter phenomenon.

Nonetheless, retaining symmetries at every step greatly simplifies the renormalization process. For simplicity then, we will insist on using a gauge-invariant regulator, so that the most general marginal and relevant local Euclidean Lagrangian has the form

$$Z_3\left(\frac{1}{4}F_{\mu\nu}^2\right) + \frac{1}{2}Z_s|D_\mu\phi|^2 + Z_f\bar{\psi}(i\gamma^\mu D_\mu - m_0)\psi + \mu_0^2\phi_i^*\phi_i$$

$$+ \frac{e_0^2 v_0^{(4)}}{2}(\phi_i^*\phi_i)^2 + \frac{1}{2\kappa_0}(\partial_\mu A_\mu)^2.$$

For any charged field with charge q, the covariant derivative is $D_\mu \equiv \partial_\mu - ie_0 q A_\mu$. We have chosen the loop-counting parameter e_0 to be such that $e_0^2/(4\pi)$ is the bare fine-structure constant, and are already working with fields that, at tree level, have canonically normalized kinetic terms. In tree approximation, all the Z factors are equal to 1, particle masses are given by $m = m_0$, $\mu = \mu_0$, and the quartic coupling is $e_0^2 v_0^{(4)}/2$. General renormalization theory leads us to believe that we can eliminate all cut-off dependence in physical amplitudes by tuning the Z factors and the bare parameters (those subscripted with a zero) as the cut-off is taken to infinity. We will attempt to show below that the divergent part of the one-loop quantum action, $\Gamma[A, \psi, \phi]$, has precisely the form given above, so that the divergences can be eliminated as claimed. We will demonstrate this below, to one-loop order, for a single spinor field. The generalization to the case of multiple charged scalar and spinor fields should be a straightforward exercise for the diligent reader.

For photons, there is a gauge-invariant generalization of the sort of regularization procedure we introduced for scalars:

$$F_{\mu\nu}^2 \rightarrow F_{\mu\nu}K(\partial^2/\Lambda^2)F_{\mu\nu}.$$

However, if we attempt to generalize this to charged fields, retaining gauge invariance, we introduce an infinite number of apparently irrelevant interactions. There are only three known classes of gauge-invariant regularization prescription. The first is lattice gauge theory, but we will insist on retaining Lorentz invariance. The second class uses

---

[11] Which include all the interactions in the classical Lagrangian. It is the failure of this condition that makes spontaneously broken non-abelian gauge theories non-renormalizable in the unitary gauge.

the fact that charged fields appear only quadratically in the QED Lagrangian.[12] We can formally integrate them out, obtaining a power of a functional determinant of a gauge-covariant operator D^2 or $\gamma_\mu D_\mu$. There are many ways to regulate the product over the gauge-invariant eigenvalues of this operator. For example, Pauli and Villars pointed out that

$$\prod_i \det(D + \mu_i^{p_i})^{\epsilon_i}$$

was finite for finite values of μ_i and appropriate choices of $\epsilon_i = \pm 1$. Here $D = -D^2$ for spin zero (where $p_i = 2$) and $i\gamma^\mu D_\mu$ (where $p_i = 1$ and $\mu_i = iM_i$). We need only a finite number of extra Pauli–Villars fields, but some of them must have wrong statistics ($\epsilon_i = 1$ for spin zero or -1 for spin $\frac{1}{2}$). The masses of the extra fields act as regulators, and are taken to infinity at the end of the calculation.

Schwinger's gauge-invariant proper-time formula uses the observation that

$$\ln \det \left[(-D^2 + m^2)/(-\partial^2 + m^2) \right] = \int_0^\infty \frac{dt}{t} [e^{-t(-D^2 + m^2)} - e^{-t(-\partial^2 + m^2)}].$$

If we cut the lower limit of the t integral off at t_0 we suppress the contribution of large eigenvalues of $-D^2$ and get a finite, gauge-invariant functional of A_μ. t_0 has dimensions of inverse mass squared and defines a cut-off. In order to use this formula in spinor electrodynamics, we must recognize that the determinant of the Euclidean Dirac operator is formally positive and independent of the sign of the fermion mass, and so is equal to the positive square root of the determinant of $(-D^2 + m^2 + i\sigma^{\mu\nu} F_{\mu\nu}/2)$. We can then apply proper-time regularization to the latter quantity.

The combination of either of these two methods with the higher-derivative cut-off of the photon propagator gives us a finite, gauge-invariant, Lorentz-invariant answer for all Green functions in spinor and scalar QED. The answer does not obey the quantum requirement of unitarity until we take the cut-off to infinity. Unfortunately, neither of these methods generalizes to non-abelian gauge theory. The only known Lorentz-invariant, gauge-invariant regulator for the non-abelian case is *dimensional regularization* (DR), and this is the method we will use. We begin by studying the wave-function renormalization of the photon.

Figure 9.6 shows the 1PI photon two-point-function graph at one loop in spinor QED.

Fig. 9.6.

Vacuum polarization in QED.

***

[12] We can eliminate the quartic scalar interaction by introducing an auxiliary field $\frac{g_0^2 v_0^{(4)}}{2} \rightarrow \frac{\sigma^2}{2} + i\sigma g_0 \sqrt{v_0^{(4)}}$.

The value of the diagram is

$$\Pi_{\alpha\beta}(p) = e_0^2 \int \frac{\mathrm{d}^d q}{(2\pi)^d} \, \mathrm{tr}\left(\frac{[\gamma_\alpha(\gamma^\mu q_\mu - im_0)\gamma_\beta(\gamma^\mu(p+q)_\mu - im_0)]}{(q^2 + m_0^2)\,[(p+q)^2 + m_0^2]} \right).$$

Our first task is to show that this formula is transverse:

$$\Pi_{\alpha\beta}(p) = \Pi\!\left(p^2\right)\!\left(\delta_{\alpha\beta}p^2 - p_\alpha p_\beta\right).$$

The less-than-alert reader will have missed the fact that, in our discussion of the gauge-invariant quantum action above, we did not allow for a one-loop correction to the gauge-fixing parameter κ_0. The fact that the 1PI photon propagator is transverse (to all orders) guarantees this. So we compute

$$p^\alpha \Pi_{\alpha\beta} \propto e_0^2 \int \frac{\mathrm{d}^d q}{(2\pi)^d} \, \mathrm{tr}\left(\frac{[p^\alpha\gamma_\alpha(\gamma^\mu q_\mu - im_0)\gamma_\beta(\gamma^\mu(p+q)_\mu - im_0)]}{(q^2 + m_0^2)\,[(p+q)^2 + m_0^2]} \right).$$

Write

$$p^\alpha \gamma_\alpha = [(p+q)^\alpha \gamma_\alpha - im_0] - (q^\alpha \gamma_\alpha - im_0),$$

in order to obtain

$$p^\alpha \Pi_{\alpha\beta} \propto e_0^2 \int \frac{\mathrm{d}^d q}{(2\pi)^d} \, \mathrm{tr}\left(\frac{q^\alpha \gamma_\alpha - im_0}{q^2 + m_0^2} - \frac{(p+q)^\alpha \gamma_\alpha - im_0}{(p+q)^2 + m_0^2} \right).$$

DR is defined so that we can shift variables of integration, so $p^\alpha \Pi_{\alpha\beta} = 0$.

Now take the trace Π_μ^μ, and use transversality to obtain

$$(d-1)p^2\Pi(p^2) = e_0^2 \int \frac{\mathrm{d}^d q}{(2\pi)^d} \, \mathrm{tr}\left(\frac{[\gamma_\alpha(\gamma^\mu q_\mu - im_0)\gamma^\alpha(\gamma^\mu(p+q)_\mu - im_0)]}{(q^2 + m_0^2)\,[(p+q)^2 + m_0^2]} \right).$$

We use the identities

$$\gamma^\alpha \gamma^\mu q_\mu \gamma_\alpha = (2-d)\gamma^\mu q_\mu,$$
$$\gamma^\alpha m_0 \gamma_\alpha = dm_0,$$

to rewrite this as

$$(d-1)p^2\Pi(p^2) = \frac{e_0^2 d_{\mathrm{spin}}}{(2\pi)^d} \int \mathrm{d}^d q \, \frac{q\cdot(p+q)(2-d) - m_0^2 d}{(q^2 + m_0^2)\,[(p+q)^2 + m_0^2]}.$$

Our next piece of trickery is

$$q\cdot(p+q) = \frac{1}{2}(p+q)^2 - \frac{1}{2}q^2 - \frac{1}{2}p^2 + q^2.$$

The first two terms cancel out after integration and shifting of variables. So we obtain

$$(d-1)p^2\Pi(p^2) = \frac{e_0^2 d_{\mathrm{spin}}}{(2\pi)^d} \int \mathrm{d}^d q \, \frac{[(d-2)/2]p^2 - dm_0^2 + (2-d)q^2}{(q^2 + m_0^2)[(p+q)^2 + m_0^2]}.$$

To evaluate the momentum integrals, we use Schwinger's parametric formula $1/A = \int_0^\infty d\alpha\, e^{-\alpha A}$ to get

$$(d-1)p^2 \Pi(p^2) = \frac{e_0^2 d_{\text{spin}}}{(2\pi)^d} \int_0^\infty d\alpha\, d\beta\, e^{-(\alpha+\beta)m_0^2} e^{-\beta p^2} \int d^d q\, e^{-(\alpha+\beta)q^2}$$

$$\times \left(\frac{d-2}{2} p^2 - dm_0^2 + (2-d) \frac{\nabla_p^2}{4\beta^2} \right) e^{2\beta p \cdot q}$$

$$= \frac{e_0^2 d_{\text{spin}}}{(2\sqrt{\pi})^d} \int_0^\infty \frac{d\alpha\, d\beta}{(\alpha+\beta)^{\frac{d}{2}}} e^{-(\alpha+\beta)m_0^2} e^{-\beta p^2}$$

$$\times \left(\frac{d-2}{2} p^2 - dm_0^2 + (2-d) \frac{\nabla_p^2}{4\beta^2} \right) e^{\frac{\beta^2}{\alpha+\beta}}.$$

In order to understand the last equality, the knowledgeable reader will have used the formula for Euler's gamma function,

$$\Gamma(t) = \int du\, u^{t-1} e^{-u},$$

as well as the identities

$$\Gamma(1+t) = t\Gamma(t),$$
$$\Gamma(t) = (t-1)!,$$

if t is an integer ≥ 1, and

$$\Gamma(t)\Gamma(1-t) = \frac{\pi}{\sin(\pi t)},$$

from which it devolves that

$$\Gamma(1/2) = \sqrt{\pi}.$$

It is now obvious that we should introduce $s \equiv \alpha + \beta$ and $x \equiv \beta/(\alpha + \beta)$, and note that $d\alpha\, d\beta = s\, ds\, dx$. We obtain the parametric formula

$$(d-1)p^2 \Pi(p^2) = \frac{e_0^2 d_{\text{spin}}}{(2\sqrt{\pi})^d} \int_0^\infty ds \int_0^1 dx\, e^{-s[m_0^2 + x(1-x)p^2]}$$

$$\times \left[\frac{d-2}{2} p^2 - dm_0^2 + (2-d) \left(\frac{d}{2s} + x^2 p^2 \right) \right].$$

Simple power counting gives a quadratic divergence in $\Pi_{\alpha\beta}$, which would be momentum-independent and correspond to a photon mass $\Lambda^2 A_\mu^2$. The transverse formula for $\Pi_{\alpha\beta}$ shows that such a term cannot be there. The only way we could get something that looks like a photon mass would be from a pole in $\Pi(p^2)$ at $p^2 = 0$. This would not be UV-divergent. In fact, in more than two space-time dimensions such a pole does not appear (and even there it appears only if $m_0 = 0$). The only way to get a massive photon is through the Higgs mechanism. In perturbation theory this can happen only if there is a scalar field with a negative mass squared in the tree-level Lagrangian. In that case we have to diagonalize the free Lagrangian properly and we find only massive tree-level excitations. It must be, then, that the formula we have written above vanishes when $p^2 = 0$. The diligent reader will verify that fact. We thus obtain

$$(d-1)\Pi(p^2) = \frac{e_0^2}{(2\sqrt{\pi})^d} d_{\text{spin}} \int_0^\infty ds\, s^{1-\frac{d}{2}}$$
$$\times \int_0^1 dx \left[dx(1-x) + (2-d)\left(x^2 - \frac{1}{2}\right) e^{-s[m_0^2 + x(1-x)p^2]} \right].$$

Poles when $d \to 4$ will come from UV-divergent integrals at $d = 4$. Ultraviolet divergences correspond to the small-s region of integration. It is now obvious that terms in our formula of order higher than p^2 in an expansion about 0 will not have divergences in the small-s region. Thus, the divergence resides in $\Pi(0)$, which means that it has the form of a correction to the tree-level relation $Z_3 = 1$. In fact

$$(d-1)\Pi(0) = \frac{e_0^2}{(2\sqrt{\pi})^d} d_{\text{spin}} m_0^{d-4} \Gamma\left(2 - \frac{d}{2}\right).$$

The peculiar factor of m_0^{d-4} in this formula points up something important. Away from four dimensions, the electromagnetic coupling is not dimensionless. DR introduces no explicit cut-off with the dimensions of mass, so the dimensions in formulae must be made up by powers of m_0 or some other mass parameter.

The pole in the 1PI photon Green function can all be attributed to the Z_3 factor, but there is an ambiguity regarding how much finite part to keep when we define Z_3. DR allows us to keep "just the pole," but when we express the answer in terms of e_0 this appears to introduce a dependence on m_0 because of dimensional analysis. However, we can introduce an arbitrary mass scale μ in place of m_0 to define the split between Z_3 and the renormalized field in the formula

$$A_\mu = Z_3^{1/2} A_\mu^R.$$

We thus write

$$(Z_3 - 1) = \frac{2}{3} \frac{\alpha_0}{\pi \mu^{4-d}} \frac{1}{d-4},$$

where we have evaluated the residue of the pole at $d = 4$. This prescription for parametrizing the renormalized theory is called minimal subtraction (MS). Note that

$$\Gamma\left(2 - \frac{d}{2}\right) \to \frac{1}{2 - d/2} - \gamma,$$

where γ is the Euler–Mascheroni constant. There is a modified prescription, called $\overline{\text{MS}}$, which keeps certain finite parts in addition to the pole, and removes many of the γ factors which would otherwise appear in renormalized formulae (see Peskin and Schroeder [33] for details).

9.8.1 Renormalization of the fermion propagator and the vertex

We now turn to the other two divergent one-loop diagrams in spinor QED, the fermion self-energy and the photon fermion vertex. The gauge-invariant form of the Lagrangian suggests that these are renormalized by the same multiplicative factor. The 1PI vertex function $\Gamma_\mu(p, p + q)$ is related to the Fourier transform of the Green function of the electromagnetic current,

$$\langle J^\mu(x)\psi(y)\bar\psi(z)\rangle,$$

by multiplying it by inverse fermion propagators on the external legs and by e_0, the bare charge. The current Green function satisfies the Ward identity

$$\partial_\mu^x\langle J^\mu(x)\psi(y)\bar\psi(z)\rangle = i\delta^4(x - y)\langle\psi(y)\bar\psi(z)\rangle - i\delta^4(x - z)\langle\psi(y)\bar\psi(z)\rangle.$$

Written in terms of the vertex function, this identity is

$$q^\mu\Gamma_\mu(p, p + q) = e_0[\gamma^\mu q_\mu + \Sigma(p + q) - \Sigma(p)],$$

where $\Sigma(p)$ is the sum of all 1PI self-energy graphs (the notation here differs a bit from standard texts because I have included the tree-level term in the definition of Γ_μ).

The vertex function itself is dimensionless, so the one-loop integral defining it is at most logarithmically divergent. Consequently, all derivatives of Γ_μ w.r.t. q are given by finite expressions in four dimensions and will not have poles at $d = 4$. The divergent part of Γ_μ is thus independent of q at one loop. A similar argument shows that it is independent of p as well. Reflection invariance of QED shows that it cannot involve γ_5, so we must have

$$\Gamma_\mu^\infty = e_0(Z_1 - 1)\gamma_\mu.$$

The Ward identity shows that this must be related to a divergent term in $\Sigma^\infty(p)$ $= (Z_2 - 1)\gamma_\mu p^\mu$, with $Z_1 = Z_2$.

We therefore turn to the self-energy diagram of Figure 9.7 to evaluate Z_2. The result will depend on the gauge parameter κ_0 in the free-photon propagator

$$\frac{(\delta_{\mu\nu} - q_\mu q_\nu/q^2)}{q^2} + \kappa_0\frac{q_\mu q_\nu}{q^4}.$$

We will notice that Landau (also called Lorentz) gauge, $\kappa_0 = 0$, has certain nice features (it avoids some infrared divergences in low orders). One way of understanding why the Landau-gauge fermion propagator is so nice is to note that it is actually the value of a

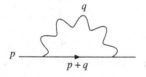

One-loop self-energy in QED.

Fig. 9.7.

fairly simple non-local gauge-invariant operator, evaluated in Landau gauge. Indeed, if we multiply the bilinear $\psi(x)\bar\psi(y)$ by

$$e^{i \int A_\mu(x)j^\mu(x)},$$

where $\partial_\mu j^\mu(z) = \delta^4(z - x) - \delta^4(z - y)$, then we get a gauge-invariant operator. On choosing $j^\mu = \partial^\mu A$, and noting that the resulting A vanishes at infinity (it is the four-dimensional analog of the field of a dipole), we see, upon integrating by parts, that the non-local exponential vanishes in Landau gauge.

We will do the calculation in Euclidean space. We write the one-loop contribution to $\Sigma(p)$ as

$$\frac{-e_0^2}{(2\pi)^d} \int d^d q \, \frac{\gamma_\alpha(\not{p} - \not{q} - im_0)\gamma_\beta}{q^2[(p-q)^2 + m_0^2]}\left(\delta^{\alpha\beta} - (1 - \kappa_0)\frac{q^\alpha q^\beta}{q^2}\right). \tag{9.1}$$

We introduce Schwinger parameters to write the denominator as

$$\int_0^\infty d\alpha \, d\beta \, e^{-(\alpha + \beta)q^2} e^{-\beta(p^2 + m_0^2)} e^{2\beta pq}.$$

Note that the extra inverse power of q^2 in the last term of the numerator turns into an extra power of α in the numerator of the parametrized integral. To do the Gaussian integral over loop momentum, we shift the integration variable to $r = q - xp$. Here we have introduced the standard one-loop passage from Schwinger to Feynman parameters: $s = \alpha + \beta$, $\beta = sx$, $d\alpha \, d\beta = s \, ds \, dx$. When we do the integrals, terms linear in r will integrate to zero, and we need to use

$$\int d^d r \, e^{-sr^2} = \left(\frac{\pi}{s}\right)^{\frac{d}{2}}$$

and

$$\int d^d r \, e^{-sr^2} r_\mu r_\nu = \frac{\delta_{\mu\nu}}{2s}\left(\frac{\pi}{s}\right)^{\frac{d}{2}}.$$

Keeping only terms even in r, the numerator is

$$\gamma^\alpha(\not{p}(1 - x) - im_0)\gamma^\beta[\delta_{\alpha\beta} - (1 - \kappa)(1 - x)s(r_\alpha r_\beta + x^2 p_\alpha p_\beta)]$$

$$+ \gamma^\alpha \not{r}\gamma^\beta x(1 - \kappa)s(r_\alpha p_\beta + r_\beta p_\alpha).$$

Note that we have inserted the renormalized gauge parameter rather than the bare one, because the correction is of higher order in e^2. After doing the r integration, terms quadratic in r acquire an extra power of s. Terms proportional to $s^{1-d/2}$ give $\Gamma(2 - d/2) \sim 2/(4 - d)$, but terms with higher powers of s give convergent integrals. Thus, the divergent part of the self-energy is given by

$$\Sigma_\infty = \frac{2}{4 - d} \int_0^1 dx \, \gamma^\mu(\not{p}(1 - x) - im_0)\gamma_\mu\left(1 - \frac{(1 - \kappa)(1 - x)}{2}\right)$$

$$+ d\not{p} \, x(1 - x)(1 - \kappa).$$

Finally, we use contraction identities for Dirac matrices and do the x integral to obtain

$$\Sigma_\infty = -\frac{e^2}{8\pi^2(4-d)}[\kappa(\not{p} - im) + 3im].$$

This can obviously be removed by rescaling the fields and renormalizing the mass according to

$$Z_2 = 1 + \kappa\frac{\alpha}{2\pi\epsilon}$$

and

$$\delta m = -3\frac{\alpha}{2\pi\epsilon}m_0.$$

There are several interesting points about these formulae. First, we find that $\delta m \propto m_0$, so the mass renormalization is multiplicative and vanishes if m_0 vanishes. This is a consequence of the extra chiral symmetry of the massless theory. Although DR doesn't really preserve conservation of the associated Noether current (this is the chiral anomaly we have discussed above), it does conserve the charge in perturbation theory. As a consequence, although m_0 has dimensions of mass, it does not have the sensitivity to the cut-off scale we expect for a relevant parameter. Note also that δm is independent of κ.

DR has the peculiar property that

$$\int d^d p \, \frac{1}{p^2 + M^2} \propto M^{d-2}$$

and thus vanishes for $M = 0$. That is, quadratically divergent integrals with only massless propagators vanish in DR, but they do depend quadratically on the masses of very heavy particles. For scalar fields, this leads to renormalizations of any mass term $\propto \Lambda^2$, coming from integrating out particles of mass near the cut-off. On dimensional grounds we might have expected the same for fermion masses. However, the chiral symmetry of massless fermion systems guarantees that the corresponding massive systems have only logarithmically divergent mass renormalization. We will return briefly to this point when we discuss the concept of *technical naturalness* below.

The second interesting point is that the fermion wave-function renormalization is κ-dependent and vanishes at $\kappa = 0$. In Chapter 6 we noted that QED has IR divergences associated with massless-photon emission in the scattering of charged particles. One can show using the RG equations of the next section and the IR freedom of QED that, in leading-order RG-improved perturbation theory, the fermion propagator has a cut rather than a pole at the physical mass, with a gauge-dependent power law. This is problematic for the LSZ formula for the S-matrix. In Landau gauge, the cut becomes a pole in this leading-order approximation. The gauge-invariant operator which is equal to the fermion field in Landau gauge includes the Coulomb field of charged particles, the IR effect that is of leading order in perturbation theory. In higher orders, we must include appropriate coherent states of transverse photons in the definition of charged-particle states, in order to account for bremsstrahlung radiation and cancel out the IR divergences.

9.9 Renormalization-group equations in QED

We have argued that, in Landau gauge, we can, order by order in the perturbation expansion, render the expressions for all Green functions finite by rescaling the fields and tuning the bare parameters m_0 and e_0 so that the renormalized parameters $m = Z_m m_0$ and e are finite. In the process, we have introduced a new mass scale, μ, *the renormalization scale*, writing the dimensional bare coupling e_0^2 as

$$e_0^2 = \mu^{4-d} e^2 Z_3^{-1}.$$

The Green functions of the bare fields, expressed as functions of e_0 and m_0, are independent of μ:

$$\mu \frac{\mathrm{d}}{\mathrm{d}\mu} \Gamma_{F,A}^{(0)}(e_0, m_0) = 0.$$

$\Gamma_{F,A}^{(0)}$ is the 1PI vertex with F fermion and A photon legs. It is a function of the momenta of the external legs, defined without the momentum-conserving delta function. It has mass dimension $4 - \frac{3}{2}F - A$. The renormalized Green function is

$$\Gamma_{F,A} = Z_2^{F/2} Z_3^{A/2} \Gamma_{F,A}^{(0)}.$$

μ-independence of the bare vertices implies a relation for the renormalized ones. It should be expressed in terms of the renormalized parameters, and we can do this using the chain rule:

$$(\mu\,\partial_\mu + \beta\,\partial_\alpha + \gamma_A \kappa\,\partial_\kappa + m\gamma_m\,\partial_m)\Gamma_{F,A} = -(F\gamma_F + A\gamma_A)\Gamma_{F,A},$$

where

$$\beta = \mu \frac{\mathrm{d}}{\mathrm{d}\mu}\alpha,$$

$$\gamma_m = \mu \frac{\mathrm{d}}{\mathrm{d}\mu}\ln Z_m,$$

$$\gamma_F = \frac{1}{2}\mu \frac{\mathrm{d}}{\mathrm{d}\mu}\ln Z_2,$$

$$\gamma_A = \frac{1}{2}\mu \frac{\mathrm{d}}{\mathrm{d}\mu}\ln Z_3.$$

We have used the symbol $\mathrm{d}/\mathrm{d}\mu$ to represent derivatives with α_0 and m_0 held fixed, while ∂_μ refers to derivatives with the renormalized parameters fixed. $\alpha_0 = e_0^2/(4\pi)$ is the bare fine-structure constant. The renormalized fine-structure constant is

$$\alpha = \mu^{d-4}\alpha_0 Z_3.$$

Thus

$$\beta = (d-4)\alpha + 2\gamma_A.$$

We have also used the result that the longitudinal part of the photon propagator is unchanged by loop corrections. Thus, the renormalization of the gauge parameter κ

comes only from the photon wave-function renormalization $\kappa_0 = Z_3\kappa$. This equation for the renormalized 1PI vertices is called the renormalization-group equation. It holds for all values of momenta, and expresses the fact that the renormalized theory has only one independent dimensionless parameter and we have artificially introduced another one through the mass scale μ. As we have emphasized, the reason why we must do this is because the theory is not scale-invariant even in the limit $m_0 \to 0$. If we had insisted on defining the renormalization scale in terms of the electron mass parameter, we would have introduced spurious infrared divergences into the $m_0 = 0$ theory.

The fact that the RG equation is true for all momenta shows us that the RG functions β and γ_i are all finite in the limit $d \to 4$. Read as a set of equations for these functions in terms of the finite proper vertices, we have an over-determined set of linear equations for β and γ_i. The fact that they have a solution is the content of the differential equation for the vertices. In perturbation theory, it is easy to see that, to kth order in the loop expansion, we can expect the functions Z_i to have poles up to order k at $d = 4$. Expressed in terms of α_0, the μ dependence of the kth order is just $\mu^{k(d-4)}$. The scaling derivative in the definition of the RG functions brings down a factor of $d - 4$, which cancels out the first-order pole. All the others must cancel out automatically, when the finite functions β and γ_i are expressed in terms of α, κ, and m! Thus, to calculate the kth-order term in any of these functions, we need only find the residue of the first-order pole, in the kth order in the loop expansion.

Finally we note that, in the MS prescription, the coefficients of these poles are independent of m_0. This follows from the fact that all of the Z_i are dimensionless and therefore $(\partial/\partial m_0)\ln Z_i$ is given by convergent Feynman integrals in $d = 4$, plus the fact that we have introduced μ rather than m_0 or m to provide the dimensions of e_0. It follows that all the RG functions depend only on the renormalized coupling α and the gauge parameter κ. In fact, since the transverse part of the photon propagator is κ-independent, γ_A and β are independent of κ. In a more general field theory, defined in perturbation theory around a Gaussian fixed point, the RG functions in the MS scheme depend on the marginal parameters, but not on the relevant ones.

If we rescale all the momenta $p_i \to e^t p_i$, then the equation of dimensional analysis is

$$(\partial_t + \mu\,\partial_\mu + m\,\partial_m)\Gamma_{F,A} = \left(4 - \frac{3}{2}F - A\right)\Gamma_{F,A}.$$

Using the RG equation, we can rewrite this as

$$(\partial_t + \beta\,\partial_\alpha + m(1 + \gamma_m)\partial_m)\Gamma_{F,A} = -\left[\left(\frac{3}{2}F + \gamma_F\right) + (1 + \gamma_A)A\right]\Gamma_{F,A}.$$

In words, this equation says that we can compensate for a change in momentum scale by changing the coupling and the mass, and rescaling the fields. Thus the solution of this equation is

$$\Gamma_{F,A}(e^t p_i, \alpha, m) = \Gamma_{F,A}(p_i, \alpha(t), m(t))e^{L(t)}.$$

On differentiating this expression w.r.t. t and insisting that it give the previous equation we find that

$$\dot{\alpha}(t) = \beta,$$
$$\dot{m}(t) = (1 + \gamma_m)m,$$
$$\dot{L}(t) = (4 - F\Delta_F - A\Delta_A),$$

with $\Delta_F = \frac{3}{2} + \gamma_F$ and $\Delta_A = 1 + \gamma_A$.

Thus, we learn that a rescaling of all momenta is equivalent to a flow of the coupling constants satisfying the above equations, combined with a scale-dependent rescaling of the Green functions, which can be interpreted as an *anomalous dimension* for the fields. Similarly, the flow equation for the mass can be interpreted as an anomalous dimension for the renormalized mass parameter. The anomalous dimensions $\Delta_{F,A,m} = E_{F,A,m} + \gamma_{F,A,m}$, where $E_{F,A,m} = (3/2, 1, 1)$ is the engineering dimension,[13] are functions of the scale-dependent fine-structure constant. If, in the asymptotic region $t \to \pm\infty$, $\alpha(t) \to \alpha^*$, a fixed point, then we conclude that the theory is asymptotically scale-invariant, with dimensions given by the anomalous dimensions evaluated at the fixed point.

Are there fixed points in QED? We can investigate this only in the vicinity of the free theory, because we are doing perturbation theory. The perturbative value of the β function in spinor QED is

$$\beta = (d - 4)\alpha + \frac{2}{3\pi}\alpha^2.$$

Adding scalars and more spinors changes the coefficient of the second term, without changing its sign. We note that for real values of d less than but close to 4 we find a zero of β at a non-zero but small value of the coupling, where we can trust perturbation theory. This will occur in any theory with couplings that are dimensionless in four dimensions but have positive coefficient for the one-loop correction to the RG function. This observation is the basis of one of the methods of calculating critical exponents for second-order phase transitions in $d = 2, 3$. One computes the fixed points and anomalous dimensions in a power series in $4 - d$, and extrapolates to the values of the dimension relevant for real condensed-matter systems. This turns out to give quite good agreement with experiment, though other methods based on field-theory calculations in integer dimensions do somewhat better. A more detailed account can be found in the books of Peskin and Schroeder [33] and Zinn-Justin [111] and the references to the original literature found therein.

As high-energy physicists, we are interested in $d = 4$. There, the RG equation has the solution

$$\alpha(t) = \frac{\alpha(0)}{1 - [2/(3\pi)]\alpha(0)t}.$$

Note that, for $t \to -\infty$, this solution remains in the perturbative regime if $\alpha(0)$ is small. On the other hand, if $t \to \infty$ the coupling becomes strong, and in fact appears

---

[13] Often the term anomalous dimension is reserved for the shift $\gamma_{F,A,m}$ rather than the dimension itself.

to reach infinity at the *Landau pole*, $t_L^{-1} = [2/(3\pi)]\alpha(0)$. Our basic definition of field theory tells us that a mathematically consistent field theory will approach its fixed point in the deep UV. Our perturbative philosophy was based on the assumption that this fixed point was Gaussian. The Landau pole tells us that this assumption was wrong. The QED coupling is marginally irrelevant, which means that the only mathematically consistent value for the renormalized coupling is zero. If we insist on a finite value for this coupling at some momentum scale $t = 0$, then the theory becomes singular at a finite scale, that of the Landau pole.

Since $\alpha(0)\sim 1/137$ in the real world, our attitude to this as physicists is somewhat different than that of the mathematical field theorist. If $t = 0$ corresponds to the atomic scale $\sim 10\,\text{eV}$ at which the fine-structure constant is measured, then the scale of the Landau pole is $\sim 10^{331}\,\text{GeV}$, which is much larger than the Planck scale $10^{19}\,\text{GeV}$ defined by quantum gravity. We certainly expect our notions of quantum field theory to break down long before we reach the scale of the Landau pole. Thus, QED is a perfectly good effective field theory at all scales at which we expect the whole notion of quantum field theory to work. In contrast with a truly irrelevant operator, like the four-fermion coupling of Fermi's theory of weak interactions, a small, marginally irrelevant coupling does not hint at new physics around the corner. The renormalized perturbation series of QED could in fact be the whole story up to the Planck scale. We know that this is not true, because, above the QCD and weak-interaction scales, physics becomes more complicated. QED is certainly incorporated into the full standard model. The point is that there is nothing in pure QED that could have led us to such a conclusion.

9.9.1 The static potential and the definition of α

The flip side of the marginal irrelevance of QED is that the coupling becomes arbitrarily weak in the IR. Indeed, an alternative name for marginally irrelevant interactions is *IR-free*. One must, however, be careful about the precise meaning of this. In the IR limit, the mass parameter also becomes large. It turns out that this makes the MS definition of the coupling very different from a physical definition in terms of real scattering amplitudes. One useful physical definition is in terms of the long-distance potential felt by a pair of oppositely charged heavy particles. If a particle is sufficiently heavy we can describe its interaction with the electromagnetic field in terms of its classical trajectory through space-time, $x^\mu(\tau)$. The current of the particle is

$$J^\mu(x) = q \int d\tau \, \frac{dx^\mu}{d\tau} \delta^4(x - x(\tau)),$$

where q is its charge. The interaction with the electromagnetic field is obtained by inserting the factor

$$e^{i \int d^4x \, A_\mu(x) J^\mu(x)} = e^{iq \int A_\mu \, dx^\mu},$$

where the second integral is a line integral along the particle path. This expression is the Wilson line. A similar formula holds in non-abelian gauge theory, where A_μ is a matrix and the exponential is replaced by a path-ordered exponential along the particle path.

It is convenient to discuss a particle–anti-particle pair as a closed Wilson line, or Wilson loop. This corresponds to pair production, propagation through space-time, and annihilation at some later time. The pair-production and annihilation events are described in an idealized manner and do not correspond to a realistic experiment. However, if we take a rectangular loop in Euclidean space with one side T much greater than the other, $T \gg R$, and Wick rotate to Lorentzian signature, then to leading order in T/R the answer will be dominated by the long period of particle–anti-particle propagation, and the unrealistic pair creation events will be a sub-leading effect.

The exact answer in Euclidean QED is given in terms of the generating function of connected Green functions, by $e^{-W[J]}$. However, in the large-R limit, the contribution of the two-point function dominates (cluster decomposition works even in this theory with massless photons). Furthermore, in the same limit, the contribution of the two-point function is dominated by its behavior at $p = 0$. If we assume that $e_0^2/(1 + \Pi(p^2))$ goes to a finite $p = 0$ limit, then the Wilson loop is given by

$$e^{-T \frac{e_0^2}{1+\Pi(0)} \frac{1}{4\pi R}}.$$

This is the Coulomb potential with physical renormalized charge

$$e_{\text{Coul}}^2 = \frac{e_0^2}{1 + \Pi(0)}.$$

We have already noted that we could have chosen to parametrize the renormalized theory by this quantity. However, we also commented that this introduced a logarithmic dependence on m_0 (and thus on m) and we preferred to introduce the arbitrary momentum scale μ to eliminate this. For finite m, this choice of parametrization cannot change the fact that e_{Coul} is a finite quantity. Thus, the physical renormalized charge defined in terms of the Coulomb potential is not equal to the IR limit of the MS coupling $\alpha(-\infty) = 0$. Physically, this corresponds to the fact that the renormalization of the electric charge due to loops of electrons goes to zero below the threshold for electron–positron pair production. Mathematically it corresponds to the fact that one must take $m(t)$ to infinity as $\alpha(t)$ goes to zero in computing e_{Coul} in the MS scheme.[14]

In the $m = 0$ theory, on the other hand, the MS RG equation can be used to show that the Fourier transform of the "Coulomb" potential is modified to $1/[q^2 \ln(q^2)]$, so that the potential falls off more rapidly than $1/R$, corresponding to vanishing charge in the IR.

---

[14] IR freedom *is* useful in the massive theory when dealing with the problem of soft-photon IR divergences. It justifies the perturbative resummation procedures one uses to extract finite answers for inclusive cross sections.

9.10 Why is QED IR-free?

I now want to give two different answers to this question. By comparing and contrasting them, we will learn a lot of things, including things relevant to non-abelian gauge theories. Consider first the Fourier transform of the two-point function of unrenormalized electromagnetic field strengths:

$$\int d^4x \, e^{-iqx} \langle F_{\mu\nu}(x) F_{\lambda\kappa}(0) \rangle.$$

This is a gauge-invariant quantity, which can be written entirely in terms of the transverse photon propagator $D_{\mu\nu}(q)$. By inserting a complete set of physical states, we can derive a Lehmann representation

$$D_{\mu\nu}(q) = \left(\delta_{\mu\nu} - \frac{q_\mu q_\nu}{q^2} \right) \left[\frac{Z_3}{q^2} + \int dM^2 \, \frac{\rho(M^2)}{q^2 + M^2} \right].$$

Note that the Z_3 appearing in this formula corresponds to the charge e_{Coul} rather than the one defined by the MS scheme. In a covariant gauge, the Hilbert space of QED is not positive definite, but the subspace generated by acting with gauge-invariant operators on the vacuum is. Thus ρ is positive. The electric and magnetic fields satisfy the canonical commutation relation

$$[F_{0i}(x,t), F_{jk}(y,t)] = i(\delta_{ij} \partial_k - \delta_{ik} \partial_j)\delta^3(x-y).$$

This leads to the sum rule

$$1 = Z_3 + \int dM^2 \, \rho(M^2),$$

which implies $0 \leq Z_3 \leq 1$. That is, the renormalized charge is always smaller than the bare charge, e_0. If we take $e_0 \to 0$ as the cut-off is taken to infinity, as would be appropriate for a marginally relevant perturbation of the Gaussian fixed point, then the renormalized charge vanishes. So the electromagnetic coupling is marginally irrelevant: QED. Note that this does not mean that there could not be *interacting* fixed-point theories containing electrodynamics. However, they cannot be accessed in perturbation theory around the Gaussian fixed point.

Furthermore, the relation $\beta = \alpha\gamma_A$ shows that at *any* fixed point the two-point function of the electromagnetic field is proportional to its value in free-field theory. The representation theory of the four-dimensional conformal group can then be used to show that all higher connected Green functions of $F_{\mu\nu}$ vanish. That is, at any interacting fixed point, the electromagnetic field decouples from the rest of the dynamics.

Let us now derive the marginal irrelevance of the electromagnetic coupling in another way. This method works only at one loop, and is based on the idea of thinking of the interacting vacuum state of QED as a *medium* filled with bare particles. The low-energy effective action of the electromagnetic field in a medium has the form

$$\frac{1}{2}\left[(e_0^2 - \chi_e)E^2 - \left(\frac{1}{e_0^2} - \chi_m \right)B^2 \right],$$

where

$$\chi_e = \left\langle \frac{\partial^2 H_m}{\partial E^2} \right\rangle,$$

and

$$\chi_m = \left\langle \frac{\partial^2 H_m}{\partial B^2} \right\rangle,$$

are the electric and magnetic susceptibilities. H_m is the Hamiltonian of the particles in the medium, and the angle brackets refer to statistical averaging over the distribution of these particles. The vacuum is a peculiar kind of medium. It is exactly Lorentz-invariant. This means that

$$e_0^2 - \chi_e = \frac{e_0^2}{1 - e_0^2 \chi_m};$$

the electric polarizability is related to the magnetic susceptibility.

We are interested in renormalizations of the electric charge due to effects going on in the UV. This means that the problem is extremely relativistic. It turns out that, because magnetism is a relativistic effect in condensed-matter physics, it is really only our intuition for magnetic systems that is relevant here. Indeed, we are familiar with two kinds of magnetic behavior of materials: *diamagnetism* with $\chi_m < 0$ and *paramagnetism* with $\chi_m > 0$. Both have their origins in quantum mechanics, because Van Leeuwen's theorem in classical statistical mechanics shows that neither can occur. Landau understood that diamagnetism arises from the quantization of particle orbits (Landau levels) in a constant magnetic field. Pauli was the first to interpret paramagnetism as a consequence of the intrinsic spins of atoms and molecules. Indeed, given the concept of such an intrinsic magnetic moment, paramagnetism *can* be understood classically. The intrinsic magnetic moments of the particles line up with the external field, enhancing it. In any given material, the overall magnetic properties are determined by a competition between these two effects. In a Lorentz-invariant medium, paramagnetism is equivalent to *anti-screening* of electric charge, an effect that is hard to understand in terms of non-relativistic electrostatics.

In order to understand what is going on in terms of scalar and spinor QED we need one more twist: for fermions the massless particles which give rise to the magnetic properties of the vacuum must be thought of as having negative energy. This can be understood in a variety of ways, the simplest of which is Dirac's picture of the free-fermion vacuum as a filled Fermi sea of negative-energy electrons. The fact that the vacuum energy has opposite sign for bosons and fermions arises from the famous minus sign for closed fermion loops in Feynman's rules. Bosonic vacuum energy is just the zero-point energy of the field oscillators, which is obviously positive. Fermionic vacuum energies have the opposite sign because commutators are replaced by anti-commutators in the normal ordering prescription. However one derives it, the effect of this is to flip the sign of any individual contribution to the susceptibility. Thus, for fermionic vacuum particles, a paramagnetic effect gives $\chi_m < 0$, while diamagnetism corresponds to $\chi_m > 0$. Our explicit calculations and/or the general result $Z_3 < 1$ which follows from unitarity then tell us that *for massless spin-$\frac{1}{2}$ particles with gyromagnetic ratio 2, paramagnetism is more important than diamagnetism.*

We can now use this result to learn something about the electromagnetism of charged spin-1 particles. The competition between paramagnetism and diamagnetism obviously depends on the gyromagnetic ratio of particles. It turns out that the only renormalizable theories of charged spin-1 particles are non-abelian gauge theories. The Lagrangian for the simplest such theory, corresponding to the group SU (2), contains an "electromagnetic" field A_μ and a charged vector field $W_\mu^\pm$ (the two fields are Hermitian conjugates of each other). In addition to the minimal coupling of A to W the electromagnetic Lagrangian is modified to

$$\frac{1}{4}(F_{\mu\nu} - e_0(W_\mu^+ W_\nu^- - W_\mu^- W_\nu^+))^2.$$

The diligent reader will verify that this gives the charged vector fields an anomalous magnetic moment,[15] in such a way that their gyromagnetic ratio is 2. Thus, in this model the winner of the competition between diamagnetism and paramagnetism is already determined by the spin-$\frac{1}{2}$ case, since the magnetic moment of spin-1 particles with $g = 2$ is larger than that of Dirac particles. Paramagnetism wins and e^2 is a marginally relevant coupling. Indeed, this hand-waving argument can be made quantitative, and reproduces the result of the rigorous calculation we will do later. The relevant formalism is the *background field gauge* and can be found in Peskin and Schroeder [33].

We can also conclude that something must go wrong with the gauge invariance of the electromagnetic field-strength tensor in this theory. Otherwise we would find a contradiction with our unitarity argument for $Z_3 < 1$. Indeed, in the non-abelian gauge theory, electromagnetic gauge invariance is part of an SU (2) gauge group. The traditional Maxwell field strength transforms under the other independent gauge transformations. Indeed, any one component of the triplet A_μ^a composed of A_μ and the real components of W_μ can be thought of as the photon, with the other two components transforming as charged fields. The field strength

$$F_{\mu\nu}^3 = F_{\mu\nu} - e_0(W_\mu^+ W_\nu^- - W_\mu^- W_\nu^+)$$

is the third component of an SU (2) triplet. It is not gauge-invariant, and produces negative norm states when acting on the vacuum in a covariant gauge.

The detailed calculation of coupling renormalization in a form that reveals the diamagnetic/paramagnetic split we have discussed here can be found in many textbooks

---

[15] Weinberg [112] proved long ago that, to lowest order in electromagnetic couplings, the only value of the gyromagnetic ratio for particles of any spin which is consistent with good behavior of high-energy cross sections is $g = 2$.

under the heading of *background field methods*. We will instead perform the calculation in terms of a simple physical quantity, the potential between static sources, or Wilson loop.

9.11 Coupling renormalization in non-abelian gauge theory

We will calculate the renormalization of the coupling by computing the one-loop correction to the potential between two static external sources. As explained in Chapter 8 and Problem 8.4, this potential is defined by

$$V(R) = -\lim_{T \to \infty} \frac{1}{T} \ln \langle W_{\mathrm{R}}(\Gamma) \rangle / d_{\mathrm{R}}. \tag{9.2}$$

Γ is the rectangular Wilson loop shown in Figure 9.8 and d_{R} is the dimension of the representation R.

In perturbation theory we can compute this in terms of the Feynman rules of Appendix D, with additional vertices for the interaction of gluons[16] and the static source. The contributions proportional to T come exclusively from the parts of the Wilson loop which point in the Euclidean time direction. Thus we have

$$g_0 \mu^{\frac{4-d}{2}} \delta_{\mu 0} \delta^3(x - R) T_{\mathrm{R}}^a,$$

for the upward-going line at R, and

$$-g_0 \mu^{\frac{4-d}{2}} \delta_{\mu 0} \delta^3(x - R) T_{\mathrm{R}}^a,$$

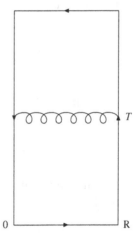

Fig. 9.8.   **Wilson loop for the static potential.**

[16] We will use the QCD terminology, *gluon*, to refer to the generic non-abelian gauge bosons of this section.

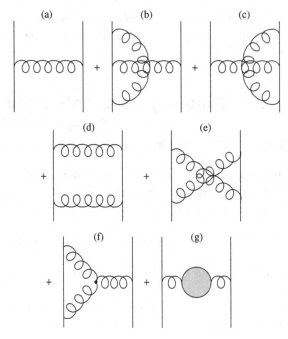

Fig. 9.9. **One-loop contributions to the potential.**

for the downward-going line at the origin. We will work in Feynman gauge, because the triple gluon vertex of Figure 9.9(f) will not contribute in this gauge. All the gluons couple to static sources, so only their 0 components appear. The triple vertex is anti-symmetric in group indices (we have a compact group) and in space indices (Bose statistics), and so vanishes in this configuration.

To complete the Feynman rules, we must remember that all of the T_R^a matrices must be path ordered around the loop. The second-order contribution comes just from the graph of Figure 9.8. Its value is

$$g_0^2 \mu^{4-d} \, \mathrm{tr}\big(T_R^a T_R^a\big) \int \mathrm{d}t \, \mathrm{d}s \, \frac{1}{4\pi^2} \frac{1}{(t-s)^2 + R^2}, \tag{9.3}$$

where the integrals go from $-T/2$ to $T/2$. The integral over $(s+t)/2$ gives a factor of T, while the integral over relative times is just

$$\frac{1}{R} \int_{-\infty}^{\infty} \frac{\mathrm{d}s}{s^2 + 1} = \frac{\pi}{R}.$$

We have evaluated the integral in four dimensions because there are no divergences in leading order. We obtain

$$V(R) = -g_0^2 \frac{\mathrm{tr} \, C_2(R)}{4\pi R},$$

where $C_2(R)$ is the Casimir operator in the R representation. Note that the Wilson loop contains the trace of this Casimir operator, but the dimension of the representation is divided out to normalize the static particle states.

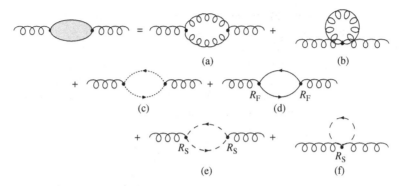

Fig. 9.10.   **One-loop contributions to gluon vacuum polarization.**

At fourth order in the coupling we have the seven graphs of Figure 9.9, the sixth of which vanishes in Feynman gauge. The gluon vacuum polarization, denoted by a shaded oval, is computed from the graphs of Figure 9.10.

The first two graphs of Figure 9.9 sum up to

$$
\left(\frac{1}{2}\right)^2 g_0^4 \mu^{2(4-d)} \int dx_1\, dx_2\, dx_3\, D(x_1 - x_3) D(x_2 - x_4)
$$
$$
\times \operatorname{tr}\!\left[(T_R^a T_R^a T_R^b T_R^b)\theta(t_1 - t_2)\theta(t_3 - t_4) + (T_R^a T_R^b T_R^a T_R^b)\theta(t_1 - t_2)\theta(t_4 - t_3) \right].
$$

All x_i variables are integrated over the full Wilson loop, from $-T/2$ to $T/2$ at **R** and then back again at the origin (we neglect the horizontal sections of the loop because they don't give contributions that grow like powers of T). The factor of $(\frac{1}{2})^2$ eliminates double counting. The θ functions refer to ordering around the Wilson loop. The first contribution comes from the graph where the internal gluon lines do not cross each other when all lines are drawn inside the loop.

If, in Figure 9.9(e), we write $\operatorname{tr}(T_R^a T_R^b T_R^a T_R^b) = \operatorname{tr}(T_R^a T_R^a T_R^b T_R^b) + \frac{1}{2} \operatorname{tr}[T_R^a, T_R^b]^2$ then we can combine the first term with the graph of Figure 9.9(d) to get precisely the square of the second-order result. The novel term thus has the form

$$
\left(\frac{1}{2}\right)^2 g_0^4 \mu^{2(4-d)} \int dx_1\, dx_2\, dx_3\, D(x_1 - x_2) D(x_3 - x_4)
$$
$$
\times \operatorname{tr}\!\left[[T_R^a, T_R^b]^2 \theta(t_1 - t_3)\theta(t_4 - t_3) \right].
$$

The group-theory factor here gives $-f^{abc}f^{abc} D(R) = -C_2(G) d_G D(R)$, in terms of the second Casimir operator of the group, its dimension, and the Dynkin index of the representation R.

There are three types of configurations of the x_i that contribute to the integral. If *all* x_i are on either the upward- or the downward-going line, then the integral has no R dependence. It contributes to the self-energy of the static sources, but not to the potential between them. If three x_i are on either the upward or the downward line, as in Figure 9.9(c), we get (note that there is a minus sign from the orientations of the different integration variables)

$$4 \int dt_1 \, dt_2 \, dt_3 \, dt_4 \, \theta(t_2 - t_3)\theta(t_3 - t_1)D(t_1 - t_2, 0)D(t_3 - t_4, \mathbf{R}).$$

All the integrals here go from $-T/2$ to $T/2$ and the θ functions are ordinary step functions. In DR, the Green functions are

$$D(t, \mathbf{R}) = \int \frac{d^d p}{(2\pi)^d} \frac{e^{-i(t p_0 + pR)}}{(p_0^2 + p^2)}$$

$$= \left(\frac{1}{2\sqrt{\pi}}\right)^d \int_0^\infty d\alpha \, \alpha^{-d/2} e^{-\frac{t^2 + R^2}{4\alpha}}.$$

The contribution from Figure 9.9(b) is similar. The two kinds of R-dependent contributions (Figures 9.9(b) and (c) and the commutator term in Figure 9.9(e)) have equal group-theoretic factors, since they are really part of the same topological configuration of the Wilson-loop diagrams, distinguished only by the fact that we have chosen to put part of the loop at infinity. The differences between them come from R dependence of the propagators, and the relative minus sign from the orientation of the static line. Their sum is proportional to ($t_{ij} \equiv t_i - t_j$)

$$T \int \left[dt_{23} \, dt_{21} \, dt_{34} \, \theta(t_{23})\theta(t_{21} - t_{23}) e^{-\left(\frac{t_{21}^2}{4\alpha} + \frac{t_{34}^2}{4\beta}\right)} \right.$$

$$\left. - dt_{31} \, dt_{21} \, dt_{34} \, \theta(t_{31})\theta(t_{21} - t_{31} + t_{34}) e^{-\left(\frac{t_{21}^2 + R^2}{4\alpha} + \frac{t_{34}^2 + R^2}{4\beta}\right)} \right].$$

This must be integrated over α and β, with weight $(\alpha\beta)^{-d/2}$.

In the first integral, we do the integral over t_{23} to get a factor of t_{21}. The latter variable is constrained to be positive. The rest of the integral is a product of decoupled Gaussians and gives us a factor $4\sqrt{\pi\alpha^2\beta}$. In the second integral we integrate over t_{31} to get a factor of $\theta(t_+)t_+$, where $t_\pm = t_{12} \pm t_{34}$. In terms of these new variables we find the sum of contributions to the Schwinger-parameter integrands of the two diagrams to be

$$T\left[4\sqrt{\pi\alpha^2\beta} e^{-\frac{R^2}{4\beta}} - dt_+ \, dt_- \, \theta(t_+)t_+ e^{-(t_+^2 + t_-^2 + 4R^2)\left(\frac{1}{16\alpha} + \frac{1}{16\beta}\right)} e^{-t_+ t_-\left(\frac{1}{8\alpha} - \frac{1}{8\beta}\right)} \right].$$

We now do the Gaussian integral over t_-, followed by that over t_+. The resulting sum of the two diagrams is

$$-\frac{g_0^4 \mu^{2(4-d)} C_2(G) d_A D(R)}{2^d \pi^{d/2}} 4\sqrt{\pi} \int d\alpha \, d\beta (\alpha\beta)^{-d/2} e^{-\frac{R^2 d}{4\beta}}$$

$$\times \alpha\sqrt{\beta}\left(1 - e^{-\frac{R^2}{4\alpha}}\sqrt{1 + \frac{\beta}{\alpha}}\right)$$

$$
= -\frac{g_0^4 \mu^{2(4-d)} C_2(\mathrm{G}) d_A D(\mathrm{R})}{2^d \pi^{d/2}} 4\sqrt{\pi} \left(\frac{R^2}{4}\right)^{\frac{7}{2}-d}
$$

$$
\times \int \mathrm{d}s\, \mathrm{d}x\, s^{\frac{5}{2}-d} [x(1-x)]^{-d/2} \mathrm{e}^{-\frac{1}{sx}} (1-x)\sqrt{x}\left(1 - \mathrm{e}^{-\frac{1}{s(1-x)}} \sqrt{\frac{1}{1-x}}\right). \quad (9.4)
$$

As usual, divergences come from the limits of the Schwinger-parameter integrals. Note, however, that the s integral converges for small s. The UV divergence comes from the region $x \sim 1$, from which we get a pole at $d = 4$. It is probing the short-distance singularity of only a single propagator. Indeed, we can do the s integral exactly and obtain

$$
-\frac{g_0^4 \mu^{2(4-d)} C_2(\mathrm{G}) d_A D(\mathrm{R})}{2^d \pi^{d/2}} 4\sqrt{\pi}\, \Gamma\left(d - \frac{7}{2}\right)\left(\frac{R^2}{4}\right)^{\frac{7}{2}-d}
$$

$$
\times \int \mathrm{d}x\, x^{d/2-3}\left[(1-x)^{1-d/2} - (1-x)^{d/2-3}\right]. \quad (9.5)
$$

The integrals can be evaluated in terms of Euler beta functions, and the first term has a pole at $d = 4$, as a consequence of the singularity of its integrand at $x = 1$.

Next we compute the diagrams which correct the internal gauge-boson line. We will do the computation in Feynman gauge, and in Minkowski space, but drop iϵ factors in propagators. In Problem 8.13 the reader will verify that the answer for the full Wilson loop is the same in any covariant gauge. The theory is invariant both under constant gauge transformations and under BRST symmetry. The first of these tells us that the gauge-boson two-point function is proportional to δ^{ab}. Given this constraint, the BRST-symmetry relation for the two-point function reduces to the same constraint as that which we found in abelian gauge theory: the two-point function is transverse. So we have

$$
\Pi_{\mu\nu}^{ab}(q) = (\delta_{\mu\nu} q^2 - q^\mu q^\nu)\delta^{ab} \Pi(q^2).
$$

Our QED calculations have shown us that dimensional regularization preserves BRST symmetry, so it is sufficient to calculate the trace of the diagrams, which gives us $\delta^{ab}(d - 1)q^2 \Pi(q^2)$. The reader is urged to carry out the full calculation of all components in a general covariant gauge, in order to convince her/himself that the result is indeed transverse. In this exercise, it will become apparent that individual graphs do not satisfy the Ward identities. They are true only for the sum of graphs at a given order.

The graphs in Figures 9.10(d)–(f) involving fermions or scalars are proportional to their values in QED. The square of the integer charge of the fields is replaced by the Dynkin index ($\mathrm{tr}(T_\mathrm{R}^a T_\mathrm{R}^b) = D(\mathrm{R})\delta^{ab}$) for the, generally reducible, representations of the gauge group in which these fields live. Thus, the divergent part of the gauge-boson two-point functions from these diagrams is

$$
\Pi_{\mathrm{F+S}}^{\mu a \nu b} = \mathrm{i}(q^2 \eta^{\mu\nu} - q^\mu q^\nu)\delta^{ab}\left(-\frac{\alpha_\mathrm{G}}{4\pi}\right)\mu^{d-4}\left[\frac{2}{3} D(\mathrm{R_F}) + \frac{1}{3} D(\mathrm{R_S})\right]\Gamma\left(2 - \frac{d}{2}\right). \quad (9.7)
$$

α_G is the analog of the fine structure constant for the non-abelian group G. We have done the computation for Weyl fermion fields in the representation R_F. For parity-invariant gauge couplings to Dirac fields, we should make the replacement $2/3 \rightarrow 4/3$. Similarly, our computation was done for complex scalar fields. If some of the irreducible components of R_S are real, and we have only one real field in this irreducible representation, the corresponding contribution is smaller by a factor of $1/2$. The reader should remember these factors of two when applying the above formula to specific theories.

The pure gauge-theory contribution to the gauge-boson two-point function consists of three graphs, Figures 9.10(a)–(c). The most complicated is the one involving triple gauge-boson vertices. It has the form

$$\Pi_{V1}^{\mu a\nu b} = \frac{(-i)^2 g_0^2 \mu^{d-4}}{2} f^{acd} f^{bcd} \int \frac{d^d p}{(2\pi)^d} \frac{N^{\mu\nu}}{p^2(p+q)^2}, \tag{9.8}$$

where the numerator factor is

$$N^{\mu\nu} = \left[\eta^{\mu\rho}(q-p)^\alpha + \eta^{\rho\alpha}(2p+q)^\mu - \eta^{\alpha\mu}(p+2q)^\rho \right]$$
$$\times \left[\delta_\rho^\nu(p-q)_\alpha - \eta_{\rho\alpha}(2p+q)^\nu + \delta_\alpha^\mu(p+2q)_\rho \right].$$

The overall factor of $1/2$ is a symmetry factor. The trace of the numerator is

$$N_\mu^\mu = -6(d-1)(p^2 + q^2 + pq).$$

The group-theory factor can be written in terms of the second Casimir operator, $f^{acd} f^{bcd} = \delta^{ab} C_2(G)$. Thus

$$\Pi_{V1\mu}^{\mu ab} = 3(d-1)g_0^2 \mu^{d-4} \delta^{ab} C_2(G) \int \frac{d^d p}{(2\pi)^d} \frac{p^2 + q^2 + pq}{p^2(p+q)^2}. \tag{9.9}$$

The diagram of Figure 9.10(b) is given by

$$\Pi_{V2}^{\mu a\nu b} = \frac{(-i)^2 g_0^2 \mu^{d-4}}{2} \int \frac{d^d p}{(2\pi)^d} \frac{\delta_{dc}\eta_{\alpha\beta}}{p^2} \tag{9.10}$$
$$\times \Big[f^{abe} f^{cde}(\eta^{\mu\alpha}\eta^{\nu\beta} - \eta^{\nu\alpha}\eta^{\beta\mu})$$
$$+ f^{ace} f^{bde}(\eta^{\mu\nu}\eta^{\alpha\beta} - \eta^{\mu\alpha}\eta^{\beta\nu})$$
$$+ f^{ade} f^{bce}(\eta^{\mu\nu}\eta^{\alpha\beta} - \eta^{\nu\alpha}\eta^{\beta\mu}) \Big].$$

We can reduce the group-theory factors and find that

$$\Pi_{V2}^{\mu a\nu b} = -C_2(G)(d-1)g_0^2 \mu^{d-4} \int \frac{d^d p}{(2\pi)^d} \frac{\delta^{ab}\eta^{\mu\nu}(p^2 + q^2 + 2pq)}{p^2(p+q)^2}. \tag{9.11}$$

Finally, we have the diagram in Figure 9.10(c), with the ghost loop,

$$\Pi_{V3}^{\mu a\nu b} = -(i)^2 g_0^2 \mu^{d-4} f^{dac} f^{cbd} \int \frac{d^d p}{(2\pi)^d} \frac{(p+q)^\mu p^\nu}{p^2(p+q)^2}, \tag{9.12}$$

where we note the absence of a symmetry factor and a minus sign, coming from the fact that the ghosts are complex scalar fermions. In contrast to physical scalars, there is no quartic coupling between the ghosts and the gauge bosons in the gauge we are using. The group-theory factor in this diagram is $-C_2(G)\delta^{ab}$.

The sum of these three diagrams gives

$$\Pi_\mu^{\mu ab} = g_0^2 \mu^{d-4} \delta^{ab} C_2(G) \int \frac{d^d p}{(2\pi)^d} \frac{N(d,x,p,q)}{p^2(p+q)^2},$$

where

$$N(d,x,p,q) = \left[3(d-1)(p^2+q^2+pq) - d(d-1)(p^2+q^2+2pq) - (p^2+pq)\right].$$

We now introduce a Feynman parameter x to simplify the denominator, and define $r = p - xq$, $U = -x(1-x)q^2$. We can then discard terms in the numerator linear in r and write it as

$$-(d-2)^2(r^2 + x^2 q^2) + (2d^2 - 5d + 4)xq^2.$$

Note that the term in the numerator quadratic in the integration variable is multiplied by $(d-2)$. The integral over this term will give rise to a pole at $d-2$, which is the signal of a quadratic divergence in DR. The fact that it is multiplied by $d-2$ indicates that this pole is absent and there is no quadratic divergence, and thus no divergent gauge-boson mass.

As usual, we do the integrals by analytically continuing to Euclidean space. I take the opportunity to record here the Minkowski-space values of two dimensionally regulated integrals which are obtained by this method:

$$\int \frac{d^d p}{(2\pi)^d} \frac{1}{(p^2 - U)^n} = \frac{(-1)^n i}{2^d \pi^{d/2}} \frac{\Gamma(n - d/2)}{\Gamma(n)} \left(\frac{1}{U}\right)^{n-d/2}.$$

$$\int \frac{d^d p}{(2\pi)^d} \frac{p^\mu p^\nu}{(p^2 - U)^n} = \frac{(-1)^{n-1} i}{2^{d+1} \pi^{d/2}} \frac{\Gamma(n - d/2 - 1)}{\Gamma(n)} \left(\frac{1}{U}\right)^{n-d/2-1}.$$

The reader should note how the argument of the numerator Euler function corresponds to the power of U. Once one has understood the methods of Euclidean field theory, it is often convenient to just remember such a table of effective Minkowski integrals.

The reader is asked to complete the computation of the renormalized static potential $V(R)$ in Problem 9.10. He/she will find that the β function for the Yang–Mills coupling is given by

$$\beta(g) = -\frac{g^3}{16\pi^2} \left(\frac{11}{3} C_2(G) - \frac{2}{3} D(R_F) - \frac{1}{3} D(R_S)\right).$$

$C_2(G)$ is the quadratic Casimir invariant of the adjoint representation, and $D(R)$ is the Dynkin index of the representation R. We have done the computation for Weyl fermions (Dirac fermions will give an extra factor of two) and complex scalars (real scalars are possible for real components of R_S and give an extra factor of $\frac{1}{2}$). As long as there are not too many matter fields, the non-abelian gauge coupling is marginally irrelevant or asymptotically free [113–118]. A definition of the coupling in terms of the potential is manifestly independent of the gauge parameter. On the other hand, Problem 9.9 shows that the first two terms in the β function are the same for all renormalization schemes whose couplings are related by power-series expansions. Thus the first two terms of the β function are universal, scheme-independent, and gauge-invariant. A general definition of the gauge coupling will not have a κ-independent β function beyond two loops.

The static potential in a massless gauge theory will satisfy the RG equation

$$(\mu \, \partial_\mu + \beta(g)\partial_g)V(R,g,\mu) = 0,$$

while dimensional analysis implies that

$$(\mu \, \partial_\mu - R \, \partial_R)V = V.$$

Together, these equations imply that

$$V = -\frac{g_0^2 C_2(\mathrm{R})}{4\pi R}.$$

The potential is essentially the Coulomb potential, multiplied by the running coupling at scale $1/R$. In perturbation theory, this relation arises because all non-scale-invariant R dependence is a function of $R\mu$, but all μ dependence comes through the bare coupling and survives in the limit $d = 4$ only when it multiplies a pole. The RG equation implies that the kth loop term in perturbation theory will be a kth-order polynomial in $\ln(\mu R)$.

9.12 Renormalization-group equations for masses and the hierarchy problem

So far we have talked mostly about the use of RG equations for Euclidean Green functions. In Lorentzian signature, various complications arise, which we will sketch in this section. First consider the RG equation for a connected two-point function

$$\mathcal{D}W_2 = -2\gamma \, W_2,$$

where $\mathcal{D}$ is the infinitesimal RG operator ($\mu\partial_\mu + \beta\partial_\alpha + m\gamma_m\partial_m$ in spinor QED) and γ is the anomalous dimension of the field in question (this argument is applicable to any operator that undergoes multiplicative renormalization). Since the RG equation is independent of momentum, it applies near a pole, where

$$W_2 \sim \frac{Z}{p^2 - M^2 + i\epsilon}.$$

Since $\mathcal{D}$ is a first-order differential operator, it produces both a single-pole and a double-pole term when acting on W_2. The two cannot cancel out, and hence must satisfy the equation separately. Thus

$$\mathcal{D}Z = -2\gamma Z$$

and

$$\mathcal{D}M^2 = 0.$$

The latter equation is particularly interesting when the theory has no relevant operators, as is true for a chirally symmetric abelian or non-abelian gauge theory, like massless QED. In massless QED, we get the equation

$$\left(\mu\,\frac{\partial}{\partial\mu} + \beta\,\partial_\alpha\right)M^2 = 0.$$

On combining this with dimensional analysis we have the solution

$$M = \mu\mathrm{e}^{-\int^\alpha \frac{\mathrm{d}x}{\beta(x)}} \to \mu c\mathrm{e}^{\frac{3\pi}{2\alpha}}.$$

This equation makes no sense unless $c = 0$. If $c \neq 0$ the theory would not be continuous at $\alpha = 0$, contradicting the assumption of perturbation theory on which we based the calculation.

This is a very general result: a marginally irrelevant perturbation cannot produce a finite mass scale. On the other hand, if the perturbation were marginally relevant, the sign of the exponent would change, and we would learn that exponentially small masses *could* be generated. This result contains the seed of one of the most important ideas in quantum field theory. It gives us an idea of how the vast hierarchy of mass scales in the real world might be generated from a system with a single intrinsic scale as large as the Planck mass. All that is necessary is that the theory have a low-energy effective approximation that is a marginally relevant perturbation of a fixed-point quantum field theory. In that case, if the underlying high-energy theory can give us an explanation of why the initial values of parameters are moderately close to the critical surface, we can explain the natural occurrence of scales exponentially smaller than the Planck scale.

The nice behavior of marginally relevant parameters is to be contrasted with what happens if we have truly relevant parameters, like scalar masses near the Gaussian fixed point. In that case, the RG tells us that we need a fine tuning of initial conditions with accuracy $(m/M_\mathrm{P})^2$ in order to explain a mass scale of order m. Most physicists consider this unnatural. By contrast, the tuning of the bare coupling of QCD at a Planck scale cut-off, which is required in order to explain the QCD scale in terms of the Planck scale, is only about one part in 25.[17] Fermion masses, although they seem like relevant perturbations, actually behave like marginal perturbations because of chiral symmetry. This means that, if the underlying theory has enough symmetry to forbid the appearance of a fermion mass in the high-energy Lagrangian, then we could explain a small value of this mass in the real world in terms of spontaneous breaking of this symmetry by the IR effects of a marginally relevant coupling.

---

[17] I am here assuming that the particle content of the minimal supersymmetric standard model is the only thing that renormalizes the QCD coupling between laboratory scales and the Planck scale. Extra matter makes the tuning of the bare coupling even less severe, as long as it preserves asymptotic freedom.

The concept of technical naturalness is based on this example. We have a parameter that could be classed as relevant, but might naturally be small as a consequence of a fundamental symmetry broken spontaneously by a marginally relevant coupling. It is then said that a theory containing a small value of this parameter is technically natural, because there will not be power-law renormalizations of its value. It becomes truly natural when we supply the underlying explanation for the symmetry, and the dynamical symmetry-breaking mechanism.

The standard model contains one parameter, the quadratic term in the Higgs potential, that is not technically natural. It receives quadratically divergent renormalizations, and it is not easy to understand why it is not of order the Planck scale. This is called the gauge hierarchy problem, since this parameter determines the scale of the massive weak gauge bosons. Proposals to solve this problem divide roughly into two classes. The first goes under the name of technicolor, and replaces the Higgs field by a new QCD-like sector with a dynamical scale of order a few TeV. This gives an elegant account of gauge symmetry breaking, but suffers from numerous phenomenological problems when we try to couple it to quarks and leptons. The second solution of the hierarchy problem is based on supersymmetry. Supersymmetry is a symmetry that relates bosons to fermions, and hence allows bosons to benefit from the chiral protection afforded to fermion masses. It is also incredibly interesting because it forces us to think about gravitational effects, and seems to be an intrinsic component of string theory, our only successful theory of quantum gravity. Generic supersymmetric extensions of the standard model, which allow for parameters that break supersymmetry in the phenomenologically necessary way, also have a variety of phenomenological problems. However, it is possible to solve these problems in specific models. The key question seems to be to understand the mechanism by which supersymmetry is broken. String theory contains hints suggesting that supersymmetry breaking is much more constrained than low-energy field theory would have us believe.

There is another resolution, or rather postponement, of the hierarchy problem in a class of models that go under the rubric "little Higgs" [119]. These models successfully postpone the hierarchy discussion to the 100-TeV scale, and declare that, since that scale will be out of experimental reach for the forseeable future, we can safely ignore it. I will leave the reader to decide on the value of these models for her/himself.

9.12.1 Marginally relevant perturbations

A remarkable theorem of Coleman, Gross, and Zee [120–122] shows that theories with marginally relevant parameters in four-dimensional space-time *always involve non-abelian gauge bosons*. Non-abelian gauge couplings are always marginally relevant unless the matter representation is too large. Some of the other couplings in the theory may also turn out to be marginally relevant, as a consequence of their interaction with the non-abelian bosons. The result we have just proved suggests strongly that non-abelian gauge interactions will be our only explanation of the large hierarchy in energy scales between the Planck mass, 10^{19} GeV, and the typical scales of particle physics. This does indeed seem to be the explanation of the scale of strong interactions. Notice that the phenomenon of confinement, which we discussed in Chapter 8, involves a dimensionful parameter, the string tension or energy per unit length of the QCD flux tube. If the QCD coupling were not marginally relevant, we would expect this parameter to be at the cut-off scale. The scale Λ_{QCD} is therefore sometimes called *the confinement scale of QCD*, though experiment shows that the actual string tension involves a scale roughly 2π larger than Λ_{QCD}.

Actually, if we follow this philosophy stringently, we are led to either omit fundamental scalar fields from our Lagrangian or insist on *supersymmetry* as a fundamental principle. Supersymmetry relates scalar fields to Weyl fermions, in such a way that chiral symmetries act on scalars. Thus, in supersymmetric theories, scalar mass parameters, like fermion masses, behave like marginally relevant operators, even though they have dimension 2. They receive only logarithmic renormalizations. Thus, it is consistent to have scalar mass parameters of order 100 GeV, even if the cut-off scale is much higher than that. Supersymmetry is broken in the real world, and its virtues can be preserved only if the superpartners of standard-model particles lie not too far above the electro-weak scale.

This kind of solution of the hierarchy problem is technical in nature. It does not yet explain the value of the electro-weak scale in the way that asymptotic freedom explains the scale of strong interactions. In order to do that, we have to make the breaking of the chiral symmetry that protects the Higgs mass into a dynamical mechanism, which depends on some new marginally relevant parameter in the Lagrangian of the world. This can be done either with or without supersymmetry. In the non-supersymmetric solution, which is called *technicolor* [123], the Higgs field is a fermion bilinear in a new QCD-like sector. These models have severe phenomenological problems. There is a variety of supersymmetric scenarios for a completely dynamical explanation of the electro-weak scale. Many of them are roughly consistent with current experimental data, but there are various causes for unease, and no uniquely beautiful model has yet emerged.

9.13 Renormalization-group equations for the S-matrix

The LSZ formula for S-matrix elements has the schematic form

$$S = \prod f_i (p_i^2 - M_i^2) Z_i^{-1/2} W_{n+m}.$$

The f_i are normalized ingoing or outgoing single-particle wave functions, which depend only on the masses, the Z_i are the residues of poles in two-point functions, and the W_n are renormalized connected Green functions. It is easy to see, as a consequence of the equations in the last subsection and of the first-order nature of $\mathcal{D}$, that S-matrix elements satisfy

$$\mathcal{D}S = 0.$$

This simply expresses the fact that the theory is over-parametrized and that the S-matrix is a physical amplitude, which does not depend on how we have chosen to define the couplings.

Unlike the case of Green functions, we cannot combine this equation with dimensional analysis to extract a momentum dependence of the scattering matrix. The point is that the S-matrix is defined on mass shell, and we cannot arbitrarily rescale the masses of particle states. These typically form a discrete spectrum. Thus, the RG has nothing so say directly about scattering-matrix elements.

However, we can combine the RG with perturbation theory in certain weak interactions to obtain information about inclusive cross sections in a strongly interacting sector of the theory. Consider for example the production of hadrons via electron–positron annihilation. To lowest order in QED, the amplitude for producing a given hadronic state X is given in terms of the matrix element $\langle 0|J_\mu(x)|X\rangle$ of the electromagnetic current $\bar{q}Q\gamma_\mu q$ in QCD. This cannot be calculated without a real solution of QCD. However, the inclusive cross section, summing over all hadronic final states, is related to the Fourier transform of the two-point function of the hadronic current, $J_{\mu\nu}(q)$. Even for large time-like q, which is relevant to high-energy e^+e^- annihilation, the RG equation does not say anything definitive about this two-point function. For example, if there are heavy quarks, with mass well above the scale Λ_{QCD} at which strong interactions are strong, there may be bound states of these quarks, which would appear as resonances in this two-point function. Their masses would satisfy the RG equation, and any function of q, M_i satisfying dimensional analysis would be an acceptable solution.

However, it is a phenomenological observation that, away from resonance singularities, such cross sections are smoothly varying functions of q. In the deep Euclidean region, where q is space-like and $\gg \Lambda_{\text{QCD}}$, we can invoke asymptotic freedom to calculate the two-point function in terms of perturbative graphs involving quarks and gluons. Analyticity, in the form of the Lehmann representation, allows us to write this explicitly known behavior in terms of integrals over the time-like region. These formulae show that, in smoothly varying regions, the time-like behavior is simply the analytic continuation of the perturbative space-like formulae, while near resonances the perturbative formulae reproduce only certain integrals over the resonance. Thus, combining knowledge of the fixed-point behavior with general principles, we can use the RG to predict things about the high-energy behavior of QCD [32].

A similar, but more complicated, analysis works for the inclusive cross sections for lepton scattering off hadrons, in the regime of high energy and momentum transfer (deep inelastic scattering). After making an angular momentum projection, one can again relate the cross section to the short-distance OPE of two currents, this time taken between single-particle hadron states. These calculations justify and extend Feynman's heuristic parton model for high-energy scattering in the strong interactions. Workers have applied the successful QCD parton model to a variety of other processes for which a rigorous analysis is harder to come by. This elaboration of perturbative QCD has been quite successful. You can read about some of the details in Peskin and Schroeder [33]. A more modern account can be found in [124].

9.14 Renormalization and symmetry

9.14.1 You don't need symmetry

Wilson's approach to the interplay of renormalization and symmetry is based on the observation that the relevant and marginal perturbations of a Lagrangian with symmetry are generally finite in number. At tree level, one can set the coefficients of all symmetry-breaking operators to zero. Even if one ignores the symmetry in constructing a cut-off version of the theory, one can still imagine tuning the coefficients of the symmetry-breaking relevant and marginal operators to restore the naive Ward identities at the level of renormalized Green functions.

Wilson's tuning procedure ignores the issue of naturalness that we raised above.[18] More interestingly, we have encountered a class of examples in which it fails. These are theories with "anomalies." Certain terms that appear in loop corrections to the classical Ward identities cannot be canceled out by adding local counterterms to the action. Anomalies are always associated with particles whose mass is zero, or can be made to go to zero by tuning the expectation value of a low-energy field. The only exception to this is classical scale invariance. In zeroth-order perturbation theory, the only requirement is that all operators in the Lagrangian be marginal. However we have seen that most marginal operators are actually marginally relevant or irrelevant. Interacting scale-invariant theories are scarce, and typically appear only at isolated points in the space of field theories.

9.14.2 But symmetries are nice

On the other hand, we have seen that invoking symmetries makes the whole process of renormalization less arduous. If we can invent a regulator that preserves a symmetry we should do it, if only because it saves work in calculation. Quite frankly, it wasn't until 't Hooft and Veltman introduced dimensional regularization that calculations in non-abelian gauge theory became sufficiently transparent for one to decide whether the theory was renormalizable. The Wilsonian approach can work, but only if one is a master calculator, who doesn't make mistakes.

More importantly, symmetries can help us to understand the smallness or absence of terms in the effective action that we might alternatively think of as finely tuned relevant

---

[18] Wilson was one of the very first advocates of naturalness as a criterion for a good theory.

parameters. This is the philosophy of "technically natural field theory" which we have adumbrated above. Indeed, it can be argued that much of the power of effective field theory comes from its exploitation of symmetry. For example, we are able to calculate a large number of low-energy scattering amplitudes involving pions in terms of a small number of parameters just by using chiral symmetry and effective field theory. Effective field theory alone would give us very little information.

9.14.3 And sometimes you get them for free

The deepest remark one can make about the relationship between symmetry and renormalization is that symmetries can be emergent. Like much else in this field, this remark is due to Wilson. To make the point in a simple context, consider a lattice version of a scalar field action, thought of as a perturbation of the Gaussian fixed point. There are lots of allowed operators (e.g. $\sum_{\mu}(\partial_{\mu}\phi)^4$) in the theory, on a cubic lattice, which break continuous rotation invariance. However, for most regular lattices, the lattice symmetries are sufficient to remove all relevant and marginal operators that break the continuous symmetry. Thus we can view rotation invariance (and, after Wick rotation, Lorentz invariance) as an accidental low-energy consequence of the fact that IR physics is dominated by a fixed point. In fact, most of the fixed points describing second-order transitions in condensed-matter physics have emergent rotation invariance, despite coming from underlying systems that have no such symmetry.

It is likely that Lorentz invariance is not accidental, but rather a consequence of the theory of quantum gravity. It is a large-distance manifestation of the underlying gauge principle of general covariance. However, there is lots of room for emergent continuous symmetries to play a role in our theory of the real world. In particular, the standard model has a host of exact and approximate global continuous symmetries like baryon number, lepton numbers, and various quark flavor numbers. There are good arguments that none of these symmetries is exact, but they might be emergent infrared symmetries of an underlying theory in which e.g. only discrete subgroups of them are exact. A detailed exploration of theories with emergent continuous symmetries is beyond the scope of our discussion here, but it is an important theme in the exploration of physics beyond the standard model.

Important examples of this kind of emergent symmetry abound in the standard model. For example, baryon and lepton numbers, and lepton flavor, are preserved automatically by the most general renormalizable Lagrangian with the standard-model gauge symmetries.[19] Furthermore, any theory of baryon-number violation can be parametrized by the coefficients of a small number of dimension-6 operators.

A more subtle emergent symmetry is the extra SU(2) in the Higgs sector of the standard model, which we called custodial SU(2). It is broken by the U(1) gauge interaction,

---

[19] Actually, B and L are broken, and only $B - L$ preserved, by an anomaly in the standard model. However, at low energies the breaking is exponentially small.

but that is the only renormalizable interaction which can break it. As a consequence, the tree-level broken-symmetry relation

$$M_W = M_Z \cos \theta_W$$

can get only finite corrections, when it is expressed in terms of the renormalized couplings.

9.14.4 Renormalization and spontaneous symmetry breaking

In the previous section we discussed the relation between renormalization and explicit breaking of symmetries. What then of theories with spontaneously broken symmetry? Our discussions of spontaneously broken symmetry and of the renormalization group show that the two subjects are really decoupled from each other. Renormalization is the procedure for deriving the effective theory of long-wavelength degrees of freedom of a system, while spontaneous symmetry breakdown is a property of the solution of that effective field theory.

In particular, if we consider the system in a large but finite volume, then there is no spontaneous breaking of symmetry. There is a unique, symmetric ground state. The renormalization procedure involves integrating out short-wavelength degrees of freedom, which, because of the approximate locality of interactions, are insensitive to the volume of the system.

A procedure for seeing this decoupling explicitly in perturbative calculations is to calculate the quantum action, rather than connected Green functions. One can then test the system for spontaneous symmetry breaking by looking to see whether there are translation-invariant solutions of the field equations

$$\frac{\delta \Gamma}{\delta \phi(x)} = 0$$

that violate symmetries of the classical action. To do this, we need only calculate the *effective potential*, which is defined by[20]

$$\Gamma[\phi_c] = \frac{1}{g^2} \int d^4x \, V_{eff}(\phi_c).$$

g^2 is the loop-counting parameter. The derivatives of V_{eff} w.r.t. ϕ_c are 1PI Green functions evaluated at zero momentum.

To develop an efficient graphical method to calculate V_{eff}, we begin with its Legendre transform, the connected generating functional in a constant classical source J, divided by the volume of space-time. $w(J)$ is computed as the logarithm of a path integral, in the usual way, except that we divide the answer by the volume, and multiply by g^2. At tree level, $w(J)$ is just the Legendre transform of the potential in the classical action. More generally, we can think of it as the energy density of the ground state in the

---

[20] We work in Euclidean space. In Lorentzian signature this equation needs a minus sign.

presence of the constant classical source. There are two interesting interpretations of V_{eff}, which follow from these definitions.

First of all, it is the answer to the following variational question: "What is the minimal value of the expectation value of the Hamiltonian of the $J=0$ theory, in states constrained to satisfy

$$\langle \psi | \phi | \psi \rangle = \phi_c?"$$

Indeed, the way to solve this constrained variational problem is to introduce J as a Lagrange multiplier, calculate the minimal energy as a function of J (which is $w(J)/\mathcal{V}_{\text{space}}$) and then impose the constraint

$$w'(J) = \phi_c.$$

The minimal expectation value of H is then obtained by subtracting the source term from w and then expressing the answer as a function of ϕ_c. This is just the Legendre transform.

Consider instead the path integral

$$\int [\mathrm{d}\phi] e^{iS} \delta \left(\int \phi - \mathcal{V}_{\text{space-time}} \phi_c \right).$$

Write the delta function as $\int \mathrm{d}J\, e^{i(\int \phi J - \mathcal{V}_{\text{space-time}} \phi_c J)}$, and perform the path integral to obtain

$$\int \mathrm{d}J\, e^{i\mathcal{V}_{\text{space-time}}(w(J) - J\phi_c)}.$$

Now do the integral over J by stationary phase in the large-volume limit. The answer is

$$e^{-i\mathcal{V}_{\text{space-time}} V_{\text{eff}}(\phi_c)}.$$

In other words, in the large-volume limit, the effective Wilsonian action for the zero-momentum mode of the field is just the effective potential.

In order to compute V_{eff} perturbatively we imagine that we have found the value of J corresponding to the expectation value of ϕ equal to ϕ_c. Then the effective potential is just $w(J) - J\phi_c$. Now shift the field in the path integral defining $w(J)$ by ϕ_c: $\phi = \phi_c + \Delta$. By definition, the field Δ has zero expectation value. The perturbative way to determine $J = \sum g^n J_n$ is to insist that it is chosen so that, in each order, the linear term from the explicit $J\Delta$ term in the action cancels out the "tadpole" graphs which give the one-point function of Δ. This then implies that *all* 1P reducible vacuum graphs vanish. We thus obtain the following prescription [125].

Define the action $S(\phi_c + \Delta) - \mathcal{V}_{\text{space-time}} \phi_c (\partial V/\partial \phi)(\phi_c)$. Compute the vacuum path integral for this action, ignoring all 1PI diagrams. This path integral is equal to

$$e^{-\mathcal{V}_{\text{space-time}} V_{\text{eff}}(\phi_c)}.$$

The one-loop term in this expression needs some care, since it is unusual to look at a Feynman diagram with no vertices. Let's consider an $O(n)$-invariant Euclidean Lagrangian

$$\mathcal{L} = \frac{1}{2}(\partial_\mu \phi^i)^2 + V(\phi^i \phi^i).$$

According to our prescription, the one-loop correction to V_{eff} is

$$V_{\text{eff}}^{(1)} = -\ln\left(\int d[\Delta] e^{-\frac{1}{2}(\Delta^i(-\nabla^2)\Delta^i + (M^2)_{ij}\Delta^i \Delta^j)}\right)$$

$$= \frac{1}{2}\ln \det(-\nabla^2 + M^2).$$

The matrix M^2 is defined by $M_{ij}^2 = (\partial^2 V / \partial\phi^i \, \partial\phi^j)(\phi_c)$.

To find the regularized form of this answer, let us divide the functional integral by the same expression with $M^2 \to M_0^2 \delta_{ij}$, where M_0 is a large mass that we think of as order the cut-off. This subtracts a ϕ_c-independent constant from V_{eff} and leads to a simpler formula at intermediate stages of the calculation. It has no relevance to any physical quantity. For any Hermitian matrix, $\text{tr} \ln A = \ln \det A$, so we can write

$$V_{\text{eff}}^{(1)} = -\frac{1}{2} \text{Tr}\left(\int \frac{ds}{s}\left[e^{-s(-\nabla^2 + M^2)} - e^{-s(-\nabla^2 + M_0^2)}\right]\right).$$

The identity which leads to this parametric formula can be verified eigenvalue by eigenvalue for the two operators. Note that it looks just like the Schwinger proper-time version of a one-loop vacuum diagram with no vertices, except that the proper-time integral has a factor of $1/s$ in it that we might not have guessed from ordinary Feynman rules. The trace $\text{Tr} \, O = \int d^d x \langle x | \text{tr} \, O | x \rangle$ is over the tensor product of function space and $O(n)$-index space; tr refers to the index trace alone. The operator $-\nabla^2 + M^2$ is diagonal in Fourier space. Thus we obtain

$$V_{\text{eff}}^{(1)} = -\frac{1}{2} V_{\text{space-time}} \, \text{tr}\left(\int \frac{d^d p}{(2\pi)^d} \frac{ds}{s}\left[e^{-s(p^2 + M^2)} - e^{-s(p^2 + M_0^2)}\right]\right)$$

$$= -\frac{1}{2(2\sqrt{\pi})^d} \text{tr} \, M^d \, \Gamma\left(-\frac{d}{2}\right).$$

Now

$$\Gamma(-d/2) = \frac{1}{(2 - d/2)(1 - d/2)(-d/2)} \Gamma(3 - d/2),$$

so the one-loop effective potential has a pole at $d = 4$. If the tree-level potential is quartic in the fields, $V = (m_0^2/2)\phi^2 + \lambda_0(\phi^2)^2$, then the residue of this pole proportional to $\text{tr} \, M^4$ is a mixture of quadratic and quartic terms. Furthermore, it is an $O(n)$-invariant function of the classical fields ϕ. This means that the pole can be removed by redefining m_0 and λ_0. The required renormalization (in the MS scheme) is identical to that which we would have to do in order to eliminate the divergent part of any one-loop Green function.

This calculation also shows us that theories with spontaneously broken symmetries are renormalized by the same manipulations as those which renormalize the corresponding symmetries in the symmetric phase. Indeed, since we are calculating the quantum action, we never have to make a choice of the vacuum we expand around. The renormalized quantum action is an $O(n)$-invariant functional of ϕ. The renormalized parameters (which, as low-energy observers, we are free to choose at will) are such that the minimum of the effective potential is asymmetric.[21] This decoupling between renormalization and spontaneous symmetry breaking should have been expected. We described spontaneous breaking in terms of inequivalent representations of the same field algebra, which were inequivalent precisely because of their different large-volume behaviors. Renormalization has to do with the local structure of the theory.

9.14.5 The Coleman–Weinberg potential

The computation of the effective potential in a general renormalizable field theory follows the pattern set by the $O(n)$ model. One simply computes the vacuum energy as a function of particle masses, where the masses are those induced by a scalar field VEV v. The resulting renormalized potential reads

$$V(\mathrm{v}) = \frac{1}{64\pi^2} \sum (2S + 1)(-1)^{2S} M_i^4(\mathrm{v}) \ln(M_i(\mathrm{v})^2/\mu^2),$$

where μ is the renormalization scale. This is evaluated in terms of renormalized couplings and satisfies a renormalization-group equation. The couplings (and Lagrangian mass parameters) come into the computation of the mass eigenvalues $M_i(\mathrm{v})$. One generally has to diagonalize a complicated mass matrix in order actually to evaluate this formula. Note that the minus sign for fermions comes about because Grassmann functional integrals give determinants rather than inverse determinants. More intuitively it is a consequence of the fact that "vacuum fermions have negative energy."

It is interesting to note that this expression is analytic in v, except at places where one or more mass eigenvalues go to zero. We have obtained an effective action for the zero-momentum modes of the fields, by integrating out the higher modes. However, if a mass is zero, there is no clear separation. This shows up as IR divergences in loop diagrams. The connection between non-analyticity of the effective potential and massless particles was the key to Wilson's analysis of critical phenomena in condensed-matter physics [126]. It is also of crucial importance in particle physics. One typical example is the Linde–Weinberg lower bound on the mass of the Higgs particle in the standard model [127–128]. If the Higgs is much lighter than the W and Z bosons, they can be integrated out and contribute only to the Higgs effective action. At tree level,

---

[21] Strictly speaking, the exact effective potential is a convex function of the field. If spontaneous symmetry breaking occurs, it is in fact constant over the region $|\phi| < \phi_0$, indicating that the true vacua are rotations of the vector $\phi_0(1, 0, \ldots, 0)$. Any point in the indicated region can be the VEV of the field in appropriate superpositions of these rotated vacua. Only the vacua with fixed rotation angles satisfy the cluster-decomposition principle. One never sees the convexity of the effective potential in simple approximations.

the ratio of masses is essentially the ratio between $\sqrt{\lambda}$ and $g_{1,2}$. However, no matter how small the Higgs quartic coupling is, compared with the gauge couplings, the gauge bosons become massless at $|H| = 0$. Their non-analytic contribution to the Coleman–Weinberg potential dominates the tree-level potential in this region, and produces a symmetric minimum if λ is very small. The condition that the Universe actually lives in the spontaneously broken vacuum state then puts a bound on m_H/m_W.

9.14.6 Renormalization of gauge theories in the Higgs phase

The interaction of renormalization and the Higgs phenomenon is more subtle than that between renormalization and spontaneously broken global symmetries. In the unitary gauge, the gauge-boson propagator does not fall off at high energies and non-abelian Higgs models are not renormalizable in unitary gauge. However, the S-matrix in this gauge is equal to that in the general R_κ gauge, which *is* renormalizable.

This peculiar situation can be understood better by noting that the elementary fields in unitary gauge are equal to gauge-invariant operators

$$B_\mu = \Omega^\dagger D_\mu(A)\Omega,$$

etc., where Ω is defined by

$$H(x) = \Omega(x)(\mathrm{v} + \Delta(x)).$$

v is the VEV of H, and Δ (the vector of physical Higgs fields) satisfies $\mathrm{v}^\mathrm{T} T_S^a \Delta = 0$. These gauge-invariant operators can be defined in any gauge, but are equal to the elementary fields of finite mass in the $\kappa \to \infty$ limit.

Thus, in R_κ gauge for general κ, the elementary fields of the unitary gauge are defined as non-polynomial functions of the elementary fields of the R_κ gauge. In perturbation theory, we are always working around the Gaussian fixed point, where the operators of fixed dimension are defined as Wick monomials. Higher-order monomials require more subtractions. Thus, it is not surprising that the Green functions of unitary gauge fields require an infinite number of subtractions, which is characteristic of a non-renormalizable theory.

On the other hand, we have argued that the scattering matrix can be computed from projected Green functions of the elementary fields, by use of the LSZ formula. These projected operators, smeared with normalizable wave packets at infinity, are BRST-invariant. Thus, the S-matrix and the Green functions of those gauge-invariant operators that can be defined as polynomials in the R_κ gauge fields behave as we expect a renormalizable theory to behave.

This remark also sheds light on the situation we encountered when computing in massive QED. There we found an apparent quadratic UV divergence, inversely proportional to the square of the gauge-boson mass, in the wave-function renormalization of the fermion field. When we used the gauge-invariant DR procedure, this divergence disappeared and was replaced by a logarithmic divergence with no singularity at $\mu = 0$. We could have done the calculation using the Stueckelberg formalism in an R_κ gauge, where the gauge-invariant regulator would have been natural. We had to use the gauge-invariant regulator in the unitary gauge in order to get a result consistent with the Higgs mechanism. Note that the unitary gauge abelian Higgs model is renormalizable, while this is not true for non-abelian theories. The difference has to do with the fact that the effective theory of an abelian NGB is free, whereas that of a non-abelian NGB is a non-renormalizable interacting theory.

9.14.7 Effective field theory and NGBs

Theories of quantum gravity do not have continuous global internal symmetries. This is a theorem in perturbative string theory [129], and follows more generally from the violation of global conservation laws by black-hole formation and evaporation. Thus, continuous global symmetries should be viewed as accidental symmetries of low-energy effective field theories. As such, we expect them to be broken by higher-dimension operators with scale no higher than the reduced Planck scale $m_P = 2 \times 10^{18}$ GeV. If G is a spontaneously broken U(1) (subgroup of a) continuous accidental symmetry group, and D is the dimension of O_D, the lowest-dimension G-violating operator allowed by exact symmetries of the underlying gravity theory, then we expect a potential for the approximate NGB field b associated with G, of the form

$$V = \frac{\langle O_D \rangle}{m_P^{D-4}} U(b/f).$$

f is the PNGB decay constant. When G is something like a global chiral symmetry of a strongly coupled gauge theory, we expect $\langle O_D \rangle \sim f^D$. The approximate NGB mass is then of order $(f/m_P)^{\frac{D-4}{2}} f$.

A context in which these effects are important is the axion solution of the strong CP problem. The axion is a pseudo-NGB, whose potential is supposed to come dominantly from a coupling $(a/f_a)G\tilde{G}$, to the gluons of QCD [130]. In most models where the axion arises from spontaneous breaking of a new strongly coupled gauge theory, the mass arising from "Planck slop" is larger than that coming from QCD, and the axion model does not work.

In general, even if we neglect symmetry-breaking terms, the low-energy effective theory for NGBs of a non-abelian group is not renormalizable. The interactions are all irrelevant operators. With our current understanding of renormalization, this should not bother us. The constant which sets the scale for all these irrelevant interactions is f, the NGB decay constant. The full effective theory contains all possible irrelevant operators consistent with the symmetries. In addition one is instructed to calculate with a cut-off $4\pi f$ (in four dimensions). To simplify matters one chooses the cut-off procedure to preserve the symmetry which gave rise to the NGB in the first place.[22]

---

[22] In the first few paragraphs of this subsection we argued that all such symmetries were approximate. The more precise meaning of the procedure we are now outlining is that the irrelevant operators which break the symmetry are scaled by m_P, or some other scale $>> f$. As a consequence they contribute much less than do the irrelevant terms we are discussing here, at energy scales below $4\pi f$.

It is then easy to verify that, as long as NGB momenta are below f, the higher-order terms in the effective Lagrangian make small corrections, in a systematic power series in p/f, to amplitudes computed using the leading-order effective action.

We can use this effective field theory to do loop calculations, and we must do so in order to capture non-analytic terms in the amplitudes which arise from loops of massless NGBs. All analytic corrections from the loops are simply absorbed into a redefinition of the coefficients of higher-dimension operators in the effective action. A systematic exposition of the rules of effective field theory for NGBs may be found in Chapter 19 of [131].

9.15 The standard model through the lens of renormalization

I would now like to return to the standard model of particle physics, and view it through the eyes of an expert on renormalization theory, since all of my readers should have brought themselves to that expert level at this point. We will take the point of view that there is a cut-off energy scale M, between 10^{16} and 10^{19} GeV, above which the concepts of local field theory fail. I will not enter into questions involving extra dimensions much larger than 10^{-16} GeV^{-1}, warped extra dimensions, or why the number of dimensions is four.

The first question one must ask oneself is why there should be any physically interesting scales below the cut-off. We have seen that, in general, one requires a fine tuning of parameters in order to achieve this. Only if there are marginally relevant parameters, at the fixed point that defines the universal IR behavior of the theory of the real world, can we achieve a ratio of scales as large as 10^{20} without some sort of tuning. It is therefore extremely interesting that, at the Gaussian fixed point, the only marginally relevant perturbations involve non-abelian gauge interactions [120–122], and that non-abelian gauge theories seem to describe the real world.

Another interesting point is that the fermion fields of the standard model form chiral representations of the gauge group. Thus, fermion mass terms are forbidden by gauge invariance, and masses can appear only through the Higgs mechanism. This argument would have led us to expect all quark and charged-lepton masses to be of order $100-200$ GeV, but only the top quark obeys this rule. We are led to suspect the existence of a symmetry explanation for quark mass hierarchies, a special role for the top in the electro-weak Higgs mechanism, or some explanation that involves extra dimensions. Chirality of the representation leads to a potential problem with anomalies. We have seen that in the standard model the anomalies between quarks and leptons are canceled out, leading to the suspicion of unification in the framework of a larger gauge group.

The force of the preceding paragraph is somewhat mitigated by the fact that the μ^2 parameter of the electro-weak Higgs mechanism is itself severely fine-tuned. Renormalization-group analysis would lead us to imagine it at the cut-off scale. There is by now a host of proposed explanations for this. They all involve new degrees of freedom at a scale within an order of magnitude of the Higgs VEV. The Large Hadron Collider (LHC) was built to explore this energy range. I have little doubt that it will discover at least some of the new physics that determines the scale of μ and, through it, much of the physics of the world as we know it. One of the most attractive candidates for the new physics is the supersymmetric extension of the standard model. In supersymmetric models, boson masses are linked to chiral fermion masses, and are effectively marginal parameters. Supersymmetry is also interesting from the point of view of the possible unification of the standard model into SU(5) or some other simple group, at high energy. Georgi, Quinn, and Weinberg [132] pointed out that one could test this hypothesis, assuming that there are no new particle states with standard-model quantum numbers[23] between laboratory scales and the scale at which the couplings unify. At the present level of experimental precision, unification fails in the standard model. The couplings don't unify, and the scale at which they come closest is low enough that the inevitable proton decay implied by a unified model should have been seen in the laboratory. The minimal supersymmetric extension of the standard model solves both of these problems.

To summarize, RG analysis and the remoteness of the energy cut-off lead us to expect a field-theory description of physics below the cut-off dominated by marginal, and at least some marginally relevant, operators. Irrelevant operators are too small in the IR to lead to most of the physics we see. Relevant operators tend to freeze out the degrees of freedom of the world near the cut-off scale. Marginally irrelevant perturbations are perfectly consistent, as long as their low-energy couplings are small enough that there are no Landau poles below the cut-off scale. However, we need at least one marginally relevant coupling to generate new low-energy scales. The special properties of anomaly-free chiral gauge theories lead one to expect them to dominate the structure of the world at scales below the Planck scale, and this seems to be the case. The relevant parameter in our current description of the Higgs mechanism leads to the belief that we will find new physics at or below the TeV scale.

There is an interesting application of the use of marginally irrelevant operators in the standard model, to find an upper bound on the Higgs mass. The Higgs particle in the standard model has a marginally irrelevant quartic coupling. At tree level, the ratio of the Higgs boson mass to the mass of the charged weak bosons is $M_H/M_W = \sqrt{\lambda}/g_2$, where g_2 is the measured SU(2) coupling strength. One would like to know, theoretically, how large the Higgs mass could be. To raise it, one has to raise the renormalized value of λ, but eventually this will bring the Landau pole down to the experimental regime. To do the necessary calculations in a reliable way, one must resort to non-perturbative (lattice field theory) methods. One finds a mass of order 750 GeV as the maximal theoretically allowed Higgs mass [133–134]. Nowadays, this is not as useful as it once was, because indirect experimental evidence from precision electro-weak data indicates a bound around 200 GeV. Furthermore, these considerations have neglected

---

[23] Actually, through one-loop order it is sufficient that any new particles that might exist are in complete multiplets of the unified group.

the large top-quark Yukawa coupling. When this is taken into account the Higgs mass can have an interesting quasi-fixed-point behavior [130]. It then seems possible (if one ignores the fine tuning necessary to get the electro-weak scale) to push the standard model to very high energies and predict a Higgs mass consistent with precision electro-weak measurements and direct experimental bounds, by assuming that the coupling is close to the quasi-fixed point.

Since the standard model is an effective field theory we can also expect to find strictly irrelevant corrections to the model. We can probe these corrections most sensitively by looking for violations of emergent symmetries such as baryon and lepton numbers and approximate flavor conservation. The results of such searches are quite interesting since they have almost all been negative. Flavor violation beyond that intrinsic to the standard model seems to require that the scale of new flavor-violating physics is above the 10–1000-TeV range (the scale for a given flavor-violating operator depends on precisely which process it mediates). Dimension-6 operators with a scale 10^{16} GeV, which could lead to proton decay, are on the edge of being ruled out unless there is extra suppression from weak couplings.

The only evidence for new physics from irrelevant corrections to the standard model is the evidence for neutrino masses coming from solar and atmospheric experiments. These can be explained by invoking a dimension-5 operator $(Hl)^2/M_S$, where the see-saw scale $M_S \sim 10^{14} - 10^{15}$ GeV. This is interestingly different than the other piece of evidence for new physics, which is the apparent unification of running couplings of all the simple factors of the standard-model group. This is far from perfect in the standard model, but the supersymmetric extension of the model has miraculously good unification at a scale of order 10^{16} GeV.

A possible reading of the RG tea leaves, then, is that the supersymmetric extension of the standard model (with the possible addition of other degrees of freedom whose standard-model couplings are quite constrained) will appear at LHC energies, between 100 GeV and 1 TeV. The effective theory of SUSY violation and the Higgs mechanism must have natural flavor conservation. Flavor physics and neutrino masses are determined by unification-scale physics, at an energy scale between 10^{14} and 10^{16} GeV. Here we may expect to see the effects of extra dimensions and/or quantum gravity. I caution the reader that this is only one of many scenarios, perhaps too conservative a choice. Whatever the future holds, it seems clear that the tools of quantum field theory will be useful for many decades of energy yet to be explored.

9.16 Problems for Chapter 9

*9.1. Write the pure SU(2) Yang–Mills theory as a theory of charged vector bosons coupled to photons. Show that there is an anomalous magnetic moment implied by the Yang–Mills Lagrangian, and that the gyromagnetic ratio of the charged bosons is 2.

*9.2. Carry out the renormalization program through one loop for a single Dirac fermion interacting with a single scalar through the interaction

$$\mathcal{L}_I = \bar{\psi}(g_S + g_P\gamma_5)\phi\psi.$$

Use a regularization scheme in which the Schwinger parameter s for each one-loop diagram is cut-off at the lower limit s_0. Evaluate the divergent parts of all one-loop diagrams and determine what other interactions need to be added. Verify that one has to add all relevant and marginal interactions consistent with the symmetries of the Lagrangian and regularization scheme. Compute the divergent parts of all one-loop diagrams including ALL of the required interactions and write the full set of RG equations for this model. You will need to invent a prescription for the finite parts of the renormalizations. When $g_P = 0$ you can also do the calculations by dimensional regularization. Find the relation between the definitions of the couplings in minimal subtraction and the prescription you invented.

*9.3. Calculate the operator product expansion (OPE) of $: \phi^4(x) :: \phi^4(y) :$, in massless free-field theory. To do this calculation you can use the result that the time-ordered product of $: \phi^4(x) :$ with any other collection of local operators is given in terms of Feynman diagrams with a four-point vertex where no self-contractions are allowed. Use this first of all to argue that the OPE is an operator expression, that is, that its structure is independent of how many extra ϕ fields there are in the Green function

$$\langle T : \phi^4(x) :: \phi^4(y) : \phi(y_1)\ldots\phi(y_n)\rangle.$$

Then calculate all of the singular terms in the OPE, including their numerical coefficients. Finally indicate the list of operators, O_n, and their Wilson coefficient functions $C_n(x-y)$ that appear in the full OPE. In this last part you need not calculate all the numerical factors in C_n, just its functional form.

*9.4. Compute the one-loop charge renormalization (photon wave-function renormalization) for the electrodynamics of charged scalar fields, with Lagrangian

$$\mathcal{L} = -\frac{1}{4}F_{\mu\nu}F^{\mu\nu} + |D_\mu\phi|^2 - m_0^2|\phi|^2.$$

Note that, unlike spinor electrodynamics, there are two graphs, and you have to sum over both of them to get a transverse answer for the photon 1PI two-point function. Use dimensional regularization for the computation.

*9.5. Show that there is a divergent graph at one loop in the above theory, with four external scalar lines. Argue that the divergence can be removed by tuning the coefficient λ_0 of a bare interaction of the form $(\lambda_0/4)|\phi|^4$. Note that this is a marginal operator consistent with the symmetries of the original Lagrangian, so general renormalization theory tells us that we should expect to have to tune it in perturbation theory around the free-field fixed point.

*9.6. Let ψ_i and ϕ_r be multiplets of fields transforming in real representations R_F and R_S of a gauge group G. Assume there is only a single invariant coupling $g_{rij}\bar{\psi}_i\phi_r\psi_j$. Using dimensional regularization, repeat Problem 9.2 including the gauge interactions.

*9.7. Consider two massless scalar fields with interaction Lagrangian

$$\mathcal{L}_I = -\frac{g^2}{4!}(\phi_1^4 + \phi_2^4) - \frac{2\lambda}{4!}\phi_1^2\phi_2^2.$$

When $g^2 = \lambda$ there is an O(2) symmetry. Renormalize this model at one loop. Argue that the inevitable mass renormalization does not affect the renormalization of the couplings at one loop. Also argue that, for any value of the couplings, the model has a discrete symmetry guaranteeing that the renormalized mass term is O(2)-symmetric. Write the RG equations for the couplings and show that if the initial values satisfy $\lambda/g^2 < 3$ then the theory becomes O(2)-symmetric in the IR limit.

*9.8. Compute the one-loop effective potential of scalar electrodynamics (Problems 9.4 and 9.5) in the Landau gauge $\partial_\mu A^\mu = 0$. Use dimensional regularization and minimal subtraction, and take the limit where the renormalized mass parameter in the minimal subtraction scheme goes to zero. The theory then has only the RG scale μ^2. Show that, in this limit, the potential has a minimum at a non-zero value of the scalar field ϕ. Show that if the renormalized scalar quartic coupling λ is of order e^4 then the minimum occurs at a field value that is neither extremely large nor extremely small, so perturbation theory should be valid. Construct the one-loop RG equations (RGEs) for e^2 and λ and show that every trajectory passes through the region where $e^4(t) \sim \lambda(t)$. Thus, whenever perturbation theory is valid, this massless limit of electrodynamics is actually in the Higgs phase. Calculate the "RG-improved" approximation to the effective potential. That is, write the solution to the RGE for V_{eff} with the boundary condition that for $|\phi| = M$, V_{eff} is given by its one-loop formula. The improved $V_{\text{eff}} = \mu^4 Z(t) V_{1-\text{loop}}(e^2(t), \lambda(t))$, where $t \equiv \ln(|\phi|/\mu)$ and $Z(t)$ is the rescaling factor that comes from solving the RGE. Compute the masses of the vector boson and the Higgs field, for every initial value of the couplings $e^2(0) = e^2, \lambda(0) = \lambda$.

*9.9. Consider two definitions of the renormalized coupling constant of some theory with a single marginal coupling. The two schemes are related by

$$g = g_1 + a_1 g_1^3 + a_2 g_1^5 + \cdots.$$

Using the renormalization group equation for g,

$$\frac{dg}{dt} = \beta(g) = b_1 g^3 + b_2 g^5 + \cdots,$$

show that the first two coefficients of the β function for g_1 are identical to $b_{1,2}$.

*9.10. Complete the computation of the static potential through one loop, and calculate the β function for Yang–Mills theory.

10 Instantons and solitons

Much of this book has been devoted to perturbation theory around Gaussian fixed points of the renormalization group. Such a perturbation theory can always be reorganized into a semi-classical approximation to the functional integral. The semi-classical expansion can also give us evidence about non-perturbative effects in quantum field theory, and this chapter is an introduction to the relevant technology. It is a very brief introduction to a very large subject. Readers who wish to purse the subject of instantons and solitons in depth should consult [135–138].

10.1 The most probable escape path

We begin by recalling the WKB approximation to the problem of quantum tunneling in a system with N degrees of freedom, q^i.

The Schrödinger equation is

$$\left[-\frac{g^2}{2} \frac{\partial^2}{\partial (q^i)^2} + \frac{1}{g^2} V(q) \right] \psi = E\psi. \tag{10.1}$$

For simplicity we have made all the variables, including the energy, dimensionless, set $\hbar = 1$, and introduced a dimensionless parameter g^2. The WKB approximation is valid when g^2 is small. The reader should verify that this Schrödinger equation would be appropriate for a system whose classical action is

$$S = \int dt \left[\frac{(\dot{q}^i)^2}{2g^2} - \frac{V}{g^2} \right]. \tag{10.2}$$

For small g^2, an approximate solution to the Schrödinger equation is

$$\psi = e^{\frac{i}{g^2} S}, \tag{10.3}$$

where

$$\left(\frac{\partial S}{\partial q^i} \right)^2 = 2(E - V). \tag{10.4}$$

The last equation is the Hamilton–Jacobi equation for the classical system with action (10.2). The integral curves of this first-order system are solutions of the classical equations of motion, and

$$\frac{\partial S}{\partial q^i} = p_i = \dot{q}_i. \tag{10.5}$$

Here $q^i(t)$ is the classical trajectory that goes through the point q^i at time t. The velocities are real only if $E > V$, but we can use this approximation to solve the Schrödinger equation in the tunneling region where $V > E$. Since the coordinates are still real, this corresponds to continuing to imaginary time.

The imaginary-time equations of motion are perfectly good differential equations. In fact, they correspond to Newtonian motion in the potential $-V$. The relation between solutions to this equation and the WKB approximation to the Schrödinger equation is the same as that between ordinary classical trajectories and the Hamilton–Jacobi equation. In particular, we have the usual result that, at any point in the tunneling region,

$$S = \pm i \int_{q_0}^{q} p_i \, dq^i, \qquad (10.6)$$

where the integral is taken over an imaginary-time classical trajectory that goes through the indicated points. This is just the Euclidean action of the trajectory. We generally take q_0 to be the turning point of the trajectory, where it hits the boundary at which real classical motion can begin. The two signs correspond to linearly independent solutions of the Schrödinger equation, one of which is exponentially falling and the other exponentially increasing, as we penetrate into the tunneling region. Boundary conditions on the other side of the tunneling barrier (or at infinity if it is not a finite barrier, but an infinitely rising wall) fix the coefficient of this growing solution to be small, so that the two are of the same order of magnitude in the middle of the tunneling region.

For small g, the wave function is exponentially small in the tunneling region, and it will be maximized in the vicinity of the path of minimal Euclidean action which traverses the barrier. In minimizing the action we have to search over all solutions that traverse the barrier between two classically allowed regions in configuration space. This includes a search over all end points of the solution after tunneling. However, once we have found the minimum, there is another solution in which we first follow the most probable escape path, and then retrace it to its origin. This solution, called *the bounce*, minimizes the action subject to the constraint that it starts and ends at the initial stationary point of the potential. There is no need to search over possible turning points in the second classically allowed region. The action of the bounce solution directly gives the logarithm of the probability for tunneling through the barrier.

10.2 Instantons in quantum mechanics

Consider one-dimensional quantum mechanics with a potential $V(x)$ that has two or more minima. We will concentrate on three cases: two degenerate minima, a periodic potential with an infinite number of degenerate minima, and two non-degenerate minima. The Euclidean (imaginary-time) action has the form

$$\frac{1}{g^2} \int dt \left[\frac{1}{2} \left(\frac{dx}{dt} \right)^2 + V(x) \right].$$

g^2 is the small dimensionless parameter which controls our semi-classical approximation. We will begin from a computation of the partition function $\mathrm{Tr}\, e^{-\beta H}$. According to Feynman, this is given by the Euclidean path integral, with periodic boundary conditions, $x(-\beta/2) = x(\beta/2)$. For large β, this behaves like $Ce^{-\beta E_0}$, where E_0 is the ground-state energy.[1] For small g^2 the path integral is dominated by the saddle point with minimum action, which is a constant solution $x(t) = x_{\min}$. Expansion of the path integral around this saddle point leads to a perturbative calculation of the ground-state energy $E = \sum g^{2n} E_n$. If the degenerate minima are related by a symmetry, as in the periodic case, or if the double-well potential has a reflection symmetry, then the terms in this series are independent of the minimum we expand around. The degeneracy is not lifted to any order in perturbation theory. We will choose the classical energy at the minimum to be zero.

Time-dependent saddle points of the Euclidean action, which we will dub *instantons*, lead to contributions to the ground-state energy that are of order e^{-S_0/g^2}. Normally, we would neglect these in comparison with the terms in the expansion, but they will give the leading contribution to the energy *splittings* between the classically degenerate levels. In order to have a finite action S_0, in the limit $\beta \to \infty$, the instanton solution must have the asymptotics $x(t) \to x_{\min}^{\pm}$ as $t \to \pm\infty$. It must interpolate between two different minima. Clearly this means that $|\mathrm{d}x/\mathrm{d}t| \to 0$ for large $|t|$.

Saddle points of the Euclidean action are given by solutions of

$$\frac{\mathrm{d}^2 x}{\mathrm{d}t^2} = V'(x),$$

which are Newton's equations in a potential $-V$, for a particle of unit mass. The Euclidean energy, $\frac{1}{2}\dot{x}^2 - V$, is conserved in this motion. On evaluating it at $t = \pm\infty$, we find

$$\dot{x}^2 = 2V,$$

which can be solved by quadratures. Notice that, for any pair of minima, there are two solutions, related by time reversal $t \to -t$. We (arbitrarily) call one of them the instanton, and the other the anti-instanton. Note that, in any relativistic quantum field theory, we always have a kind of time-reversal symmetry, namely TCP. Thus, in this context, every instanton will always have an anti-instanton.

Given the form of the solution, we can now understand the term instanton. Expand the equation of motion around a minimum $x(t) = x_{\min} + \delta$. Then

$$\ddot{\delta} = V''(x_{\min})\delta.$$

Since V'' is positive the solutions are rising and falling exponentials. We have to choose the falling exponential at both $t \to \pm\infty$, in order to obey the finite action boundary conditions. Thus, the instanton $x(t)$ differs from the classical ground state only in

---

[1] We will not deal with the sub-leading terms in β. From them, one can extract information about excited states. See the book by Zinn-Justin [111].

the local vicinity of some point t_1 (which may be chosen arbitrarily because of time-translation invariance). It thus has a particle-like aspect, which explains the suffix *on*, but localized in time rather than space, which is indicated by the prefix *instant*. In higher-dimensional field theories instantons will be localized in both time and space. We can also reinterpret d-dimensional instantons as localized $(d+1)$-dimensional static solutions. The latter are known as *solitons*, and actually do define particle states in the quantum theory.

Time-translation invariance implies that, in the $\beta \to \infty$ limit, we have a continuous set of solutions, labeled by t_1. We can write them as $x(t) = X(t + t_1)$, where $X(t)$ is the solution whose maximum deviation from $x_{\min}^{\pm}$ occurs at $t = 0$. Obviously, we have to integrate over this whole set of saddle points, because they have the same action. In other words, we have a *saddle line* rather than a saddle point in our multi-dimensional integral over all functions. The measure of functional integration is defined by expanding $x(t) = \sum c_n x_n(t)$ in a complete set of orthonormal functions on the real line. We then integrate over c_n with uniform measure. Our line of saddle points is a curved line through this infinite-dimensional Cartesian space. In principle we have to decompose the measure into integration along this curve and along all directions perpendicular to it. The measure of integration is no longer Cartesian, and must be computed. However, in the Gaussian approximation, as we will see in a moment, this subtlety can be ignored.

The Gaussian approximation expands the action to quadratic order around the saddle-point configuration and does the resulting Gaussian integral. The function $\dot{X}(t)$ is an obvious zero mode of the Gaussian fluctuation operator

$$\frac{\delta^2 S}{\delta x(t)\delta x(s)} = \left[-\frac{\mathrm{d}^2}{\mathrm{d}t^2} + V''(X(t))\right]\delta(t-s),$$

because the whole one-parameter family of functions $X(t+t_1)$ has the same action. Our formula for the solution shows that $\dot{X}$ does not change sign. It is therefore the lowest eigenmode of this Sturm–Liouville operator. All of the positive modes are orthogonal to it. Thus, infinitesimally close to our line of saddle points, the measure on the orthogonal directions is uniform (this is geometrically obvious for finite-dimensional integrals – draw a picture). The non-uniformity of the measure becomes important in higher orders in the expansion, which we will not have the opportunity to explore.

So far our calculation of the instanton correction to the partition function takes the form

$$Z = Z_0\left[1 + \beta e^{-\frac{S_0}{g^2}} \det^{-\frac{1}{2}}\left(\frac{-\partial_t^2 + V''(X(t))}{-\partial_t^2 + V''(x_{\min})}\right)\right].$$

Here Z_0 is the Gaussian approximation to the partition function in the expansion around $x_{\min}$. This is equal to $\det^{-\frac{1}{2}}(-\partial_t^2 + V''(x_{\min}))$, which is why we have this operator in the denominator of the second term. In field theory, this one-loop correction to the vacuum energy is infinite. However, because the instanton is a smooth function, the very high eigenvalues of the two fluctuation operators are the same. The ratio of the two differential operators is a Fredholm integral operator, and has finite determinant. In higher-dimensional field theory, there are additional UV divergences in the determinant. We can study them using the Schwinger proper-time formula for functional determinants, which we introduced in Chapter 9. As usual, they come from short proper-time intervals, and correspond to renormalizations of local functionals of the classical background field. These divergences are precisely removed by the renormalization of parameters which defines the theory.

When g^2 is small, this expression is very small, but also almost infinite, because β is taken to ∞. This sort of divergence is familiar from the perturbative expansion of the partition function in terms of Feynman diagrams. From that context, we know how to deal with it. In fact, vacuum diagrams with n disconnected parts behave like β^n for large β. Feynman rules tell us that the sum of all these disconnected diagrams is just the exponential of the sum of connected diagrams. The latter sum has a single overall factor of β, which is what we would expect for the logarithm of the partition function that has the form $\mathrm{e}^{-\beta E_0} d_{\mathrm{g}}$ (d_{g} is the number of degenerate ground states of the system), for large β. In field theory, the factor of β is multiplied by a factor of spatial volume, and the quantity we are computing is the vacuum energy density, the coefficient of space-time volume in the logarithm of the partition function.

To see the analogous exponentiation of instanton contributions to the partition function, we have to introduce some approximate solutions to the Euclidean equations of motion. Consider

$$x_{\mathrm{c}}(t) = \sum_{k=1}^{n} X_k(t + t_k),$$

where

$$|t_i - t_j| \gg V''(x_{\min}).$$

For each k, X_k is one of the instanton solutions we have already discussed. For the double-well potential we must enforce a rule, namely that each instanton X_+ is followed by an anti-instanton, X_-, in order for $x(t)$ to be a smooth function. The instanton looks like a step function. Furthermore, in this model, in order to approximate functions that were periodic before taking the large-β limit, we must restrict the total numbers of instantons and anti-instantons to being the same. On the other hand, for the periodic potential, an instanton taking minimum a into minimum $a + 1$ can be followed either by an instanton going to $a + 2$, or by an anti-instanton going back to a.[2]

---

[2] The question of whether we must have $N_+ = N_-$ for the periodic potential seems to depend on the period of the x variable. If it is the same as the period of the potential then there is no restriction on $N_\pm$. However, even if the x variable runs over the whole real line, we can compute $Z(\theta) = \sum_N \mathrm{e}^{iN\theta}\, \mathrm{Tr}[\mathrm{e}^{-\beta H} T^N]$, where T is the discrete translation which leaves the potential invariant. This computation allows arbitrary $N_\pm$ but with a phase $\mathrm{e}^{i(N_+ - N_-)\theta}$. The same phase can be inserted into the periodic x computation by adding the term $i \int dt\, \dot{x}$ into the Euclidean action. The two computations are mathematically identical. For non-periodic x the interpretation is the partition function summed over states with a fixed eigenvalue of T (Bloch waves), whereas for periodic x it corresponds to a change in the action.

These functions satisfy the equations of motion, up to exponentially small terms, as a consequence of the exponential falloff of $X_k(t)$. They have action $\approx nS_0$. The parameters t_k are approximate collective coordinates for this solution. They do not change the action much. Thus, one should integrate over them. One might worry that these integrals cover points where the $|t_i - t_j|$ are not large, but we will see that these regions give only exponentially small corrections to the calculations we are about to do. There is, however, one restriction on the integral over the t_k. Field configurations in which the centers of two or more instantons (or anti-instantons) are permuted correspond to the same field configuration. Thus, we can integrate freely over all collective coordinates if we divide the result by $N_+!N_-!$ for a configuration containing N_+ instantons and N_- anti-instantons.

If we take into account the contribution of all of these configurations to the path integral, we obtain[3]

$$Z = Z_0 \sum_{N_+,N_-=0}^{\infty} \frac{(\beta D^{-1/2})^{N_+ + N_-}}{N_+!N_-!} e^{-\frac{(N_+ + N_+)S_0}{g^2}}$$
$$= Z_0 e^{2\beta D^{-1/2} e^{-S_0/g^2}},$$

which corresponds to a shift in the ground-state energy of

$$\delta E_0 = -2D^{-1/2} e^{-S_0/g^2}.$$

If, following the prescription in the footnote above, we insert the phase $e^{i\theta(N_+ - N_-)}$ into the sum, then $2 \to 2\cos\theta$ in these formulae. The evaluation of the partition function looks like that for the configuration integral in the statistical mechanics of a classical one-dimensional gas consisting of two species of particles with identical fugacities. This approximation is therefore called the *dilute-gas approximation*. Note that the factorial factors in the denominator insert what is called *correct Boltzmann counting* into the classical statistical formulae. This factor, inserted to avoid the Gibbs paradox, is usually explained by appealing to quantum mechanics. Here we see that it would also follow from a model in which particles were localized classical field configurations.

10.2.1 Computation of determinants

Now let us explain the factor D in the above expression. Let's start from the ratio of determinants

$$R = \frac{\det'[-\partial_t^2 + V''(x_c(t))]}{\det[-\partial_t^2 + V''(x_{\min})]}.$$

The prime on the determinant in the numerator means that we omit the approximate normalizable zero modes corresponding to translating individual instantons, as well as the exact overall translation zero mode. The determinant in the denominator appears

---

[3] For simplicity, I deal with the case of a periodic potential, with no restriction on the equality of instanton and anti-instanton numbers. The reader is invited to fill in the details for the double well.

because we have pulled a factor of Z_0 out in front of our expression for the partition function. We have chosen the classical action of the constant saddle points to vanish, so Z_0 is just given by the inverse square root of the determinant of the fluctuation operator around these configurations.

Recall how this works. Given a saddle point $x_S(t)$, we expand the functional integration variable as $x(t) = x_S(t) + g \sum c_n \delta_n(t)$, where δ_n constitute the complete set of orthonormal eigenfunctions of the operator $-\partial_t^2 + V''(x_S(t))$, which appears in the quadratic part of the action for the fluctuations. The inverse of this operator is the propagator that appears in the Feynman rules for calculating higher orders in the g expansion. The Gaussian integral over the c_n gives us a constant (g-independent) pre-factor. We define the measure as

$$\prod \frac{dc_n}{\sqrt{2\pi}},$$

so that the Gaussian integral gives

$$\prod \lambda_n^{-1/2}.$$

This expression has two possible problems. It usually diverges in the very-large-λ_n regime, and it is infinite if there is a normalizable zero mode. We have, for the moment, solved the second problem by simply omitting this mode. The first is solved by a trick invented by Fredholm in the nineteenth century. The operators $-\partial_t^2 + U(t)$, for any smooth $U(t)$ approaching a constant at infinity, all have the same high-energy spectrum. In modern RG language this is the statement that a smooth potential is a relevant perturbation of the free Hamiltonian $-\partial_t^2$. As a consequence, the high-energy divergence in the determinant cancels out in the ratio.

The reason that we have omitted the (approximate and exact) zero modes is that we have already extracted the integral over these directions in field space when we integrated over collective coordinates. We just have to get our normalizations right. The reason why there is a normalizable zero mode is that the action is invariant under time translation. Thus for any saddle point $x_c(t)$ there is actually a line of saddle points $x_c(t + t_1)$. For the multiple-instanton approximate saddle points there are correspondingly multiple lines, since we can translate each instanton independently as long as they are far apart. The line of saddle points is not a straight line w.r.t. any orthonormal basis. However, right near the line of saddles there is obviously an orthonormal basis of small fluctuations, with one direction going along the line.[4] Thus

$$\delta_0 = \frac{\dot{x}_c}{S_0^{1/2}}$$

is the normalized zero mode. The reader can verify that it indeed satisfies the zero-eigenvalue equation. We have written the normalization factor in a way that exploits the classical equation $\dot{x}_c^2 = 2V$. This function is a sum of functions concentrated near

----

[4] In higher orders in perturbation theory around the instanton we have to get the correct measure in the vicinity of the curved line of saddles. This is done by an analog of the Faddeev–Popov trick.

the centers of all the instantons and anti-instantons. We can find the approximate zero modes corresponding to relative translation of a single instanton by simply dropping all terms in δ_0 except the one near that instanton.

To relate the measure dt_1 of the collective coordinate to that of the coefficient of the zero mode, dc_0, we simply insist that these two variations of $x_c(t)$ are the same,

$$g\delta_0\, dc_0 = \dot{x}_c\, dt_1.$$

Thus the correct measure, $dc_0/\sqrt{2\pi}$, is

$$\frac{dc_0}{\sqrt{2\pi}} = \sqrt{\frac{S_0}{2\pi g^2}}\, dt_1.$$

A similar formula is valid for all the approximate collective coordinates. We arrange the integral so that we integrate independently over each of the instanton positions, so the appropriate S_0 is that of a single instanton.

The zero mode of the single instanton is obviously the lowest eigenvalue, because the corresponding wave function $\dot{x}$ is monotonic and has no nodes. Since, in the Schrödinger equation analogy, the potential, $U(t) = V''(X(t))$, goes to a constant at infinity, all the rest of the eigenstates are scattering states. Thus for the single instanton

$$R = e^{\mathrm{tr} \int d\epsilon\, \ln \epsilon\, \rho(\epsilon)},$$

where $\rho(\epsilon)$ is the density of scattering states.

As noted above, the Schrödinger operator $-d^2/dt^2 + V''(x_c(t))$ is a non-negative operator. We know that it has $N_+ + N_-$ normalizable (approximate) zero modes. The rest of its spectrum consists of positive-energy scattering states. Since the instantons and anti-instantons are widely separated, we can calculate the phase shift by a multiple-scattering expansion: it is the sum of phase shifts for scattering by individual instantons. The Fredholm determinant is given by

$$\frac{\det'[-d^2/dt^2 + V''(x_c(t))]}{\det[-d^2/dt^2 + V''(x_{\min})]} = e^{\int dE\, \rho(E)\ln E}.$$

Using the relation $\rho(E) = d\delta(E)/dE$, between the density of states and the phase shift, combined with the multiple-scattering expansion, we see that the dilute-instanton-gas determinant is just the product of individual instanton and anti-instanton determinants. Thus the full one-loop expression for the contribution of the dilute gas of instantons and anti-instantons is just the exponential of the single instanton plus anti-instanton contributions.

10.3 Instantons and solitons in field theory

An *instanton* is a stable, finite-action, solution to the Euclidean field equations of a d-dimensional quantum field theory. For renormalizable quantum field theories, these equations can also be viewed as the equations for stable, finite-energy, static solutions to the field equations of the same theory in $d + 1$ dimensions. With this interpretation, the solution is called a *soliton*.[5] Solitons are particles with mass of order $1/g^2$, where g^2 is the semi-classical expansion parameter. There are also more general periodic solutions of the $(d + 1)$-dimensional equations that have such a particle interpretation.

Another kind of generalized soliton is an infinite extended object of finite energy per unit volume, like a flux tube or domain wall. We have already encountered flux tubes in our discussion of confinement. Indeed, the most important solitons in high-energy quantum field theory are the monopoles and flux tubes associated with U(1) gauge fields. Various other solitons are useful in applications to condensed-matter physics.

The fact that a soliton's mass per unit volume goes to infinity in the semi-classical limit means that part of the semi-classical expansion is an expansion in the recoil of the soliton. This makes the expansion particularly intricate, and we do not have space for a proper discussion of it here. Consequently, we will begin our discussion with the instanton interpretation of static classical solutions, for which we can give a fairly complete sketch of the semi-classical expansion. We will then outline the principal facts about solitons in gauge theories.

We have emphasized repeatedly that the Schrödinger picture is an awkward way to think about quantum field theory. Fortunately, the connection between tunneling and Euclidean field equations gives us a way to compute tunneling corrections to Green functions quite directly. The discussion in the previous subsection provides us with a conceptual framework for understanding what an instanton *is*, but to understand what it *does* we turn to the Euclidean path integral for Green functions. We will treat two cases, the instanton for decay of a metastable ground state in scalar field theory and the eponymous instanton for pure Yang–Mills theory. In the first case, we are doing an expansion of a formal functional integral of $e^{-S[\phi^i]/g^2}$ with the boundary conditions that $\phi^i(x) \to v^i$ at Euclidean infinity. v^i is a local minimum of the field potential. Finite-action stationary points[6] will lead to exponentially small corrections to Green functions.

It is intuitively appealing, and rigorously proven [139], that the minimal action solution has O(d) invariance. Thus, it is a function only of the Euclidean distance r from some point of origin. Translation invariance assures us of the existence of a d-parameter family of solutions with different origins and equal action. We will come back to the consequences of this degeneracy forthwith. The minimal-action solution is thus a minimum of

$$\int d^d r \, r^{d-1} [(d\phi^i/dr)^2 + V(\phi^i)].$$

The instanton is a particular path $\phi^i(r)$ in field space, so there is an equivalent single-field problem (with a modified potential) that solves for the field-space path length as

---

[5] This is an abuse of the mathematician's term soliton, which refers to special solutions of integrable field theories.

[6] Finite action actually refers to the action *difference* between the solution in question and the constant solution $\phi^i = v^i$.

a function of r. For simplicity then, we will restrict our attention to a one-dimensional field space, with the knowledge that the generalization is straightforward.

The variational equations are

$$\phi_{rr} + \frac{d-1}{r}\phi_r - V'(\phi) = 0.$$

These are the Newtonian equations for a particle moving under the influence of time-dependent friction, in a potential $U = -V$. U has two maxima, which we call the false vacuum v_F and the true vacuum v_T. The boundary conditions are that $\phi \to v_F$ as r goes to infinity and that $\phi_r(0) = 0$, in order to have finite action. Notice that this spherically symmetric solution is automatically what we have called *the bounce* above. The Green functions we calculate with the functional integral are *ground-state expectation values*, and contain two factors of the exponential suppression of the ground-state wave function in the tunneling region.

The variational equations always have a finite-action solution. The free boundary condition is the value of ϕ at $r = 0$. The "particle" obviously has trajectories starting on the v_T side of the minimum of U that do not make it to v_F, even in infinite time. Simply start with a Euclidean energy $\frac{1}{2}\phi_r^2 - V$, which is less than or equal to the value of the potential at v_F. To see that there are solutions that overshoot v_F in finite time, start with ϕ near v_T. If the initial condition is close enough to the maximum, the trajectory will remain near the maximum until very large r, where the friction term is negligible. At this point, energy is approximately conserved, and overshoot is guaranteed.

As the initial position is varied continuously between undershoot and overshoot solutions, we will find a unique solution that settles in to v_F in infinite time. Near $r = \infty$ the corrections $\phi - v_F$ fall off like e^{-mr}, where m is the mass of small oscillations near the metastable minimum. Thus the action is finite. The leading semi-classical approximation to connected Green functions consists of saturating the functional integral with the degenerate classical solutions $I(|x - a|)$, where I is the spherically symmetric instanton solution we have just discussed. The answer is

$$e^{-\frac{S[I]}{g^2}} \int d^d a\, I(x_1 - a) \ldots I(x_n - a) = W_n(x_1 \ldots x_n).$$

Note that it is translation-invariant.

This simple result for connected Green functions follows from a slightly more elaborate analysis, called the dilute-gas approximation (DGA) for full Green functions. Consider instanton contributions to the partition function. The single instanton corrects the partition function by an additive term of the form

$$V e^{-\frac{S[I]}{g^2}},$$

which is exponentially small but multiplied by an infinite factor of the Euclidean space-time volume. Clearly, this infinity is similar to the overall volume factor in vacuum Feynman diagrams, and we expect it to exponentiate. To see this consider a field configuration $I_{DGA} = \sum I(x - a_i)$. When $|a_i - a_j|$ is large for each pair, this is an approximate solution of the Euclidean equations, because the instanton field falls rapidly at infinity.

Its action is $nS[I]$ and it has n collective coordinates a_i that should be integrated over space-time. However, this multiple integration overcounts configurations that differ only by a permutation of the a_i. There is only one field configuration but many disjoint integration regions related by permutation. Thus, the total contribution of these dilute-gas configurations to the partition function is

$$\sum_{n=0}^{\infty} \frac{(V e^{-S[I]/g^2})^n}{n!} = e^{V e^{-S[I]/g^2}}.$$

Thus, the DGA gives an exponentially small correction to the vacuum energy density. When applied to the calculation of Green functions the DGA gives disconnected contributions, in which some finite number of instantons will affect connected clusters of fields, while most of the instantons in the gas cancel out against the denominator. Connected Green functions are, in the DGA, a one-instanton effect.

Like all leading-order semi-classical results, the overall normalization of the DGA answer is not determined until we integrate over small fluctuations around the instanton. This must be done with care. We have integrated over a d-dimensional sub-manifold of field space, parametrized by the *collective coordinate*, a, the instanton center. We must rewrite the integration measure in terms of coordinates transverse to this sub-manifold, and consider only fluctuations in the transverse direction. In principle, this introduces a Jacobian determinant reminiscent of the Faddeev–Popov determinant of gauge theory. However, most of the effects of that determinant show up only at higher orders in the semi-classical expansion. All that is left is the instruction to leave out the zero modes of the fluctuation operator

$$-\nabla^2 + V''(I)$$

when calculating its determinant.

It is easy to see, from translation invariance or explicit substitution, that the d functions $\partial_i I$ are zero modes of the fluctuation operator. Furthermore, they are normalizable, because $\partial_i I = (x_i/r)I_r$ falls off exponentially at infinity. We have run quickly through the discussion of the DGA and collective coordinates for field theory, because it precisely parallels the quantum-mechanical arguments of the previous section.

10.4 Instantons in the two-dimensional Higgs model

Before beginning the discussion of instantons and vacuum structure in higher-dimensional field theory, I want to point out one more way in which the solutions we will discuss apply to physical systems. The classical partition of a general system is the product of a simple term coming from the kinetic energy plus a configuration integral

$$\int d^n q \, e^{-\beta V(q)}.$$

For a $(d+1)$-dimensional relativistic field theory, the *potential* in this formula is just the Euclidean action of the same field theory in dimension d. So the partition function which gives the vacuum energy of the d-dimensional field theory is also the classical partition function of the system in $d + 1$ dimensions. In this way of thinking about things, the coupling g^2 plays the role of inverse temperature, with dimensional factors supplied by relevant parameters (in the RG sense) of the field theory. This should be reminiscent of the way our instanton sums in quantum mechanics reduced to a problem in classical statistical mechanics.

The utility of these formulae goes beyond relativistic field theory, because the coarse-grained, long-wavelength behavior of condensed-matter systems, especially near second-order phase transitions, is dominated by classical fluctuations of an order parameter field, whose effective energy functional is rotation-invariant. Thus, the study of the relativistic Higgs model that we present here is relevant also to the statistical mechanics of planar superconductors. We will also touch briefly on the Kosterlitz–Thouless phase transition in planar XY magnets.

Our strategy for finding finite-action instanton solutions in higher-dimensional field theory will focus on the notion of *topological charge*, a generalization of the $\int \mathrm{d}t\, \partial_t x^i$ of quantum instantons. That is, we break the space of Euclidean field configurations up into classes with different behavior at infinity. Continuous changes of the fields bounded in space cannot change the class. If we can find any finite-action configuration in the class, and the class is different from that of the constant classical background, then we should be able to find a non-trivial finite-action solution by minimizing over variations within the class.

In d-dimensional Euclidean space, infinity has the topology of a $(d-1)$-sphere S^{d-1}. This indicates the relevance of the space of maps $\Pi_{d-1}(X)$, from the sphere to the space X of behaviors at infinity of finite-action field configurations. Since finite-action field configurations are defined relative to a classical vacuum, X is the space of classical vacua. For $d=2$, we have maps from the circle, and the space of classical vacua should have a non-trivial circle in it. This implies a continuous degeneracy, which is usually an indicator of a classical U(1) symmetry, which is spontaneously broken. This is indeed the case for the model we will study.

Let χ be the periodic, angle-valued Goldstone field, with period 2π. Configurations with non-trivial winding number have

$$\int \mathrm{d}\theta\, \partial_\theta \chi = 2\pi n,$$

where the integral is taken around the circle at infinity. By Stokes' theorem, we can deform this circle into the finite interior, until we hit singularities of χ. If χ is the phase of a complex field, $\phi = \rho e^{i\chi}$, then singularities are to be expected, and don't lead to divergences, at the zeros of ρ. On the other hand, ρ should go to a non-zero constant ρ_0 at infinity. The contribution of the kinetic term of χ to the action at large radius is thus bounded by

$$S \geq \frac{\rho_0^2}{g^2} \int_{r_+}^{\infty} r\, \mathrm{d}r \left(\frac{\partial_\theta \chi}{r} \right)^2 = (2\pi n)^2 \frac{\rho_0^2}{g^2} \int_{r_+}^{\infty} \frac{\mathrm{d}r}{r}.$$

The action is thus logarithmically divergent.

Normally, we will throw out configurations with infinite action, when the origin of the infinity is infrared. However, in this case, we would miss something interesting if we did so. The field equations for χ are linear, so consider a configuration consisting of a sum of $n = \pm 1$ solutions for χ. Let the separation between the two centers with $\rho(R_{\pm}) = 0$ be R. When R is large we have an approximate solution of the field equations with action $2S_0 + \ln(R/r_+)$. The contribution to the partition function from such configurations is

$$Z_{\text{Inst}} \propto V \int d^2 R \, e^{-2S_0/g^2} e^{-\frac{1}{g^2} \ln(|R|/r_+)}.$$

The factor of V comes from the integration over the center of mass of the instanton–anti-instanton configuration, while the integral over R describes the relative coordinate. It's clear that the integral converges for small g^2, but that there is a critical value of g^2 at which the integral diverges. This is a signal of a phase transition in which a gas of tightly bound instanton–anti-instanton pairs is replaced by a plasma of separated instantons and anti-instantons. The behavior near the transition can be described exactly by converting the sum over instantons into a sine-Gordon field theory (we will see an example of the same technique in a three-dimensional example below), and using the renormalization group. This phase transition was first understood by Kosterlitz and Thouless.

Another strategy, which makes the instanton action finite, is to gauge the global U(1) symmetry. That is, instead of regarding it as a symmetry acting to transform physical states into each other, we regard it as a redundancy. Two field configurations related by a gauge transformation $\phi(x) \rightarrow e^{im\lambda(x)}\phi(x)$ (the field ϕ has charge $m \in Z$) are supposed to be the same physical state of the system. In order to do this, we have to introduce a gauge potential $A_\mu(x)$, which transforms as $A_\mu \rightarrow A_\mu + \partial_\mu \lambda$ and make the replacement $\partial_\mu \phi \rightarrow D_\mu \phi \equiv (\partial_\mu - imA_\mu)\phi$. The Lagrangian is

$$-\frac{1}{4\mu^2} F_{\mu\nu}^2 + |D_\mu \phi|^2 + \mu^2 V(|\phi|).$$

Note that, in this formula, A_μ has mass dimension 1 and ϕ has dimension 0. g^2 is dimensionless, as are the parameters in the potential V. The potential has a global minimum at $|\phi| = 1$ and a maximum at $|\phi| = 0$. The perturbative spectrum in the expansion around the minimum is exposed by introducing the gauge-invariant variables $\rho = |\phi|$ and

$$B_\mu = A_\mu - \frac{1}{m} \partial_\mu \chi.$$

We find a massive vector field,[7] with mass μ, and a massive scalar with mass $(\mu/2)V''(1)$.

The natural gauge-invariant order parameter for this system is the Wilson loop

$$W[C] = e^{ik \int_C A_\mu \, dx^\mu}.$$

---

[7] In two space-time dimensions, the particle described by the Proca equation has only one internal state.

If we choose parameters in the potential such that $|\phi| = 0$ is the global minimum, this order parameter has the following behavior for large loops C:

$$W[C] \sim \begin{cases} e^{-A[C]}, & k \neq 0 \bmod m, \\ e^{-P[C]}, & k = 0 \bmod m, \end{cases}$$

where $P[C]$ is the perimeter of C and $A[C]$ the area bounded by C. This is interpreted by thinking of a Wilson loop with a long rectangular section with

$$T \gg R \gg 1/m_{\min}.$$

That is, both sides are much larger than the Compton wavelength of the lightest particle in the spectrum. This corresponds to creating a pair of extremely heavy external sources with charge $\pm k$, moving them a distance R apart, and letting them sit for a time T before re-annihilating them. Then

$$W[C] \sim e^{-E(R)T},$$

where $E(R)$ is the minimum energy state in the presence of the static, separated, particle–anti-particle pair. The area law corresponds to the unscreened one-dimensional Coulomb potential $E(R) \sim |R|$, while the perimeter law gives us the self-energy of charges screened by the dynamical charge-μ particles in the system.

By contrast, the perturbative calculation in the Higgs phase leads to screening of all external sources, even $k \neq 0$ modulo m. In the interpretation of the Higgs model as the statistical mechanics of a planar superconductor, this screening is interpreted as screening of magnetic flux by the Meissner effect. The Wilson loop measures the Aharonov–Bohm effect experienced by transporting a particle of charge k around a closed loop.

Now let us return to the question of instanton solutions in the Higgs phase. First consider the following configuration of the fields:

$$\int_\infty d\theta \, \partial_\theta \chi = 2\pi n,$$

$$A_\mu = \frac{1}{m} \partial_\mu \chi,$$

$$\rho = 1.$$

Formally, since A_μ is a gradient, $F_{\mu\nu} = 0$, and the action of this configuration is the same as that of the classical vacuum with $\rho = 1$ and $\chi = A_\mu = 0$.

This is misleading, since

$$\int A_\mu \, dx^\mu = \int F_{\mu\nu} \, dx^\mu \, dx^\nu = \frac{2\pi n}{m}.$$

The line integral is taken over any contour that one can reach by continuous deformation of the circle at infinity, without hitting a singularity of χ. If χ has a singularity only at one point (call it the origin), then

$$F_{\mu\nu} = \frac{2\pi n}{m} \delta^2(x) \epsilon_{\mu\nu},$$

and the action is really infinite. As before, we can regularize the singularity by considering a configuration in which $\rho(r)$ has a zero at the origin, and goes asymptotically to 1 at infinity. Thus, we make an ansatz

$$A_\mu(x) = \epsilon_{\mu\nu} x^\nu a(r),$$

and

$$\phi_a(x) = x^a \frac{\rho(r)}{r}.$$

Here ϕ_a are the real and imaginary parts of the complex field ϕ. Imposing $\rho(0) = 0$ and the boundary condition that as $r \to \infty$ we approach the singular configuration $\rho = 1$, $mA_\mu = \partial_\mu \chi$, $\chi = n\theta$, makes it easy to see that we can make the action finite. Therefore, there is a minimum-action configuration with the same boundary conditions. It is somewhat more difficult to prove, but nonetheless true, that the minimum is achieved within the symmetric ansatz we have chosen. If we define the fluctuations $\delta a = a - a_\infty(r)$, $\delta\rho(r) = \rho - 1$, then they satisfy the linearized equations for large r. The fact that all gauge-invariant fields are massive in the Higgs vacuum shows that the fluctuations fall off *exponentially*. Thus, the conditions for the dilute-instanton-gas approximation are satisfied.

As in the Bloch-wave problem, it is convenient to define a partition function with an extra phase

$$e^{i\frac{\theta}{2\pi} \int F}.$$

The instantons have two translational zero modes and a positive fluctuation determinant. Thus, the dilute-gas partition function is

$$Z = e^{2LT \frac{N}{2\pi g^2} D^{-1/2} e^{-S_0/g^2} \cos\left(\frac{\theta}{m}\right)}.$$

Here $\sqrt{N}$ is the normalization factor for the single-instanton zero modes $\partial_\mu X_I$, and we have used the fact that a single instanton has flux $1/m$ for a charge-m Higgs field. LT is the volume of Euclidean space-time.

It is easy to evaluate the Wilson-loop expectation value, by noting that

$$\int A_\mu \, dx^\mu = \int_A F,$$

where the volume integral runs over the area interior to the loop. Thus the charge-k Wilson loop simply shifts the θ parameter to $\theta + 2\pi k$, within this area. When $LT \gg A \gg \mu^{-2}$, we can break the dilute-gas sum in the numerator up into the product of a sum for which all instantons are inside the loop and one for which they are all outside the loop. These sums can each be evaluated using the formula for the partition function for the appropriate volume and θ angle. Thus

$$\langle W \rangle \sim e^{2A \frac{N}{2\pi g^2} e^{-S_0/g^2} \left[\cos\left(\frac{\theta + 2\pi k}{m}\right) - \cos\left(\frac{\theta}{m}\right)\right]}.$$

This falls like the area, unless k is a multiple of m. Thus, instantons restore the confining Coulomb potential, which seemed to be screened in the Higgs phase. Note, however, that the strength of the potential is exponentially small for small g. This is quite different from the phase where there is no expectation value of the Higgs field. There the strength of the Coulomb potential is of order g^2.

In the interpretation of our partition function and Wilson loop in terms of the Aharonov–Bohm phase in a superconductor, the failure of screening in the presence of instantons is an indication that the Meissner effect disappears at finite temperature. Two-dimensional superconductors have finite-energy excitations that are "Abrikosov flux dots." At finite temperature the system is a dilute gas of dots and anti-dots. An external magnetic field penetrates the system by preferentially aligning the dots in the region where the field exists.

10.5 Monopole instantons in three-dimensional Higgs models

In three Euclidean dimensions, topological charges are classified by $\Pi_2(M_V)$, the mappings of the 2-sphere into the space of vacuum field configurations. The space of gauge field vacua is always the gauge group, and $\Pi_2(G)$ vanishes for any Lie group. Thus, we will have instantons only for models that involve scalar fields, and only if the space of minima of the scalar potential has non-vanishing Π_2. The angular-momentum part of the kinetic term has the form

$$\int dr \, \Phi^A L^2 \Phi^A,$$

so angular dependence leads to a linearly divergent action. We conclude, as in two dimensions, that finite-action instantons will occur only if the space of scalar vacua has a gauge equivalence on it. That is, locally it takes the form $G/H \times X$, where G is the gauge group and H the subgroup which preserves a point in the vacuum manifold. In physics language, the gauge group G is in the Higgs phase, with H the unbroken subgroup. In our discussion below of magnetic monopoles as solitons, we will show that $\Pi_2(G/H) = \Pi_1(H)$. The most interesting case will be that in which H is a product of U(1) groups, and the simplest example is a single U(1) and $G = SU(2)$. In this case, G/H is just S^2 and the topological charge is just the winding of the sphere on itself. The map of charge 1 is just the identity map, while that of charge -1 is the orientation-reversing map of the sphere on itself, $n^a \to -n^a$, in the representation of S^2 as the space of unit 3-vectors.

Thus, the simplest three-dimensional model with finite-action instantons is the Georgi–Glashow model of SU(2) gauge theory broken to U(1) by a Higgs field in the three-dimensional adjoint representation. The action is

$$S = \int d^3x \, \frac{1}{4g^2} \left[(F_{\mu\nu}^a)^2 + \frac{1}{2}(D_\mu^{ab}\phi_b)^2 + V(\phi^a\phi^a) \right].$$

We count the mass dimension of both the gauge potential and the scalar field as 1, in which case g^2 also has mass dimension 1. These are not the scaling dimensions of the fields in a renormalization-group analysis. Restricting our attention to renormalizable potentials, V is a polynomial of order ≤ 6 whose minimum is at non-zero ϕ^a. In order to get explicit solutions, we will restrict ourselves to the case $V = 0$, a situation that is natural if there is enough supersymmetry. None of the qualitative results we obtain will depend on this restriction. The vacuum expectation value of $|\phi^a|$ is denoted v.

Our analysis begins from the remark that

$$\int d^3x \, \frac{1}{4}[F_{\mu\nu}^a \pm \epsilon_{\mu\nu\lambda}(D_\lambda\phi)^a]^2 \geq 0.$$

Thus, if the potential is zero,

$$g^2 S \pm \frac{1}{2}\int d^3x \, F_{\mu\nu}^a \epsilon_{\mu\nu\lambda}(D_\lambda\phi)^a \geq 0.$$

It follows that

$$S \geq |M|\frac{v}{g^2},$$

where the magnetic charge M is defined by

$$Mv = \frac{1}{2}\int d^3x \, F_{\mu\nu}^a \epsilon_{\mu\nu\lambda}(D_\lambda\phi)^a.$$

v is the vacuum expectation value of the dimension-1 Higgs field. Now we can use the Leibniz rule for covariant derivatives, the Bianchi identity for the gauge field strength, and the fact that covariant derivatives of singlets are ordinary derivatives, to show that

$$M = \frac{1}{v}\int_{S^2} \phi^a F^a,$$

where $F^a = \frac{1}{2}F_{\mu\nu}^a \, dx^\mu \, dx^\nu$ is the field-strength 2-form.

The name magnetic charge comes from the fact that we have an unbroken U(1) gauge theory. If we interpret our solutions as static solutions of the $(3+1)$-dimensional version of the theory, then M measures the magnetic flux of the unbroken gauge field through the sphere at infinity. These solutions are called 't Hooft–Polyakov monopoles.

The inequality is saturated by solutions of the first-order equations

$$\frac{1}{2}\epsilon_{\lambda\mu\nu}F_{\mu\nu}^a = \pm(D_\lambda\phi^a).$$

In the supersymmetric context which justifies setting $V = 0$ in the quantum theory, these are the Bogomol'nyi–Prasad–Sommerfield (BPS) equations which say that the solution preserves half the supercharges.[8] This accounts for the first-order nature of the equations: SUSY variations are of first order in derivatives.

In the theory with no potential there are n monopole solutions. These arise because the massless scalar field which parametrizes the radius of ϕ^a gives rise to attractive long-range forces that cancel out the magnetic repulsion of the monopoles. For our purposes we will be more interested in the non-BPS solutions consisting of equal numbers of monopoles and anti-monopoles. The properties of these can be completely understood in terms of the BPS solution with unit magnetic charge. It is natural to make a spherically symmetric ansatz,

$$\phi^a = \frac{x_a}{r} f(r),$$

$$A_{\mu a} = \epsilon_{\mu a \nu} \frac{x_\nu}{r} a(r).$$

Note that f and a have mass dimension 1.

We have

$$D_\lambda \phi^a = P_{a\lambda} \left(\frac{1}{r} - a \right) f + \frac{x_a x_\lambda}{r^2} f',$$

where P_{ab} is the projector orthogonal to x_a/r.

The field strength is given by

$$2B_\lambda^a \equiv \epsilon_{\mu\nu\lambda} F_{\mu\nu}^a = 2P_{\lambda a} \left(-\frac{a}{r} - a' \right) + 2\frac{x_a x_\lambda}{r^2} \left(a^2 - \frac{2a}{r} \right).$$

The Bogomol'nyi equation

$$B_\lambda^a = \pm D_\lambda \phi^a$$

leads to

$$\pm \left(\frac{1}{r} - a \right) f = - \left(\frac{a}{r} + a' \right)$$

and

$$\pm f' = a^2 - \frac{2a}{r}.$$

If we introduce

$$q \equiv (1 - ra),$$

then the first equation reads $f = q'/q$, and the equations are solved by

$$q = \frac{rv}{\sinh(rv)},$$

where v is the vacuum expectation value of $|\phi^a|$.

---

[8] This is most easily understood in the soliton interpretation of our solutions in $3 + 1$ dimensions. The extended supersymmetry algebra

$$[Q_\alpha^i, \bar{Q}_\beta^j]_+ = (\gamma^\mu)_{\alpha\beta} P_\mu \delta^{ij} + \epsilon^{ij} M$$

has degenerate representations when the matrix $\gamma^\mu P_\mu + \epsilon M$ has zero eigenvalues. This is a condition on the mass of particles, which for solitons is the static energy. The energy becomes Euclidean action in the three-dimensional instanton interpretation of the solutions.

We will do the calculation of the effects of instantons in the low-energy effective theory below the mass of the massive gauge bosons and scalars. In a generic theory with monopole instantons, this low-energy theory consists of a three-dimensional U(1) gauge theory, coupled to a dilute gas of point-like instantons. If we think of the vector dual to F_{ij} as an *electric field* in the statistical mechanics of three-dimensional electrostatics, then our problem is equivalent to the statistical mechanics of a Coulomb gas. In that language, the phenomenon we are about to expose is called *Debye screening*. For a given configuration of monopoles and anti-monopoles, the field strength is given by

$$F_{\mu\nu}^{\text{mon}} = \epsilon_{\mu\nu\lambda}\,\partial_\lambda\phi_0,$$

where

$$\nabla^2\phi_0 = 4\pi \sum [\delta^3(x - x_i) - \delta^3(x - y_i)].$$

x_i and y_i are the positions of the monopoles and anti-monopoles, respectively. We have to sum over all monopole and anti-monopole numbers, with weight

$$\frac{\left(D^{-1/2}\sqrt{\dfrac{\mathcal{N}}{2\pi(gR_c)^2}}^3\, e^{-\frac{S_0}{(gR_c)^2}}\right)^{N_+ + N_-}}{N_+!N_-!},$$

and integrate over all the positions. $\mathcal{N}$ is the integral of the square of the translational zero modes (summed over all field components). This quantity, as well as D and S_0, are the contributions of the microscopic theory to the instanton amplitudes. The R_c in this formula is $1/v$. In principle, there are renormalization corrections to this effective field theory formula, but if gR_c is small, then they are small, and we can neglect them.

Different underlying theories, with different scalar potentials, or other massive fields, will change these parameters, but will not otherwise affect the infrared dynamics we study. The parameters should be tuned to absorb infinite effects in the low-energy theory, such as the Coulomb self-energies of the monopoles, and restore the finite values of these self-energies in the underlying theory. Additional massless fields, such as those guaranteed in a supersymmetric theory, can change the IR dynamics. Typically the change has to do with the particular Green functions to which the instantons contribute, rather than a drastic change in the instanton statistical mechanics.

The fluctuating U(1) gauge dynamics can be represented by a vector potential in a particular gauge. It is more convenient, however, to represent it by a fluctuating field strength $F_{\mu\nu}$. This is achieved by replacing the Euclidean Maxwell action by $(g^2/4)\mathcal{F}_{\mu\nu}^2 + 2i\mathcal{F}_{\mu\nu}(\partial_\nu A_\mu) + $ (gauge fixing) (note that $\mathcal{F}$ has dimension 1). Integrating over $\mathcal{F}_{\mu\nu}$ restores the Maxwell form, while integrating over A_μ gives us a functional delta function setting

$$\partial_\nu \mathcal{F}_{\mu\nu} = J_\mu.$$

The current is zero in the denominator of the functional integral formula, and represents the electrically charged particles (in the sense of $(2 + 1)$-dimensional electrodynamics) coupled to the gauge field. We will take it to be the current of a single Wilson loop.

The constraint is solved by setting $\mathcal{F}_{\mu\nu} = F^0_{\mu\nu} + \epsilon_{\mu\nu\lambda}\,\partial_\lambda\phi$, where F^0 is any special solution of the constraint equation. ϕ is a dimensionless scalar field. Different choices of F^0 are absorbed into the ϕ functional integral. Later, we will make a semi-classical approximation for ϕ, and it is convenient to choose F^0 to simplify the semi-classical analysis. We will choose it to be zero in the denominator and, in the numerator, a surface delta function on the minimal-area surface bounded by the loop.

The utility of the dual formulation is that it is easy to include the monopole instantons as sources of the fluctuating field ϕ. If we add the term $i\sum[\phi(x_i) - \phi(y_j)]$ to the denominator functional integral, integration over ϕ gives rise to the repulsive and attractive long-range Coulomb forces among monopoles and anti-monopoles. It also leads to infinite self-energies, which are absorbed into S_0. Note that the factor of i in the action is required in order to get the right sign for the Coulomb energies and make like charges repel and opposite charges attract.

We can now sum over the monopoles for fixed ϕ and obtain a correction to the effective action

$$\delta S = 2D^{-1/2}\left(\frac{\mathcal{N}}{2\pi(gR_c)^2}\right)^{3/2} e^{-\frac{S_0}{(gR_c)^2}}\cos\phi,$$

a result first obtained by Polyakov (though the equivalent physics was done long ago by Debye). The most significant aspect of this result is that the ϕ field has obtained an exponentially small mass.

Now consider trying to evaluate the functional integral over ϕ by finding a classical solution that minimizes the action. In the denominator, the appropriate solution is just $\phi = 0$. In the numerator, if we ignore the cosine term, we could try to set

$$\partial_\lambda\phi = \frac{1}{2}\epsilon_{\mu\nu\lambda}F^0_{\mu\nu}.$$

This cannot work everywhere, because of the current source on the Wilson loop. However, it fails only in the vicinity of the loop and we get an action proportional to the circumference of the loop. The large-distance behavior of the loop is in this case determined by the quadratic fluctuations of ϕ around the classical solution, which give rise to the logarithmically rising Coulomb potential of $(2 + 1)$-dimensional electrodynamics.

For simplicity, think about a loop in the x, y plane, with $z = 0$. The solution discussed above has $\phi = 0$ for negative z. It jumps to $\phi = \frac{1}{2}F^0_{xy}$, for positive z, whenever (x, y) is in the interior of the loop. In general, this will lead to an infinite contribution from the cosine term, of the form $A[\infty]$, where A is the area of the loop. We have to let ϕ go back to zero to avoid the infinity, but this leaves over a finite contribution proportional to A for the minimal-action configuration. The proportionality constant is exponentially small.

Note, however, that, when the required jump in ϕ is exactly an integral number of periods of the cosine, we can eliminate the area term entirely. By arithmetic that we will do carefully when we study monopoles as solitons, we find that the area law disappears precisely for Wilson loops corresponding to the charges carried by massive W bosons. Thus we get a linear confining potential only between charges corresponding to half-integral spin representations of SU(2).

The answer to the question of what classical configurations these monopole instantons are tunneling between is quite subtle. To find it we compactify the two space dimensions on a rectangular torus of radius R. The gauge-invariant magnetic field $\phi^a F_{12}^a$ has configurations with non-vanishing quantized flux wrapping the torus,

$$\int_{T^2} \phi^a F^a = N,$$

and these configurations have energy $\sim 1/R^2$. Thus, in the large-R limit, the system has degenerate states with different values of magnetic flux, and the monopoles are the instantons which tunnel between these states.

10.6 Yang–Mills instantons

There are two possible Lorentz-invariant, gauge-invariant, dimension-4 terms in the Euclidean Lagrangian of pure non-abelian gauge theory. These are

$$\mathcal{L} = \frac{1}{2g^2} \operatorname{tr} F_{\mu\nu}^2, \tag{10.7}[9]$$

and

$$\mathcal{L}_\theta = \frac{\theta}{32\pi^2} \operatorname{tr} F_{\mu\nu}(*F)^{\mu\nu}, \tag{10.8}$$

where

$$(*F)_{\mu\nu} = \frac{1}{2}\epsilon_{\mu\nu\lambda\kappa}F^{\lambda\kappa}. \tag{10.9}$$

The second of these is actually a total derivative. This is easy to see for the abelian case. For the non-abelian case, we use the language of matrix-valued differential forms. The gauge potential $A = iA_\mu\,dx^\mu$ is a Lie-algebra-valued 1-form. The field strength F is given by

$$F = \mathrm{d}A - A^2.$$

Note that F and A are defined to be anti-Hermitian. Furthermore, because of the anti-symmetry of the multiplication of forms, A^2 involves only the commutator of the matrices A_μ. Finally

$$\mathrm{d}F = -\mathrm{d}A\,A + A\,\mathrm{d}A = [A, F].$$

---

[9] The trace is taken in the fundamental representation with the convention $\operatorname{tr}(\lambda_a\lambda_b) = \frac{1}{2}\delta_{ab}$.

Introduce the Chern–Simons form

$$C = \mathrm{tr}\left(A\, \mathrm{d}A - \frac{2}{3} A^3 \right).$$

Then

$$\mathrm{d}C = \mathrm{tr}\left[(\mathrm{d}A)^2 - \frac{2}{3}(\mathrm{d}A\, A^2 - A\, \mathrm{d}A\, A + A^2\, \mathrm{d}A) \right]$$

$$= \mathrm{tr}[\mathrm{d}A\, F - 2\, \mathrm{d}A\, A^2],$$

where we have used cyclicity of the trace and graded anti-commutativity of differential forms. Noting that $\mathrm{tr}\, A^4 = -\mathrm{tr}\, A^4 = 0$, we conclude that

$$\mathrm{d}C = \mathrm{tr}\, F^2 \qquad (10.10)$$

is proportional to the θ term in the Lagrangian. The second term in the action is therefore an integral over the boundary of Euclidean space-time of the Chern–Simons form. The condition of finite action is that $A \to U^\dagger\, \mathrm{d}U$, a pure gauge, on the boundary. The function U defines a map from the 3-sphere at infinity in R^4 to the gauge group. Consider the case of SU(2), where the group manifold is itself S^3. There are obviously different topological classes of map of the sphere into itself, which are multiple wrappings. They are characterized by the winding number n, which is an arbitrary integer. Thus, there are different classes of finite-action gauge-field configurations on R^4, classified by the winding number of the pure gauge transformation, which they approach at infinity. The winding number n is called the topological charge of the configuration. For more general gauge groups it can be shown that every non-trivial map of S^3 into the group can be deformed into a map into some SU(2) subgroup, so the topological classification is the same.

Now note the obvious inequalities

$$\int \mathrm{d}^4 x\, \mathrm{tr}(F_{\mu\nu} \pm *F_{\mu\nu})^2 \geq 0. \qquad (10.11)$$

On expanding out the square and collecting terms we see that this bounds the action of any configuration from below,

$$S \geq \frac{8\pi^2}{g^2}|n|,$$

where n is the topological charge. Equality is achieved only for self-dual and anti-self-dual configurations

$$F = \pm *F. \qquad (10.12)$$

Solutions of these first-order equations are automatically solutions of the Euclidean Yang–Mills equations, because they minimize the action for fixed topological charge.

A basis of (anti-)self-dual tensors in four dimensions is given by the 't Hooft symbols

$$\eta^a_{\mu\nu} = \delta_{\mu 0}\delta_{va} - \delta_{v0}\delta_{\mu a} + \epsilon_{\mu\nu a}, \qquad (10.13)$$

$$\bar{\eta}^a_{\mu\nu} = \delta_{\mu 0}\delta_{va} - \delta_{v0}\delta_{\mu a} - \epsilon_{\mu\nu a}. \qquad (10.14)$$

These appear in the product rule for Euclidean Weyl matrices

$$\sigma^\mu \sigma^\nu = \delta^{\mu\nu} + \eta^{\mu\nu a}\sigma_a, \tag{10.15}$$

$$\bar{\sigma}^\mu \bar{\sigma}^\nu = \delta^{\mu\nu} + \eta^{\bar{\mu}\nu a}\sigma_a.$$

It is natural to search for a minimal-action solution that is maximally symmetric, via the ansatz

$$A_\mu = g_1^{-1}(\hat{x})\partial_i g_1(\hat{x})f(x^2). \tag{10.16}$$

g_1 is the winding number 1 mapping of the 3-sphere into SU(2),

$$g_1 = \hat{x}_\mu \sigma^\mu, \tag{10.17}$$

where $\hat{x}$ is the unit vector in the x direction. In Problem 10.5 the reader will show that this gauge configuration is invariant under simultaneous space-time and gauge rotations, and that its field strength is self-dual if

$$f = \frac{x^2}{x^2 + \rho^2}.$$

ρ is an arbitrary positive parameter, called the scale size of the instanton.

In order to find a tunneling interpretation for this Euclidean solution we consider Yang–Mills theory on the Euclidean manifold $R \times S^3$. Note that this manifold is conformally equivalent to R^4, so we can use the conformal invariance of the classical Yang–Mills equations to construct the instanton on this manifold from the solution we have just found. Using the fact that $F \wedge F = dC$, we find that the topological charge of the instanton for this manifold is equal to the difference of the winding numbers of the pure gauge configurations, which it approaches as $t \to \pm\infty$.

10.6.1 Hamiltonian formulation of Yang–Mills theory in temporal gauge

It is always possible to use the gauge freedom of a non-abelian gauge theory to set $A_0^a = 0$. When this is done, the non-abelian electric field is just

$$\vec{E}^a = \frac{1}{g^2}\,\partial_0 \vec{A}^a.$$

The Lagrangian has a standard canonical form and can be quantized in a straightforward manner, with E_i^a and A_i^a as canonical conjugates. However, we must also impose the condition that follows from varying the original action with respect to A_0^a. It is easy to see that this is

$$G^a(x) = D_i^{ab}E_i^b - \rho^a = 0.$$

This is the non-abelian form of Gauss' law, and ρ^a is the time component of the Noether current for matter fields, which follows from the global G symmetry. This condition should be understood as follows. The system with $A_0^a = 0$ is invariant under *time-independent* gauge transformations, and

$$G(\omega) = \int \mathrm{d}^3 x\, \omega^a(x) G^a(x)$$

is the generator of these transformations when ω^a vanishes sufficiently rapidly at infinity that integrations by parts are permitted.

It is easy to verify that each $G(\omega)$ commutes with the Hamiltonian, so that we have an infinite-dimensional symmetry group of the mechanical system we have defined. The statement that $G(\omega) = 0$ is imposed on the allowed physical states of this theory. It is somewhat analogous to the restriction to BRST-invariant states in covariant quantization.

An interesting twist on this formalism occurs if there is a topological term in the Yang–Mills action, and G is broken down to H $=$ U(1) (or a product of U(1) groups). The topological term does not affect the equations of motion, but it does change the canonical momentum, and thus the generator of gauge transformations. In electrodynamics, it is easy to see that the constraint equation is now

$$\partial_i E_i - \frac{\theta}{2\pi}\, \partial_i B_i - \rho = 0,$$

where ρ is the electric charge density of matter fields. We see that, in the presence of θ, magnetic monopoles will carry (generally irrational) electric charges [140].

In the classical approximation, ground states of the theory are found by looking for minima of the energy. The classical states of minimum energy are static, pure gauge-field configurations. When space has the topology S^3 (equivalently, when it has the topology R^3 but we impose falloff conditions at infinity), the space of such field configurations is classified by the topological winding number n. The instanton represents a tunneling process between two classical vacuum states with winding numbers differing by one. This is closely analogous to the tunneling between different wells of an infinite periodic potential in one dimension. The true semi-classical vacuum state in that case is a Bloch wave, characterized by a Bloch momentum, which runs between 0 and 2π. In an analogous fashion, semi-classical Yang–Mills theory would appear to have a continuous set of vacua characterized by a parameter θ that is periodic with period 2π. In fact this is true, even though the semi-classical approximation is valid only for Green functions at short distances, in the confining phase of the theory. The parameter θ is just the coefficient of the $F \wedge F$ term in the Lagrangian, which we introduced above. A simple example of the correspondence between vacuum parameters and terms in the Lagrangian is treated in Problem 10.2

10.6.2 Bosonic zero modes

There is also an anti-instanton solution of the anti-self-duality equations in which the winding-number-1 gauge transformation is replaced by the transformation of opposite winding number. Atiyah and Hitchin were able to find all solutions of the self-duality equations [141]. These exact solutions are useful primarily in supersymmetric gauge theories, where they allow us to calculate certain correlation functions exactly. Their study goes beyond the scope of this book.

For our purposes, it is more interesting to look at approximate solutions of the second-order equations consisting of a dilute gas of widely separated instantons and anti-instantons. In this context, widely separated means separations large compared with all of the parameters ρ_i of the instantons and anti-instantons. The existence of the parameter ρ is a consequence of the classical conformal invariance of the Yang–Mills equations. Given any solution with finite action, we can scale the coordinates to get a new solution with the same action. Note that the same is not true for finite special conformal transformations, because they map finite points to infinity. When we consider small fluctuations of the instanton, these symmetries show up as zero modes of the fluctuation operator. The zero modes corresponding to special conformal transformations are not normalizable and the corresponding directions in field space are not included in the functional integral.

The procedure for dealing with the scale zero mode is similar to that for translations. One extracts a scale collective coordinate and does the integral over it exactly. The classical measure for this integration is the scale-invariant measure

$$\frac{\mathrm{d}\rho}{\rho},$$

but there are generally quantum corrections to this. In Yang–Mills theory, the dominant effect is the replacement of g^2 in the instanton action by the running value of the renormalized coupling $g^2(\rho)$ [142]. In asymptotically free theories g goes to zero for zero scale size, which tends to make the integral over ρ converge near 0. On the other hand, this leads to an enhanced divergence at large ρ. Instanton calculations of the short-distance behavior of Green functions are under control, because the ρ integrals get cut off at scales of order the distance between operators. However, instanton technology, just like perturbation theory, fails to capture the large-distance behavior of asymptotically free theories.

In a purely bosonic gauge theory, the result of the dilute-gas computation of Green functions is

$$\langle O_1(x_1) \ldots O_n(x_n) \rangle_{\mathrm{c}} = D^{-1/2} \int \mathrm{d}^4 a \int \frac{\mathrm{d}\rho}{\rho} \int \mathrm{d}\Omega$$

$$\times g^{-(5+d_{\mathrm{G}}-3)}(\rho) \mathrm{e}^{-\frac{8\pi^2}{g^2(\rho)}} O_1^{(\mathrm{c})}(x_1 + a) \ldots O_n^{(\mathrm{c})}(x_n + a). \tag{10.18}$$

The integrals are over the position, scale size, and embedding of the instanton configuration in the gauge group. $O_k^{(\mathrm{c})}$ is the classical value of the operator O_k in the instanton configuration of fixed position, size, and gauge orientation. The factor $D^{-1/2}$ is the determinant computed from the non-zero modes of fluctuation around the instanton. It has been computed by 't Hooft [142]. The inverse powers of g come from the normalization of the zero mode integrals, as explained above.

10.6.3 Fermion zero modes and anomalies

In gauge theories with fermions, an extremely interesting phenomenon occurs in connection with zero modes of the massless Dirac operator in an instanton background. Consider the eigenvalue equation for the Euclidean Dirac operator:

$$i \, \slashed{D} \psi = \lambda \psi. \tag{10.19}$$

$\gamma_5 \psi$ is then an eigenfunction as well, with eigenvalue $-\lambda$. Now consider modes with $\lambda = 0$, and the way in which they behave under continuous deformations of the Yang–Mills background. In general, we would expect such a deformation to change the eigenvalue, but we can only change pairs of eigenvalues away from zero.

For $\lambda = 0$, the eigenfunctions may be chosen to have a definite chirality, $\gamma_5 \psi = \pm \psi$, because γ_5 anti-commutes with the Dirac operator and preserves the zero eigenspace. This is not true for $\lambda \neq 0$, since we have seen that multiplication by γ_5 reverses the sign of the eigenvalue. As a consequence, the wave function of a non-zero eigenvalue is a linear combination of the two different chiralities. Thus, left- and right-handed zero modes must be lifted in pairs.

This means that the difference between the number of left- and right-handed zero modes (the so-called *index* of the Dirac operator) is a topological invariant of the gauge field, unchanged by continuous deformations. We might be tempted to think that this is related to the topological invariant we have just been discussing. A famous mathematical theorem, due to Atiyah and Singer [143–146], tells us that this is the case. A physics proof of this equation follows from the anomaly equation for the axial U(1) current of the massless fermions

$$\partial_\mu \bar{\psi} \, \gamma^\mu \gamma_5 \psi = \frac{1}{32\pi^2} F^a_{\mu\nu} (*F)^{\mu\nu}_a. \tag{10.20}$$

Consider the Euclidean fermion functional integral in a background gauge field with finite action and topological charge. The functional integral of the left-hand side of the anomaly equation is

$$\sum \partial_\mu \bar{\psi}_n \, \gamma^\mu \gamma_5 \psi_n, \tag{10.21}$$

where the sum is over all normalized eigenfunctions of the Dirac operator,

$$\slashed{D} \psi_n = \lambda_n \psi_n. \tag{10.22}$$

Since the gauge potential falls off at infinity, the finite-λ_n eigenfunctions fall off exponentially at infinity. Thus, if we integrate (10.21) over all of Euclidean space, only the zero modes contribute, and the integral just counts the difference between the numbers of right- and left-handed zero modes. On the other hand, the right-hand side integrates to the topological charge. So we find that the topological charge is in fact equal to the difference between the numbers of right- and left-handed zero modes.

Actually, this calculation is valid only when the fermions are in the fundamental representation. For a general fermion representation, R, if we choose a representative of the topological class for which the gauge field sits in an SU(2) subgroup, then R will break up into a direct sum of spin-j representations of SU(2). The anomaly of the spin-j representation is $2\sum_{m=-j}^{j} m^2$ times larger than that of the spin-$\frac{1}{2}$ representation, so we can use this formula to count the number of zero modes for any representation.

Bosonic zero modes give apparently infinite contributions to the path integral. We have to integrate over the corresponding collective coordinates exactly in order to eliminate or interpret the infinity. By contrast, fermion zero modes tell us that the single-instanton contribution to the partition function vanishes. Instead, the single instanton contributes to the expectation value of operators that can "absorb" the zero modes. The simplest operator, with minimal number of fields, which can do the job, is called the 't Hooft operator associated with the instanton. The dilute-instanton-gas approximation for theories with fermion zero modes corresponds to the lowest-order perturbation theory of a modified theory in which the 't Hooft operator is added to the Lagrangian with coefficient De^{-S_I}, where D includes the determinant of non-zero modes, and inverse coupling factors for bosonic zero modes. These expressions must be integrated over bosonic collective coordinates.

The anomaly equation tells us that the U(1) axial symmetry is broken by instantons. This shows up in the structure of the 't Hooft operator, which explicitly violates the symmetry. In general, if there are k simple factors of the gauge group, we may expect to have k independent symmetries of the classical Lagrangian, which are broken by these non-perturbative effects.

10.7 Solitons

Every instanton solution that we have discussed can also be viewed as a static solution of the equations of motion of the same field theory in one more dimension. That is, the Euclidean action in d dimensions is the same as the energy for static solutions of the $(d+1)$-dimensional Lorentzian field theory with the same field content. In those cases in which our field theory contains vector fields, we are looking at static solutions in the $A_0 = 0$ gauge. The spherically symmetric solutions look like point objects, and we might imagine that they represent some new class of particles in the theory, called *solitons*. We note, for future reference, that the static solutions can also be promoted to a theory in more spatial dimensions. They are solutions that are independent of some of the coordinates, and are thus non-trivial on p-dimensional hyperplanes. The string-theory-inspired name for such extended solutions is *p branes*, with particles corresponding to $p = 0$, strings to $p = 1$, and membranes to $p = 2$. By abuse of language, all of these solutions representing extended objects are called solitons.[10]

In order to prove that 0-brane solitons are indeed particles, we begin by showing that they do indeed correspond to points in the spectrum of the quantum Hamiltonian. The derivation applies to general periodic solutions of the classical equations, and is

---

[10] This is an even further abuse of the mathematical term, which referred only to solutions in integrable field theories.

basically the old Bohr–Sommerfeld quantization rule. We consider the path-integral representation of the trace of the resolvent of the Hamiltonian:

$$\mathrm{Tr}\left(\frac{1}{z-H}\right) = \int [\mathrm{d}q(t)] e^{iT[q(t)]z} e^{\frac{iS[q(t)]}{g^2}},$$

where the integral is over all periodic paths

$$q(t + T[q(t)]) = q(t).$$

We have used notation appropriate for a single quantum degree of freedom, but it should be obvious that the formulae generalize to any number of degrees of freedom, including field theory. Notice that we are doing a *Lorentzian* functional integral here. The derivation of this formula is easy and is left as an exercise.

When g^2 is small, we approximate the functional integral by a stationary phase, saturating the integration by classical solutions. The action of a classical solution is given by

$$\frac{S}{g^2} = \int \mathrm{d}t \, \frac{L}{g^2} = \int p\dot{q} \, \mathrm{d}t - E,$$

where we note that the classical energy $\propto 1/g^2$. We consider periodic solutions satisfying the Bohr–Sommerfeld condition

$$\int p \, \mathrm{d}q = 2\pi k,$$

where the integral is taken over a period. For any such solution, there will be an infinite number of others, given by multiple traverses of the same trajectory. These have period nT and action nS, where T and S are the period and action for the single-pass trajectory. On summing up the contributions from this infinite class of trajectories, we get

$$\mathrm{Tr}\left(\frac{1}{z-H}\right) \sim \frac{1}{(1 - e^{i(z-E)})},$$

which has a pole at $z = E$. Thus, in the semi-classical approximation, the trace of the resolvent has a pole at the energy of each periodic classical solution. The usual quantization of small oscillations around the minimum of the potential (which gives rise to particles in perturbative field theory) is a special case of this rule.

The fact that static spherically symmetric solutions are particles now follows from Lorentz invariance. This is easiest to prove for $1+1$ dimensions, for which the derivation is not complicated by non-covariant choices of gauge for the static solutions. We simply note that Lorentz covariance of the theory implies that $\phi_j(\Lambda x)$ is a solution for every Lorentz matrix Λ. If the original solution was localized around the point x, the boosted solution is localized around $x - vt$. Furthermore, its energy and momentum are related by the usual relativistic dispersion relation $E = \sqrt{p^2 + m^2}$, where m is the energy E_c of the static classical solution.

The fact that our solitonic particle has a mass of order $1/g^2$ resolves a puzzle about how the soliton can be localized in its rest frame. The uncertainty in its velocity, for some small uncertainty in its position, is of order $g^2/\Delta x$. Thus, for small g it can be well localized both in velocity and position. These remarks lead us to expect that the semi-classical expansion is also a non-relativistic expansion for the motion of the soliton. In particular, in this expansion, the number of solitons is conserved, rather than just the difference between soliton and anti-soliton numbers.

The latter remark relieves a tension that someone schooled in the tenets of effective field theory might have been feeling about the whole notion of solitons. Recall that all quantum field theories are just low-energy approximations to something else. Why should we trust the predictions of quantum field theory for these states whose energy goes to infinity in the semi-classical approximation? The answer goes back to the Wilsonian procedure for integrating out degrees of freedom. This is usually done by integrating out heavy fields in the Euclidean functional integral. Even without thinking about solitons, we can imagine an absolutely stable particle of large mass. We consider states containing an arbitrary number of these massive particles, interacting with themselves and with light degrees of freedom, under conditions in which no heavy particles are created and all energies and momentum transfers are much smaller than the heavy mass. The generalization of effective field theory to such situations involves light fields interacting with the coordinates describing the motion of the heavy particles. In general it will contain additional renormalization constants relating to the heavy-particle properties.

We will now show that a similar procedure works for solitons. The major difference is that the quantum theory of solitons requires no extra renormalization constants. Their properties are completely determined by the parameters of the low-energy effective theory in which they appear as classical solutions. The key observation, motivated both by our remarks about Lorentz transformations and by the spatial translation invariance of the field theory, is that the function

$$\phi^i = \phi_i^{\mathrm{c}}(x - X(t)),$$

where $X(t)$ is slowly varying, and ϕ_i^{c} is the static classical solution, has energy

$$E = E_{\mathrm{c}} + \frac{1}{2} \int \mathrm{d}x \left(\frac{\mathrm{d}\phi_i^{\mathrm{c}}}{g\,\mathrm{d}x}\right)^2 \dot{X}^2.$$

This is the energy of a non-relativistic particle, with mass

$$m = \int \mathrm{d}x \left(\frac{\mathrm{d}\phi_i^{\mathrm{c}}}{g\,\mathrm{d}x}\right)^2.$$

If we recall the classical equations of motion

$$-\frac{\mathrm{d}^2}{\mathrm{d}x^2}\phi_i^{\mathrm{c}} + \frac{\partial V}{\partial \phi_i} = 0$$

and the manipulations that lead to *Euclidean energy conservation* when we considered the same solution as an instanton, we find that

$$\left(\frac{d\phi_i^c}{dx}\right)^2 - 2V$$

is x-independent. We can evaluate the constant at infinity, where both terms in the expression vanish. The reader should verify that this shows that the mass appearing in the non-relativistic kinetic energy of the soliton is just the static energy, as we expect from Lorentz invariance.

The strategy for getting a perturbative expansion of soliton interactions is simply to expand the functional integral around the sub-manifold of field space parametrized by $\phi_i^c(x - X(t))$ for slowly varying X and then to do the X functional integral. The rationale is like that of the Born–Oppenheimer approximation in molecular physics. The time scales for X motion (in the soliton rest frame) are much longer than the time scales involved in perturbative particle motion and interaction. Therefore, in the spirit of effective field theory, we can first solve for the particle scattering in a fixed soliton background, then use this to compute the effective action for $X(t)$, and finally do the quantum mechanics of $X(t)$.

The sub-manifold of slowly moving soliton configurations is a curved sub-manifold of field space. Therefore, in order to isolate it we have to compute the measure in the vicinity of this sub-manifold. The procedure for doing this order by order in perturbation theory is analogous to the Faddeev–Popov procedure for integrating along a curved gauge slice in non-abelian gauge theory. It has been worked out in great detail by Gervais, Jevicki, and Sakita [147]. Fortunately, to leading order, every curved manifold is straight. When we expand around the static soliton by writing

$$\phi_i(t, x) = \phi_i^c(x) + g \sum c_n \delta_n(x, t),$$

we find a quadratic action for fluctuations, whose equations of motion are

$$([\partial_t^2 - \partial_x^2]\delta_{ij} + U_{ij}(x))\delta_j(t, x) = 0.$$

$U_{ij} = \partial^2 V / \partial \phi_i \partial \phi_j$, and is time-independent. On writing $\delta_n = e^{i\omega_n t}\hat{\delta}_n$, we find that ω_n^2 are the eigenvalues of the Schrödinger operator we encountered when implementing small fluctuations around quantum instantons. We know that the spectrum of this operator consists of a normalizable zero mode, proportional to $d\phi_i/dx$, and a continuum of scattering states. The zero mode is obviously the part of the field in the direction parametrized by $X(t)$. Thus, to leading order in the expansion about the soliton we simply find free soliton motion, accompanied by scattering of the perturbative excitations from the static soliton. Soliton recoil appears only at next order in the perturbation expansion.

It's interesting to note that these amplitudes for particle scattering by solitons are independent of g, whereas scattering amplitudes for ordinary particles are of order g^2. If we look in the two-soliton sector, then we can compute the S matrix for soliton–soliton scattering by solving the classical equations with initial conditions corresponding to two incoming, widely separated solitons. The phase shift is computed in terms of the classical action of this solution, and is of order $1/g^2$. So solitons are strongly coupled to themselves, and have order-1 couplings to elementary particles.

We can understand these results intuitively by thinking about the soliton states in terms of the elementary-particle Fock space. In terms of fields with canonically normalized kinetic terms, which create elementary-particle states with amplitude 1, the classical soliton field is of order $1/g$. A state with the field shifted to the classical soliton value is

$$e^{i \int dx \, \pi^i \phi_i^c} |0\rangle,$$

and the average number of particles in such a state is of order $1/g^2$. So we can view the various types of scattering amplitudes as coherent sums over scattering of elementary particles off the elementary constituents of the soliton. This picture should not be pushed too far, but it gives the right order of magnitude in powers of g for each type of soliton scattering amplitude.

10.8 't Hooft–Polyakov monopoles

We now skip to the case of most interest, the static 't Hooft–Polyakov solutions of $(3+1)$-dimensional gauge theories. Let us study their topological charge with a little generality. We start with a general gauge group G and a Higgs field in a representation R of G. Recall that the finite-energy (formerly finite-action) condition is that the Higgs field approaches some minimum of the potential on the sphere at infinity. It must satisfy

$$\phi(t_1, t_2) = g(t_1, t_2)\phi_0,$$

where ϕ_0 is a fixed vector in the representation, whose stability subgroup is H. It is the value of $\phi(t_1, t_2)$ at a point on the sphere, which we call the North Pole. This defines a continuous mapping of S^2 into G/H, an element of $\Pi_2(G/H)$. $t_{1,2}$ are coordinates on the sphere. Continuity of $\phi(t_1, t_2)$ does not imply that $g(t_1, t_2)$ is continuous on the sphere, and indeed it is not, whenever ϕ is topologically non-trivial. Without loss of generality we can map the sphere onto the square, with the entire boundary of the square identified with the North Pole. ϕ thus takes on the value ϕ_0 everywhere on the boundary. Let (t_1, t_2) be the Cartesian coordinates on the square, running from 0 to 1.

The gauge potential at infinity is related to ϕ by the equation

$$D_i \phi = 0.$$

There are many gauge-equivalent solutions of this equation. One is chosen by setting

$$g(t_1, t_2) = P e^{i \int_C A},$$

where, for each point (t_1, t_2), C is the horizontal line starting at the left boundary of the square and going to that point. With this definition, $g = 1$ on the left, top, and bottom boundaries of the square, but may be non-trivial on the right boundary. The condition that ϕ be continuous is that

$$g(1, t_2)\phi_0 = \phi_0,$$

for all t_2, i.e. $g(1, t_2) \in H$. This then defines a map of the circle into H, or a member of $\Pi_1(H)$. It is easy to verify that the map we have just constructed, namely that of $\Pi_2(G/H)$ into $\Pi_1(H)$, is a group homomorphism.

Conversely, given any element of $\Pi_1(H)$ that can be continuously distorted into the identity when $\Pi_1(H)$ is embedded in $\Pi_1(G)$, we can construct $g(t_1, t_2)$ in terms of the homotopy which distorts that element into the identity. Finally, the image of the identity in $P_2(G/H)$, the mapping $g(t_1, t_2) = 1$, is clearly the identity in $\Pi_1(H)$. Our map is a group homomorphism, so this means that it is a one-to-one and onto mapping (an isomorphism) between $\Pi_2(G/H)$ and the subgroup of $\Pi_1(H)$ which maps into the identity in $\Pi_1(G)$. In particular, if G is simply connected we have an isomorphism between $\Pi_1(H)$ and $\Pi_2(G/H)$. The most interesting case is when $\Pi_1(H)$ is an integral lattice, i.e. when H is a product of U(1) factors. In this case, as in our SO(3)/U(1) example, the topological charge of solitons is precisely the magnetic charge of the multiple Maxwell fields in the low-energy effect theory below the scale of the Higgs VEV. One way to get such a pattern of symmetry breaking for any group is to introduce a Higgs field in the adjoint representation. A general VEV for an adjoint Higgs breaks the group to its Cartan torus.

Every Lie group G has a simply connected covering group $\bar{G}$, such that $G = \bar{G}/\Pi_1(G)$. Familiar examples are $\bar{SO}(3) = SU(2)$ and more generally $\bar{SO}(n, m) = Sp(n, m)$. These examples illustrate a general pattern: the covering group has extra representations, the spinors, with the property that the eigenvalues of Cartan generators (the weights of the representation) are fractions of the weights in bona-fide representations of G. A more general class of examples is obtained by taking any simply connected Lie group G modulo its center Z_G. Only representations invariant under the center are representations of G/Z_G.

The pure Yang–Mills Lagrangian contains only adjoint fields. Let us add Higgs fields only in representations of G/Z_G. Then monopole charges are restricted. The gauge group of the theory could be either G or G/Z_G, but the Lagrangian does not distinguish between them, so monopole charges are restricted to be in the subgroup of Π_1 (H) which is trivial inside of $\Pi_1(G/Z_G)$. So, in our SO(3)/U(1) example, the monopole charges must all be even.

The physical meaning of all of this has to do with the Dirac quantization condition. We will describe it explicitly only for the case of SO(3)/U(1), but the reader should be able to generalize it to all other cases. As we have discussed in Chapter 8, the low-energy effective theory is Maxwell's electrodynamics coupled to both electric and magnetic charges. This raises a well-known problem, which was first solved by Dirac. An electrically charged particle couples to the vector potential

$$\int A_\mu \, \frac{\mathrm{d}x^\mu}{\mathrm{d}\tau},$$

but the vector potential cannot be globally defined in the presence of magnetic charge, because

$$\epsilon^{\mu\nu\lambda\kappa}\, \partial_\nu F_{\lambda\kappa} = J_m^\mu.$$

Dirac solved this problem for point-like monopoles by taking a particular solution of this equation with F non-zero only along a 2-surface running from the world line of the monopole to infinity, and then introducing a vector potential to describe the rest of the electromagnetic field. Essentially he described a monopole as a semi-infinite solenoid. But we would like physics to be independent of the choice of where we place this fictitious solenoid, if we want the description of the interaction of monopoles and charges to be local. Classically there is no problem, since the charged particle will never hit the infinitely thin solenoid, but quantum mechanically we must make sure that there is no Aharonov–Bohm phase when the charged particle follows a trajectory encircling the solenoid. As we proved in Chapter 8, this leads to the Dirac quantization condition

$$(e_i g_j - e_j g_i) = 2\pi n_{ij}. \tag{10.23}$$

We have written the condition for arbitrary pairs of particles, assuming that each pair has both electric and magnetic charge. Each n_{ij} must be an integer.

Another illuminating way of deriving this condition is to compute the angular momentum of the electromagnetic field of the particle pair. It's easy to verify that this contains a term

$$\delta L = \frac{e_i g_j - e_j g_i}{4\pi}\hat{r},$$

where $\hat{r}$ is the unit vector pointing between the pair. Quantization of angular momentum in half-integer units now implies that n_{ij} is an integer. We will see a striking consequence of this fact a bit later on.

For solitonic monopoles in theories in which U(1) arises from spontaneous breakdown of a simple group, both electric and magnetic charges are already quantized. Monopole quantization follows from the topological conditions we have been discussing, while electric charge quantization follows from the fact that the weights of representations of compact Lie groups are discrete. The restriction we discussed above, namely that the magnetic charge be in the subgroup of $\Pi_1(H)$ which maps to the identity in $\Pi_1(G/Z_G)$, is the statement that the monopole satisfies the Dirac quantization condition *for electric charges in all representations of* G, including those which are not representations of G/Z_G. For SO(3)/U(1) the monopole charge is twice what it needs to be to satisfy the Dirac quantization condition with particles in tensor representations of SO(3). If this were not so we would run into an inconsistency when we tried to add fields in the spinor representations of SU(2) to the Lagrangian.

The discussion of charge quantization in the context of non-trivial classical solutions of the field equations leads us to ask about electric charge carried by classical field configurations. Why can't such classical charges take on arbitrary continuous values? To answer this, we first choose a gauge, $A_0^a = 0$. In this gauge we canonically quantize the spatial components of the gauge potential, and then impose the constraint equation derived by varying A_0^a as a constraint on physical states:

$$D_i^{ab} \Pi_b^i + \epsilon^{abc}(\pi_b \phi_c - \pi_c \phi_b)|\Psi_{\text{phys}}\rangle = 0.$$

$P_b^i = (1/g^2)\partial_t A_b^i$ is the canonical conjugate to the vector potential and π_a the canonical conjugate to the Higgs field. The operator which must vanish on physical states is the generator of time-independent local gauge transformations. The corresponding generator of global U(1) transformations is

$$Q = \int_{S^2} \hat{\phi}^a n_i \Pi_a^i.$$

In this gauge, all charge-carrying configurations have time-dependent vector potentials at infinity. Julia and Zee [148] found the corresponding configurations with both electric and magnetic charge in a different, static gauge.

In $A_0 = 0$ gauge, the solutions have the form of a time-dependent gauge transformation of the monopole solution. For small deviations, the behavior near infinity is

$$\delta A_i^a = \delta_i^a \frac{\theta(t)}{r^2},$$

while the leading behavior of the Higgs field is unchanged. This is a normalizable zero mode of the monopole solution and the collective coordinate $\theta(t)$ must be quantized. However, since it represents a U(1)-gauge rotation θ is a periodic variable. Its canonical momentum, the electric charge Q, is therefore quantized. The minimal unit is the charge on the adjoint representation. Quantization of θ thus leads to a discrete spectrum of dyonic (or dual-charged) excitations of the monopole. The spacing between these levels is of order $(\Delta Q)^2$ in units of the mass of the massive gauge bosons.

A surprise is in store [149–150] when we carry out the collective coordinate quantization for monopoles in theories in which there are fields in representations that transform under the center of SU(2). The unit of charge quantization is half what it was in the SU(2)/Z_2 theory. As noted above, if we calculate the angular momentum of some of the dyonic bound states of this system, we will find half-integer values. Thus, we have constructed spin-$\frac{1}{2}$ particles from a purely bosonic system!

10.9 Problems for Chapter 10

*10.1. Discuss vacuum decay for a scalar field theory in d dimensions. That is, we have a single scalar field, with the same kind of 2-minimum potential we discussed in the case of quantum mechanics. The Euclidean path integral is over all field configurations $\phi(x)$ that approach ϕ_f, the higher minimum of the potential, at infinity (set the potential equal to zero there). Assume that the instanton configuration is SO(d)-invariant. Find the field equation for it. If you call the Euclidean radius t, you will find that the equation is identical to a mechanics problem in the upside-down potential, with a time-dependent friction term $(d-1)/t$.

Using simple energetic considerations, argue that there is a finite-action bounce solution. Write down the equations for small fluctuations around the bounce. Since the bounce is spherically symmetric, they can be separated by expanding in d-dimensional spherical harmonics. How many exact zero modes of the solution exist, following from symmetries? In what representation of SO(d) do they transform? Try to argue that there is exactly one negative-eigenvalue fluctuation mode. (Don't try too hard, the general result for a large class of quantum systems can be found in [151]. If you can do it easily for this special case, do it.) The answer to most of this exercise can be found in the Coleman book [138]. Try to do it yourself, but you can go there for hints if you get stuck.

*10.2. Prove Derrick's theorem. Consider a Lagrangian of the form $(\nabla \phi^i)^2 - V(\phi)$ in $d > 2$ dimensions. Consider any finite-energy non-singular field configuration and show that you can lower its energy by considering $\phi(ax)$ with appropriate a. This shows that, without further constraints, there are no stable solitons in such a theory. Generalize this to an arbitary number of scalars. Now consider a complex scalar field with a charge quantum number. The potential is a function of $\phi^* \phi$. Show that time-dependent fields of the form $e^{i\omega t}\phi(x)$, where x are the space variables, carry an electric charge $Q = i \int d^{d-1}x(\phi^* \, \partial_t \phi - \phi \, \partial_t \phi^*)$. Evaluate the energy of such configurations. Show that Derrick's theorem no longer applies. In the appendix to the chapter on lumps in Coleman [138], you will find an example of a theory of this type that has stable charged solitons. Such objects are called Q-balls.

*10.3. Consider the quantum Lagrangian

$$L = \frac{1}{2}(\dot{\phi})^2 - V(\phi),$$

where V is periodic with period 2π. There are two different interpretations of this system. In the first, ϕ lives on a circle. In the second, ϕ lives on the real line, in the presence of a periodic potential. In the second interpretation the system is invariant under translation of ϕ by 2π and contains a global symmetry operator T that commutes with the Hamiltonian. We can diagonalize the Hamiltonian in the sector satisfying

$$T\psi(\phi) = \psi(\phi + 2\pi) = e^{i\theta}\psi(\phi),$$

where θ is called the Bloch momentum. In the first interpretation, translating ϕ by 2π does nothing, and wave functions are required to be periodic. In either case, we are allowed to add the term $\delta L = \theta\dot{\phi}$ to the Lagrangian, which is a total derivative, violating time-reversal invariance, which is analogous to the $F\tilde{F}$ term in gauge theories. Show that, in either interpretation, there are finite-action Euclidean instantons that have value 2π times an integer for this topological term. Show that, in the case that ϕ lives on the real line, the path integral with non-zero θ corresponds to computing the ground-state expectation values in the lowest energy state with Bloch momentum θ.

*10.4. The instanton solutions of a quantum-mechanics problem are solitonic particle solutions of field theory with the same Lagrangian in $1 + 1$ dimensions. Show that, in any higher dimension, the same solutions can be interpreted as *domain walls* separating two ground states of the theory. Similarly, show that the instantons of the $(1+1)$-dimensional Higgs model can be thought of as particles in the $(2+1)$-dimensional version of the model and as strings or (Abrikosov–Nielson–Olesen) vortices in the $(3 + 1)$-dimensional version. This problem requires no additional calculation.

*10.5. Prove that the instanton solution of four-dimensional gauge theory is invariant under rotations, in the sense that all gauge-invariant functions of the field are rotation-invariant.

Concluding remarks

If you've successfully worked your way through this short but arduous journey to the world of quantum field theory, you should be exhilarated! The landscape is full of fabulous beasts (most of which seem to go by the last name -on) and elegant formulae. Most surprisingly, all of this elegance seems to give us a remarkably precise and general description of many facets of our world, from phase transitions in humdrum materials to the interiors of stars and the intergalactic medium. In this concluding section I want to emphasize again a few general lessons, and chart out for you the parts of the quantum-field-theory landscape we have NOT explored.

The first of the important lessons that you should take away from this book is the beautiful unification of the classical theories of fields and particles that is forced on us by combining relativity and quantum mechanics in a fixed space-time background. The second is the unification of the methods of quantum field theory and classical statistical mechanics, which is provided by the Euclidean path-integral formulation of field theory.

Next I would ask you to remember the difference between a symmetry and a gauge equivalence and the different meanings of the idea of *spontaneous symmetry breakdown* in the two cases. Spontaneous breakdown of a global symmetry is related to locality. A quantum field theory is *defined* by its behavior at short distances, but there may be different infrared realizations of the same short-distance operator algebra and Hamiltonian. Sometimes these are related by a symmetry transformation. If it is a continuous group of transformations, any one infrared sector of the theory contains massless excitations called Nambu–Goldstone particles.

By contrast, no physical quantity transforms under a gauge equivalence, which is merely a convenient redundancy of our description of a physical system. What we call a spontaneously broken gauge symmetry is a particular phase, the Higgs phase, of gauge theories. In this phase there are no massless particles associated with the generators of the gauge equivalence. Gauge theories can also have another massive phase (which is distinct from the Higgs phase if the Higgs fields are invariant under a non-trivial subgroup of the center of the gauge group) called the *confining* phase, which is related to the Higgs phase by electric–magnetic duality. In the real world the non-abelian electro-weak gauge group is in the Higgs phase, and the color group of QCD is in its confining phase. The U(1) group of electromagnetism is in yet a third phase, called the Coulomb phase. This is a special case of a more general possibility in which the gauge theory behaves like a conformal field theory in the infrared.

Perhaps the most important topic in quantum field theory is the theory of renormalization. As formulated by Wilson, Kadanoff, and Fisher, this is a very general way of understanding how physical processes at different length scales are related to each other. From the philosophical point of view, this is the reason why we are able to do physics at all. If we had to uncover the correct theory of the smallest length scales and highest energies in order to talk about what we see in our laboratories, there would be no hope for a theory of physics at all. The renormalization group (RG) explains to us why it is that we can do long-distance physics without knowing about short distances. The key concept is that of the RG fixed point, parametrized by a small number of relevant and marginal perturbations. All UV theories containing a given IR set of degrees of freedom flow to the vicinity of the fixed point whose basin of attraction they lie in. They are differentiated only by the values they determine for the marginal and relevant parameters. This gives rise to universality classes of behavior, a post-diction spectacularly confirmed by observations of critical phenomena in condensed-matter physics.

Given this broad-brush picture of what we *have* covered in this book, we now turn to the subjects we have omitted. The most important of these are supersymmetry, finite-temperature field theory, and field theory in curved space-time. Supersymmetry is (in dimensions higher than 2) the only allowed extension of the space-time conformal group. Remarkably, it joins together bosons and fermions as a single entity. It also seems to be deeply connected to the quantum theory of gravity. At the technical level, supersymmetry allows us to make many exact statements about quantum field theory that are not possible with ordinary field theories. There are several good reviews of SUSY field theory [152–156], but, if I may be allowed an opinion, as yet no comprehensive treatment of all modern developments.

Finite-temperature relativistic field theory is primarily applicable in cosmology and astrophysics, though the techniques are similar to those of the non-relativistic theory, which has wide applications in condensed-matter physics. Equilibrium calculations are quite similar to what we have discussed: to calculate averages in the canonical ensemble, instead of using the vacuum density matrix, one simply performs the Euclidean path integral on a space with one compactified dimension, whose length is the inverse temperature. Good reviews of this subject can be found in [157]. Non-equilibrium field theory is much more difficult, and has been the focus of a lot of recent work. As far as I know, there is no definitive modern summary, but the reader can consult [158].

Quantum field theory in curved space-time is relatively easy to define for space-times whose complexification has a real Euclidean section. Again, one simply performs the path integral on the appropriate Euclidean manifold, and analytically continues the answer. More general space-times with no time-like isometries can also be dealt with at a certain level of generality [159]. However, the most interesting thing about this subject is that it contains the seeds of its own demise, and shows us that we need to replace quantum field theory with a quantum theory of gravity. I am referring to the phenomenon of Hawking radiation. Within the context of field theory in curved space-time, it seems to imply that the formation and evaporation of black holes violates the unitary evolution postulate of quantum mechanics. It is only with the advent of string theory, the first real theory of quantum gravity, that we have begun to understand how this paradox is resolved.

Other subjects to which we have given short shrift are perturbative QCD [124] and weak interactions [160–161]. These subjects are nicely treated in the textbook by Peskin and Schroeder [33]. Lattice field theory has developed into a computer-intensive subfield. Good reviews of the subject can be found in [162]. Appendix A contains references to a number of books on the vast subject of quantum field theory in statistical physics. Finally, I want to mention the use of two-dimensional conformal field theory in statistical physics and string theory. This is a subject of great beauty and utility [163–164].

I hope you've come to the end of this trip with an appreciation of the beauties of quantum field theory and a hunger to know more about it. As you can see, there are lots of directions to follow from this point. The methods of field theory and the concept of renormalization have become so pervasive that it is probably beyond the capabilities of any single person to be an expert in the entire field. But every single avenue you can follow is interesting and exciting. Even those of you who prefer to explore the mysteries of quantum gravity, where the paradigms of field theory appear to fail, will continue to use many of the ideas and techniques of field theory. Indeed, we have found that, in many cases, the dynamics of gravity in some class of space-times is completely equivalent to that of a quantum field theory living on an auxiliary space [165–167].

Appendix A Books

This is a brief guide to other books on quantum field theory. The standard modern textbook is *An Introduction to Quantum Field Theory*, by Peskin and Schroeder [33]. I recommend especially their wonderful Chapter 5, and all of the calculational sections between 16.5 and 18.5, as well as Chapters 20 and 21. Every serious student of QFT should work out the final project on the Coleman–Weinberg potential, which can be found on page 469. Another standard is Weinberg's three-volume opus [131]. Here I recommend the marvelous sections on symmetries and anomalies in Volume II. The technical discussions of perturbative effective field theory are invaluable. The section on the Batalin–Vilkovisky treatment of general gauge equivalences is also useful. Volume I should probably be read *after* completing a first course on the subject. It presents an interesting but idiosyncratic approach to the logical structure of the field. Volume III on supersymmetry is full of gems. In my opinion, it is flawed by an idiosyncratic notation and a tendency to obscure relatively simple ideas in an attempt to give absolutely general discussions. Finally, let me mention a relatively new book by M. Srednicki [168]. I have not gone through it thoroughly, and I do not agree with the author's ordering of topics, but the pedagogical style of the sections I have read is wonderful. It is clear that everyone in the field will turn to this book for all those nasty little details about minus signs and spinor conventions. I think there is also a chance that it will replace Peskin and Schroeder as a standard textbook.

I have also enjoyed using the books by Bailin and Love [169] and Ramond [170] in my many years of teaching the subject. The books by Itzykson and Zuber [171] and Zinn-Justin [111] are more monographs/encyclopediae than textbooks, but they contain a wealth of detail on specific subjects in field theory that can be found nowhere else. I mention especially Zinn-Justin's discussions of the large-order behavior of perturbation theory, of the use of field theory for calculating critical exponents, and of instantons. Other treatments of the field theory/statistical physics interface can be found in the book by Drouffe and Itzykson [172], the marvelous book by Parisi [173], and the book by Ma [174]. Schwinger's source theory books [175] also belong in the category of non-textbooks, which contain scads of invaluable information about field theory.

Among older field-theory books, the second volume of Bjorken and Drell [176] contains lots of useful information, like explicit forms for the spectral representations for higher spin. The books of Nishijima [177] (Chapters 7 and 8), Bogoliubov and Shirkov [178], and even Schweber [179] will reward the really serious student of the subject.

Finally, I want to mention various shorter documents that I think are essential reading for students of quantum field theory. The most important is Kogut and

Wilson [126], still the best introduction I know of to Wilson's profound ideas about renormalization. Next is the 1975 Les Houches lecture-note volume *Methods in Field Theory* [180], every chapter of which is a gem. Coleman's book [138] is a collection of lectures on a variety of topics in field theory. I've drawn on it heavily for the material about instantons and solitons, but the other lectures are also worth reading. The contributions of Adler, Weinberg, Zimmermann, and Zumino to the 1970 Brandeis Summer School Lectures,[1] and of Weinberg to the 1964 Brandeis volume, are also worth reading [181]. Much of Weinberg's material reappears in his textbook [131]. Finally, let me mention the reprint volume *Selected Papers on Quantum Electrodynamics* [182] edited by Schwinger. The contributions of Feynman and Schwinger in particular should be read by every student of field theory.

There are lots of other books on field theory, and I apologize to those authors I haven't mentioned. I've emphasized those texts I've found most useful in my own career. Others will have different favorites.

---

[1] These lectures also contain a marvelous introduction to string theory by Mandelstam.

Appendix B **Cross sections**

Here we give the standard formulae for the differential cross section of a reaction in which two incoming particles produce an arbitrary final state

$$d\sigma = \frac{1}{2\omega_1\omega_2|v_1 - v_2|} \prod_f \frac{1}{(2\pi)^3} \frac{d^3 p_f}{2\omega(p_f)} |\mathcal{M}|^2 (2\pi)^4 \delta^4\left(P_1 + P_2 - \sum p_f\right).$$

$|v_1 - v_2|$ is the relative velocity of the two particles in the laboratory frame. Similarly the differential decay rate of an unstable particle is

$$d\Gamma = \frac{1}{2M} \prod_f \frac{1}{(2\pi)^3} \frac{d^3 p_f}{2\omega(p_f)} |\mathcal{M}|^2 (2\pi)^4 \delta^4\left(P - \sum p_f\right).$$

Appendix C Diracology

Here we collect a variety of identities for Dirac matrices and spinors. The basic commutation relations are

$$[\gamma^\mu, \gamma^\nu]_+ = 2\eta^{\mu\nu}. \tag{C.1}$$

The Weyl representation of these relations is

$$\gamma^\mu = \begin{pmatrix} 0 & \sigma^\mu \\ \bar{\sigma}^\mu & 0 \end{pmatrix}, \tag{C.2}$$

where $\sigma^\mu \equiv (1, \sigma)$, $\bar{\sigma}^\mu = (1, -\sigma)$, and σ is the usual 3-vector of Pauli matrices. In this representation

$$\gamma_5 \equiv i\gamma^0\gamma^1\gamma^2\gamma^3 = \begin{pmatrix} -1 & 0 \\ 0 & 1 \end{pmatrix}.$$

These matrices obviously satisfy

$$(\gamma^\mu)^\dagger = \gamma^0\gamma^\mu\gamma^0. \tag{C.3}$$

We will work only with representations related to the Weyl representation by unitary (rather than special linear) transformations, so this relation will always be true.

A convenient representation is the Dirac representation, in which

$$\gamma^0 = \begin{pmatrix} 1 & 0 \\ 0 & -1 \end{pmatrix}.$$

It is related to the Weyl representation by $\gamma_D^\mu = S_D^\dagger \gamma_W^\mu S_D$ with

$$S_D = \frac{1}{\sqrt{2}}(1 - i\sigma_2) \otimes 1.$$

The Majorana representation is related to the Dirac representation by $\gamma_M^\mu = S_M^\dagger \gamma_D^\mu S_M$ with

$$S_M = \frac{1}{\sqrt{2}}(\sigma_3 + \sigma_2) \otimes 1.$$

In the Majorana representation the Dirac equation is satisfied separately by the real and imaginary parts of the field.

The space of all 4×4 matrices is spanned by the anti-symmetrized products of Dirac matrices, $\gamma^{\mu_1 \cdots \mu_k}$, with $0 \leq k \leq 4$. But note the relations

$$\gamma^{\mu\nu\lambda\kappa} = \epsilon^{\mu\nu\lambda\kappa} \gamma_5,$$
$$\gamma^{\mu\nu\lambda} = \epsilon^{\mu\nu\lambda\kappa} \gamma_\kappa \gamma_5,$$

and

$$\gamma^{\mu\nu} = \frac{1}{2} \epsilon^{\mu\nu\lambda\kappa} \gamma_{\lambda\kappa} \gamma_5.$$

Calculations of spin-averaged cross sections and closed fermion loops lead to traces of products of Dirac matrices:

$$\mathrm{tr}(\gamma^{\mu_1} \ldots \gamma^{\mu_k}).$$

These can be calculated using anti-commutation relations and cyclicity of the trace, or by the tensor method. The latter comes from the observation that the result must be a numerical Lorentz tensor (and thus built from $\eta_{\mu\nu}$ and $\epsilon_{\mu\nu\lambda\kappa}$), which is invariant under cyclic permutation of the indices. Note also that the result is zero for k odd, because $\gamma_5^2 = 1$ and γ_5 anti-commutes with all the γ^μ. For example

$$\mathrm{tr}(\gamma^\mu \gamma^\nu) = A \eta^{\mu\nu},$$

and we compute $A = 4$ by hitting both sides with $\eta_{\mu\nu}$. A similar manipulation shows that

$$\mathrm{tr}(\gamma^\mu \gamma^\nu \gamma_5) = 0,$$

$$\mathrm{tr}(\gamma^\mu \gamma^\nu \gamma^\lambda \gamma^\kappa) = C(\eta^{\mu\nu}\eta^{\lambda\kappa} + \eta^{\mu\kappa}\eta^{\lambda\nu}) + D\eta^{\mu\lambda}\eta^{\nu\kappa} + B\epsilon^{\mu\nu\lambda\kappa}.$$

Taking all the indices different, the RHS is just B and the LHS is $\propto \mathrm{tr}\,\gamma_5 = 0$, so $B = 0$. On hitting both sides with $\eta_{\mu\nu}$, we get

$$5C + D = 16.$$

Doing the same with $\eta_{\mu\lambda}$ gives

$$\mathrm{tr}(\gamma^\mu \gamma^\nu \gamma_\mu \gamma^\kappa) = -8\eta^{\nu\kappa} = (2C + 4D)\eta^{\nu\kappa}.$$

Here we have used an identity

$$\gamma^\mu \gamma^\nu \gamma_\mu = -2\gamma^\nu,$$

which we will derive in a moment. We conclude that $C = -D = 4$. Similar manipulations allow us to calculate any trace with relative ease.

To evaluate $\gamma^\lambda \gamma_{\mu_1} \ldots \gamma_{\mu_k} \gamma_\lambda$, we proceed by induction, using the anti-commutation relations

$$\gamma^\lambda \gamma_\mu \gamma_\lambda = -4\gamma_\mu + 2\gamma_\mu = -2\gamma_\mu,$$

$$\gamma^\lambda \gamma_\mu \gamma_\nu \gamma_\lambda = -\gamma_\mu(-2\gamma_\nu) + 2\gamma_\nu \gamma_\mu = 4\eta_{\mu\nu},$$

$$\gamma^\lambda \gamma_\mu \gamma_\nu \gamma_\kappa \gamma_\lambda = -4\gamma_\mu \eta_{\nu\lambda} + 2\gamma_\nu \gamma_\kappa \gamma_\mu,$$

$$\gamma^\lambda \gamma_\mu \gamma_\nu \gamma_\kappa \gamma_\alpha \gamma_\lambda = \gamma_\mu(4\gamma_\nu \eta_{\kappa\alpha} - 2\gamma_\alpha \gamma_\kappa \gamma_\nu) + 2\gamma_\nu \gamma_\kappa \gamma_\alpha \gamma_\mu.$$

The reader is encouraged to continue computing these trace and contraction identities to higher orders.

The wave functions which convert fermion creation operators into local fields are solutions of the momentum-space Dirac equations

$$(\not{p} - m)u(p,s) = (\not{p} + m)v(p,s) = 0.$$

They satisfy the completeness relations

$$\sum_s u(p,s)\bar{u}(p,s) = (\not{p} + m),$$

$$\sum_s v(p,s)\bar{v}(p,s) = (\not{p} - m),$$

where the right-hand sides are proportional to the projection operators on solutions of the respective equations. Note that many books use a different normalization. We have followed the convention that, for fields of any spin, the Fourier transform is $(2\omega_p(2\pi)^3)^{-\frac{1}{2}}$ times a creation or annihilation operator, times a covariant wave function.

In the Weyl basis, the solutions have the form

$$u(p,s) = \begin{pmatrix} \sqrt{p \cdot \sigma}\, \chi(s) \\ \sqrt{p \cdot \bar{\sigma}}\, \chi(s) \end{pmatrix},$$

$$v(p,s) = \begin{pmatrix} \sqrt{p \cdot \sigma}\, \eta(s) \\ -\sqrt{p \cdot \bar{\sigma}}\, \eta(s) \end{pmatrix}.$$

Here χ and η are normalized two-component spinors corresponding to the choice of s.

In a convenient convention

$$\eta(s) = 2si\sigma_2 \chi^*(s),$$

where $s = \pm 1/2$ is either the component of spin in some fixed direction, $\hat{n}$, or the helicity.

Appendix D Feynman rules

In this section we list the Feynman rules in both Euclidean and Lorentzian signature, for all renormalizable interactions in four space-time dimensions.

D.1 Propagators

$$\frac{-\mathrm{i}\left(\eta_{\mu\nu} - \dfrac{p_\mu p_\nu (1-\kappa)}{p^2 - \kappa m_A^2}\right)}{p^2 - m_A^2 + \mathrm{i}\epsilon} \qquad \frac{\delta_{\mu\nu} - \dfrac{p_\mu p_\nu (1-\kappa)}{p^2 + m_A^2}}{p^2 + m_A^2}$$

D.1.1 Gauge boson propagator in R$_\kappa$ gauge

m_A^2 is the matrix $v^{\mathrm{T}} T^a T^b v$ in the adjoint representation of the gauge group, and v is the expectation value of the Higgs fields. For gluons in QCD, we have $m_A^2 = 0$ and we denote gluon lines by

Scalars are denoted by dashed lines (with arrows on them if the scalars are complex).

$$\frac{\mathrm{i}}{p^2 - m_S^2 + \mathrm{i}\epsilon} \qquad \frac{1}{p^2 + m_S^2}$$

D.1.2 Scalar propagator in R$_\kappa$ gauge

The scalar mass-squared matrix is

$$m_S^2 = \kappa T^a v (T^a v)^{\mathrm{T}} + m_H^2,$$

where the first term is a sum of dyadics and operates in the subspace of would-be NGBs, while the second term operates in the orthogonal subspace of physical Higgs bosons. Everything is written in terms of real scalar fields.

$$\frac{i}{p^2 - \kappa m_V^2 + i\epsilon} \qquad \frac{1}{p^2 + \kappa m_V^2}$$

D.1.3 Ghost propagator in R_κ gauge

The ghosts are complex scalars in the adjoint representation. Closed ghost loops add an extra minus sign.

The propagator for Dirac fermions is

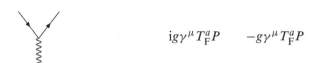

$$\frac{i}{\not{p} - m_F + i\epsilon} \qquad \frac{1}{\not{p} - im_F}$$

D.1.4 Dirac fermion propagator

The fermion mass matrix can be an arbitrary combination of 1 and γ_5, with coefficients that are complex matrices in internal index space. The momentum in the propagator goes in the direction of the particle number, which for charged particles is minus the electric charge. Unpaired Weyl fermions are treated by using a Dirac field with couplings such that the right-handed components are free. Closed fermion loops get an extra minus sign.

D.2 Vertices

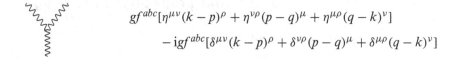

$$ig\gamma^\mu T_F^a P \qquad -g\gamma^\mu T_F^a P$$

D.2.1 Fermion gauge vertex

The projection operator P is either the unit matrix, or the projector on left-handed Weyl spinors.

$$gf^{abc}[\eta^{\mu\nu}(k-p)^\rho + \eta^{\nu\rho}(p-q)^\mu + \eta^{\mu\rho}(q-k)^\nu]$$
$$- igf^{abc}[\delta^{\mu\nu}(k-p)^\rho + \delta^{\nu\rho}(p-q)^\mu + \delta^{\mu\rho}(q-k)^\nu]$$

D.2.2 Three-gauge-boson vertex

All three momenta are ingoing. The second line is the Euclidean vertex.

$$gV_{\mu\nu\rho\sigma}^{abcd}$$

D.2.3 Four-gauge-boson vertex

The Euclidean vertex has $-\mathrm{i} \to 1$ and $\eta^{\mu\nu} \to \delta^{\mu\nu}$.

$$V_{\mu\nu\rho\sigma}^{abcd} = -\mathrm{i}g^2\Big[\, f^{abe}f^{cde}(\eta^{\mu\rho}\eta^{\nu\sigma} - \eta^{\nu\rho}\eta^{\mu\sigma})$$
$$+ f^{ace}f^{bde}(\eta^{\mu\nu}\eta^{\rho\sigma} - \eta^{\nu\rho}\eta^{\mu\sigma})$$
$$+ f^{ade}f^{bce}(\eta^{\mu\nu}\eta^{\rho\sigma} - \eta^{\mu\rho}\eta^{\nu\sigma})\,\Big]$$

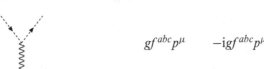

$$gf^{abc}p^{\mu} \qquad -\mathrm{i}gf^{abc}p^{\mu}$$

D.2.4 Ghost gauge vertex

$$-\mathrm{i}gM \qquad gM$$

D.2.5 Yukawa vertex

For each scalar field, the matrix M is a combination of 1 and γ_5, with coefficients that are matrices in the space of fermion internal indices.

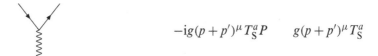

$$-\mathrm{i}g(p+p')^{\mu}T_{\mathrm{S}}^{a}P \qquad g(p+p')^{\mu}T_{\mathrm{S}}^{a}$$

D.2.6 Scalar gauge vertex

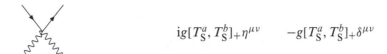

$$\mathrm{i}g[T_{\mathrm{S}}^{a}, T_{\mathrm{S}}^{b}]_{+}\eta^{\mu\nu} \qquad -g[T_{\mathrm{S}}^{a}, T_{\mathrm{S}}^{b}]_{+}\delta^{\mu\nu}$$

D.2.7 Scalar–scalar gauge vertex

For purely scalar interactions one generally includes a $1/k!$ in the Lagrangian, for a term that contains k identical scalar fields. The vertices are then ig in Lorentzian signature and $-g$ in Euclidean signature, where g is the coefficient of the inverse factorials. Graphs are counted by their symmetry numbers. To compute symmetry numbers one either figures out the order of the group of geometrical symmetries of the graph, or goes through the combinatorics of Wick's theorem. See D.4.

In our computation of Wilson loops, we have described Feynman rules for the coupling to static non-abelian sources.

D.3 External lines

Here we give the rules for invariant amplitudes. Refer to Appendix B for the additional factors for S-matrix elements and cross sections. The external lines in a diagram look exactly like the internal lines except that they have one free end. For scalars the rule that goes with an external line is just a 1. For vectors, we have a factor of $\epsilon^\mu(k)$ for incoming particles and $(\epsilon^\mu(k))^*$ for outgoing particles. For massive particles with momentum in the 3-direction, the (linear) polarization vectors are

$$\epsilon^\mu_{1,2}(k) = (0, \hat{e}_{1,2}, 0),$$

where $\hat{e}_{1,2}$ are two-dimensional unit vectors in the 1- and 2-directions. Helicity (circular polarization) states are the complex linear combinations with $2^{-\frac{1}{2}}(\hat{e}_1 \pm \hat{e}_1)$. The other polarization is called longitudinal and has the form

$$\epsilon^\mu_3 = \left(\frac{p_3}{m}, 0, 0, \frac{-\omega_p}{m} \right).$$

If we rotate the momentum to another direction, each of the three polarizations rotates like a vector. We do not rotate the polarization indices, because we keep the spin quantization axis along the 3-direction.

For external Dirac fermions we have the rules

$$u(p, s) - \text{incoming fermion},$$

$$\bar{u}(p, s) - \text{outgoing fermion},$$

$$v(p, s) - \text{outgoing anti-fermion},$$

$$\bar{v}(p, s) - \text{incoming anti-fermion}.$$

The spinor solutions of the momentum-space Dirac equation were discussed in Appendix C.

D.4 Combinatorics

Here are a few examples of Feynman graphs in a theory with $\mathcal{L}_I = (\lambda/4!)\phi^4$, along with the symmetry numbers by which they are multiplied. The denominators in the combinatoric factors are a $1/2$ from the exponential in perturbation theory, and $1/4!$ from each vertex, coming from the definition of the coupling. The numerators count the number of ways a given diagram is obtained from Wick's theorem, written in a way that makes the counting clear.

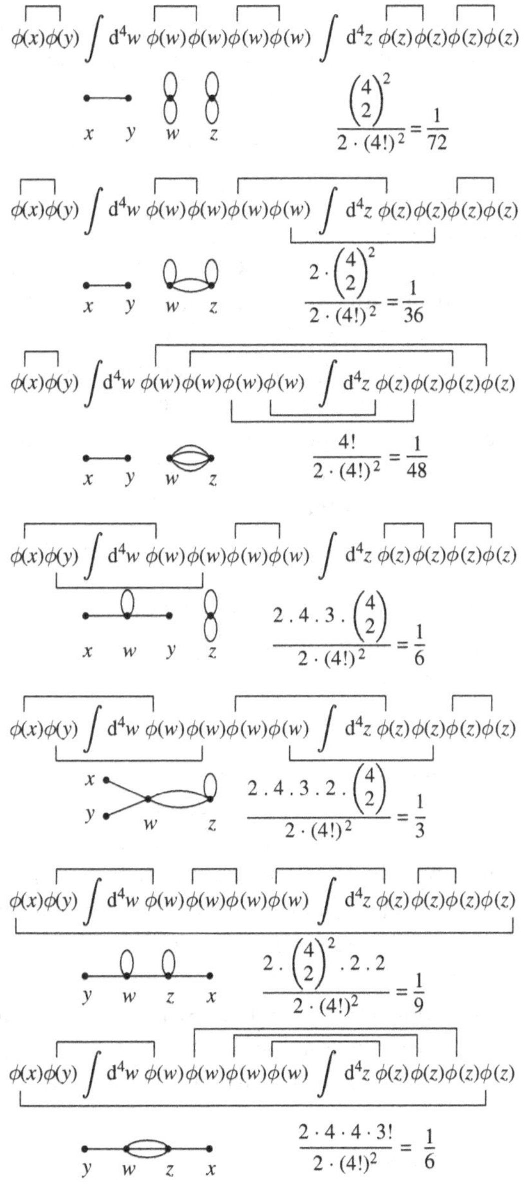

$$\phi(x)\phi(y) \int \mathrm{d}^4w\; \phi(w)\phi(w)\phi(w)\phi(w) \int \mathrm{d}^4z\; \phi(z)\phi(z)\phi(z)\phi(z)$$

$$\frac{\binom{4}{2}^2}{2 \cdot (4!)^2} = \frac{1}{72}$$

$$\phi(x)\phi(y) \int \mathrm{d}^4w\; \phi(w)\phi(w)\phi(w)\phi(w) \int \mathrm{d}^4z\; \phi(z)\phi(z)\phi(z)\phi(z)$$

$$\frac{2 \cdot \binom{4}{2}^2}{2 \cdot (4!)^2} = \frac{1}{36}$$

$$\phi(x)\phi(y) \int \mathrm{d}^4w\; \phi(w)\phi(w)\phi(w)\phi(w) \int \mathrm{d}^4z\; \phi(z)\phi(z)\phi(z)\phi(z)$$

$$\frac{4!}{2 \cdot (4!)^2} = \frac{1}{48}$$

$$\phi(x)\phi(y) \int \mathrm{d}^4w\; \phi(w)\phi(w)\phi(w)\phi(w) \int \mathrm{d}^4z\; \phi(z)\phi(z)\phi(z)\phi(z)$$

$$\frac{2 \cdot 4 \cdot 3 \cdot \binom{4}{2}}{2 \cdot (4!)^2} = \frac{1}{6}$$

$$\phi(x)\phi(y) \int \mathrm{d}^4w\; \phi(w)\phi(w)\phi(w)\phi(w) \int \mathrm{d}^4z\; \phi(z)\phi(z)\phi(z)\phi(z)$$

$$\frac{2 \cdot 4 \cdot 3 \cdot 2 \cdot \binom{4}{2}}{2 \cdot (4!)^2} = \frac{1}{3}$$

$$\phi(x)\phi(y) \int \mathrm{d}^4w\; \phi(w)\phi(w)\phi(w)\phi(w) \int \mathrm{d}^4z\; \phi(z)\phi(z)\phi(z)\phi(z)$$

$$\frac{2 \cdot \binom{4}{2}^2 \cdot 2 \cdot 2}{2 \cdot (4!)^2} = \frac{1}{9}$$

$$\phi(x)\phi(y) \int \mathrm{d}^4w\; \phi(w)\phi(w)\phi(w)\phi(w) \int \mathrm{d}^4z\; \phi(z)\phi(z)\phi(z)\phi(z)$$

$$\frac{2 \cdot 4 \cdot 4 \cdot 3!}{2 \cdot (4!)^2} = \frac{1}{6}$$

Fig. D.1.   Combinatoric factors for some two-loop diagrams.

For physicists, a Lie group is a group whose elements depend smoothly on a finite number of parameters ω_a. Mathematically this means that the group is also a smooth differentiable manifold and that the action of the group on itself (by left or right multiplication) is a diffeomorphism. Much of the structure of Lie groups can be extracted by studying their behavior near the identity element, where we write[1]

$$U(\omega) \approx 1 + \mathrm{i}\omega_a T^a.$$

A representation of the group is a homomorphism of the group into the group of invertible linear transformations on a vector space. The representation is faithful if the only group element represented by the operator 1 is the identity of the group. In physics we deal with two kinds of non-faithful representation. The first is the trivial, one-dimensional representation, where every element is mapped into the identity. The second is exemplified by the three-dimensional vector representation of SU(2), where the Z_2 center of the group is represented by the identity. In this case, no transformation near the identity is mapped to the identity, so the representation is locally faithful.

In a faithful representation, the requirement that the group multiplication close is equivalent to the fact that the matrices T^a form a Lie algebra:

$$[T^a, T^b] = \mathrm{i} f_c^{ab} T^c.$$

The constants f_c^{ab} are called the structure constants of the group. The Jacobi identity for double commutators constrains these constants. The constraint is equivalent to the statement that the matrices

$$(T_A^a)_c^b \equiv -\mathrm{i} f_c^{ab}$$

form a representation of the same Lie algebra. This is called the adjoint representation.

A Lie group is semi-simple if it contains no abelian invariant subgroup. An example of a Lie group that is not semi-simple is the Poincaré group. Translations are the abelian invariant subgroup. We know how to construct induced representations of this group, starting from representations of the abelian subgroup, and we have used this technique to construct the states of free particles, in the text. Mathematicians classify groups that are direct products of $U(1)^k$ and a semi-simple group as non-semi-simple, but in physics it is convenient to lump these into the semi-simple category. Every semi-simple group can be written as a direct product of simple factors $G_1 \otimes G_2 \otimes \ldots \otimes G_n$. A simple group is one that cannot be further factored in this way.

---

[1] We use conventions suited for thinking about unitary representations of the group.

A final step in classification is to distinguish between compact and non-compact groups. The reference is to the range of the parameters, or what mathematicians call the group manifold. In physics we deal with the non-compact Poincaré group and sometimes its completion to the conformal group $SO(2, d)$. Apart from that, all of the Lie groups which appear in field theory are compact. It is fortunate that Cartan was able to classify all compact simple groups. They fall into three infinite families (the rotation, unitary, and unitary symplectic groups in n real, n complex, and $2n$ real dimensions) plus five exceptional groups G_2, F_4, $E_{6,7,8}$. The subscripts on these names refer to the rank of the group, the maximal number of commuting generators. The rank of $SU(N)$ is $N-1$, the number of traceless diagonal Hermitian matrices. That of $SO(n)$ is determined by the maximal number of orthogonal 2-planes in n-dimensional space, which is k for $n = 2k$ or $2k+1$. The rank of $Sp(2k)$ (often called $Sp(k)$) is k.

Choose any maximal commuting set of generators. This is called a Cartan subalgebra, and the generators are called Cartan generators H_i, $i = 1, \ldots, r$. The eigenvalues of the Cartan generators in a representation R of the group form a set of r-dimensional vectors called the *weights* of the representation. The lattice formed by adding together weights of all possible representations is called the weight lattice. For the special case of the adjoint representation, the weights are called roots. The root lattice is obtained by adding together all possible roots.

As noted above, a Lie group is called compact if its parameter space is compact as a manifold. For example, $SU(2)$ is the set of matrices $n_4 + i n_a \sigma_a$, with $n_4^2 + n^2 = 1$ and all components real. This is the same as the sphere S^3 in four-dimensional Euclidean space, which is compact. The group of all $N \times N$ unitary matrices is compact for any N. As a manifold it is equivalent to N complex, mutually orthogonal unit vectors (its columns), with some identifications under permutations. Any Lie group that has a faithful finite-dimensional unitary representation is thus compact.

If R is any such representation of a compact group G, and T_R^a the generators in that representation, then $\text{tr}(T_R^a T_R^b) = D(R)\delta^{ab}$. The trace defines a scalar product in the space of Hermitian generators, and we can always define orthogonal linear combinations. The coefficient $D(R)$ is called the Dynkin index of the representation. If we want the generators in any representation to have the same commutation relations, then the Dynkin indices will differ. We usually normalize the Dynkin indices by

$$\text{tr}(T_F^a T_F^b) = \frac{1}{2}\delta^{ab},$$

in the smallest representation of the group, often called the *defining* or *fundamental* representation (though representations corresponding to these words are not always identical). This also normalizes the structure constants, via

$$\text{tr}[T_F^a, T_F^b]T_F^c = \frac{i}{2}f^{abc}.$$

This shows that for compact groups the structure constants are totally anti-symmetric. In any other representation

$$\text{tr}[T_R^a, T_R^b]T_R^c = iD(R)f^{abc}.$$

If we replace the commutator by an anti-commutator in the last formula we get

$$\mathrm{tr}[T_{\mathrm{R}}^a, T_{\mathrm{R}}^b]_+ T_{\mathrm{R}}^c = \mathrm{i}A(\mathrm{R})d^{abc}.$$

$A(\mathrm{R})$ is called the anomaly coefficient of the representation and appears in the ABJS anomaly equations, which are discussed extensively in the text. There is a unique totally symmetric invariant in the product of three adjoint representations of a compact simple group, so d^{abc} is representation-independent. The same fact tells us that it is sufficient to compute the trace of T^3 for any generator for which it is non-zero in order to compute $A(\mathrm{R})$.

The proof of the uniqueness of d^{abc} follows from the theory of tensor products of representations. Given two representations with representation matrices $M_1(g)$ and $M_2(g)$, we can make a new one by taking the tensor product $M_1 \otimes M_2(g)$. This representation is called $R_1 \otimes R_2$. The Peter–Weyl theorem tells us that every finite-dimensional representation is a direct sum of irreducible representations, so

$$R_1 \otimes R_2 = \bigoplus R_k.$$

The theory of which representations appear in this sum is the generalization of the Clebsch–Gordan series for angular momentum. It is most easily worked out in terms of properties of the weights of the representations. The product of two adjoint representations always has another adjoint in it, because of the structure constants. That is, if V_a and W_a both transform like adjoints, so does $f_{abc}V_b W_c$. The existence of d^{abc} implies a second copy of the adjoint representation in the product of two adjoints. For SU(N) its existence is guaranteed by the fact that the representation of the generators in the fundamental representation is a basis for all traceless Hermitian matrices. Thus

$$T^a T^b = \frac{1}{2N}\delta^{ab} + \mathrm{i}f^{abc}T^c + d^{abc}T^c,$$

since the commutator is anti-Hermitian and the anti-commutator Hermitian. It turns out that, for all other compact simple Lie algebras, there is only one adjoint in the product of two adjoints, so the d^{abc} symbol vanishes.

A *direct product* of two groups $\mathrm{G}_{1,2}$ is the group formed by pairs (g_1, g_2) and the obvious multiplication law. In a matrix representation the direct product matrices are tensor products acting on independent indices in the representation space. The Lie algebra of a direct product is a direct sum of Lie algebras, and the generators have the form

$$1 \otimes t^a \omega_a + T^I \Omega_I \otimes 1,$$

where the lower-case and capital letters refer to the generators and infinitesimal parameters of the two different groups.

A *semi-direct product* group G has two subgroups $\mathrm{G}_{1,2}$ that do not commute with each other. Conjugation of one group by an element of the other $\mathrm{G}_1 \to g_2^{-1}\mathrm{G}_1 g_2$ (and vice versa) leads to a different, though isomorphic, subgroup. The canonical geometrical example is the group of rotations and translations, or Lorentz transformations and translations, in a Euclidean or Lorentzian signature space. The commutation relations are

$$[J_{\mu\nu}, J_{\rho\sigma}] = i[\eta_{\mu\rho}J_{\nu\sigma} - \eta_{\nu\rho}J_{\mu\sigma} - \eta_{\mu\sigma}J_{\nu\rho} + \eta_{\nu\sigma}J_{\mu\rho}],$$
$$[J_{\mu\nu}, P_{\rho}] = i(\eta_{\mu\rho}P_{\nu} - \eta_{\nu\rho}P_{\mu}),$$
$$[P_{\mu}, P_{\nu}] = 0.$$

In words, these equations tell us that P transforms as a vector, and J as a tensor of second rank under rotations (Lorentz transformations). The second equation also tells us how angular momentum and boosts respond to translations. These equations are valid in any dimension and for any signature of the metric $\eta_{\mu\nu}$.

Further material on Lie groups and algebras can be found in [1–5].

In addition to the big omissions I mentioned in the text, there is a host of subjects in field theory that I have neglected. I will mention them here, without detailed reference. In the days of Spires and Google detailed reference hardly seems necessary: all you need is a name. The large-N approximation, particularly the connection between large-N matrix field theories and string theory, is a big lacuna. I've barely mentioned the field of computational lattice gauge theory, which is slowly achieving a quantitative understanding of the hadron spectrum and other low-energy properties of QCD. The use of field theory techniques in condensed-matter physics produced the theory of superconductivity, of critical phenomena, and of the quantum Hall effect as well as a variety of other phenomena. The books by Parisi, Ma, Drouffe and Itzykson, and Zinn-Justin provide an entrée into this vast body of knowledge. Another big lacuna is the study of integrable two-dimensional theories and exact S-matrices. There has been a variety of attempts to study the high-energy fixed-momentum-transfer region of scattering amplitudes (the Regge region) by summing selected classes of diagrams in field theory. This has led to an effective field theory approach called Reggeon calculus as well as to a very interesting set of equations called the BFKL equations (this is not quite the Regge region). Then there is the use of two-dimensional conformal field theory to do perturbative string theory. Polchinski's volumes and the new book by the Beckers and Schwarz have excellent discussions of that. The next thing that comes to mind is heavy-quark effective field theory. The use of field theory as an exact formulation of quantum gravity (matrix theory and AdS/CFT) has been referred to only in passing, so I reiterate it here. The study of field theory in dimensions higher than four has mostly been done in connection with the Kaluza–Klein program, supergravity and string theory. Another topic I've neglected is what is colloquially known as "light-cone quantization" (it's actually *light-front quantization*). This is a fascinating approach that should make field theory amenable to the tricks of condensed-matter physicists, but it has not had as much impact as one would have thought. An old topic, which has fallen out of fashion, but still has useful lessons, is *dispersion theory*, which tries to extract general properties of scattering amplitudes from minimal principles of unitarity, crossing symmetry, Lorentz invariance, and a somewhat vaguely defined notion of analyticity. Axiomatic and constructive and algebraic quantum field theory are more rigorous approaches, but have not produced results significant to the wider community of field theorists for a long time. There are repeated attempts to use the techniques of field theory, in particular the renormalization group, to study turbulence. This project has not really made a breakthrough yet, but many people expect one. Finally let me note that path integrals, instantons, and lattice gauge theory have made their appearance in fields like neurobiology, quantum computing, and finance. Field theory is a general tool for studying complicated systems with many degrees of freedom, so it will almost certainly find future applications that we cannot yet dream of.

References

[1] H. Georgi (1999), *Lie Algebras in Particle Physics*, 2nd edn. (New York, Perseus Books).

[2] R. Gilmore (1974), *Lie Groups, Lie Algebras, and Some of Their Applications* (New York, Wiley).

[3] H. J. Lipkin (1966), *Lie Groups for Pedestrians*, 2nd edn. (Amsterdam, North Holland).

[4] R. N. Cahn (1984), *Semi-simple Lie Algebras and Their Representations* (New York, Benjamin-Cummings).

[5] Particle Data Group (W.-M. Yao *et al.*) (2006), *J. Phys. G: Nucl. Part. Phys.* **33**, 1

[6] W. Pauli (1940), *Phys. Rev.* **91**, 716.

[7] S. Weinberg (1966), *The Quantum Theory of Fields*, Vol. I (Cambridge, Cambridge University Press), Section 5.7.

[8] G. Lüders and B. Zumino (1958), *Phys. Rev.* **110**, 1450.

[9] J. Schwinger (1951), *Proc. Nat. Acad. Sci.* **37**, 452.

[10] F. J. Dyson (1949), *Phys. Rev.* **75**, 1736.

[11] R. F. Streater and A. S. Wightman (1964), *PCT, Spin Statistics and All That* (New York, Benjamin).

[12] B. Simon (1974), *The $P(\phi)_2$ Euclidean Quantum Field Theory* (Princeton, MA, Princeton University Press).

[13] G. 't Hooft and M. Veltman (1973), CERN preprint-73-09.

[14] G. Källen (1953), *On the Magnitude of the Renormalization Constants in Quantum Electrodynamics, Kongelige Danske Vidensk. Selskab.*, Vol. 27, no. 12.

[15] H. Lehmann (1954), *Nuovo Cim.* **11**, 342.

[16] S. Weinberg (1996), *The Quantum Theory of Fields*, Vol. I (Cambridge, Cambridge University Press), Chapter 4.

[17] R. F. Streater and A. S. Wightman (1964), *PCT, Spin Statistics and All That* (New York, Benjamin).

[18] M. Goldberger and K. L. Watson, *Collision Theory* (New York, Dover Press).

[19] R. G. Newton, *Scattering Theory of Waves and Particles* (New York, Dover Press).

[20] H. Lehmann, K. Symanzik, and W. Zimmermann (1955), *Nuovo Cim.* **1**, 205.

[21] R. Haag (1958), *Phys. Rev.* **112**, 669.

[22] D. Ruelle (1962), *Helv. Phys. Acta.* **35**, 34.

[23] R. Eden, P. Landshoff, D. Olive, and J. C. Polkinghorne (1966), *The Analytic S-Matrix* (Cambridge, Cambridge University Press).

[24] A. Proca (1936), *Comptes Rendus Acad. Sci. Paris*, **202**, 1490.

[25] G. 't Hooft (1976), *Phys. Rev.* D **14**, 3432 [Erratum *Phys. Rev.* D **18**, 2199 (1978)].

[26] J. C. Pati and A. Salam (1974), *Phys. Rev.* D **10**, 275 [Erratum *Phys. Rev.* D **11**, 703 (1975)].

[27] R. N. Mohapatra and J. C. Pati (1975), *Phys. Rev.* D **11**, 566.

[28] R. N. Mohapatra and J. C. Pati (1975), *Phys. Rev.* D **11**, 2558.

[29] G. Senjanovic and R. N. Mohapatra (1975), *Phys. Rev.* D **12**, 1502.

[30] I. I. Bigi and A. I. Sanda (2000), *CP Violation* (Cambridge, Cambridge University Press).

[31] T. Takagi (1927), *Jap. J. Math.* **1**, 83.

[32] E. C. Poggio, H. R. Quinn, and S. Weinberg (1976), *Phys. Rev.* D **13**, 1958.

[33] M. Peskin and D. Schroeder (1995), *An Introduction to Quantum Field Theory* (New York, Addison Wesley).

[34] T. Kinoshita and D. R. Yennie (1990), *Adv. Ser. Direct. High Energy Phys.* **7**, 1.

[35] P. P. Kulish and L. D. Faddeev (1970), *Theor. Math. Phys.*, **4**, 745.

[36] Y. Nambu (1960), *Phys. Rev. Lett.* **4**, 380.

[37] Y. Nambu and G. Jona-Lasinio (1961), *Phys. Rev.* **122**, 345.

[38] Y. Nambu and G. Jona-Lasinio (1961), *Phys. Rev.* **124**, 246.

[39] J. Goldstone (1961), *Nuovo Cim.* **19**, 154.

[40] J. Goldstone, A. Salam, and S. Weinberg (1962), *Phys. Rev.* **127**, 965.

[41] S. L. Adler (1965), *Phys. Rev.* B **137**, 1022.

[42] S. Weinberg (1996), *Quantum Theory of Fields*, Vol. II (Cambridge, Cambridge University Press), Chapter 19.

[43] J. Schwinger (1951), *Phys. Rev.* **82**, 664.

[44] S. Mandelstam (1962), *Annals Phys.* **19**, 1.

[45] K. G. Wilson (1974), *Phys. Rev.* D **10**, 2445.

[46] C. Becchi, A. Rouet, and R. Stora (1976), *Annals Phys.* **98**, 287.

[47] S. Weinberg (1996), *Quantum Theory of Fields*, Vol. II (Cambridge, Cambridge University Press), Chapter 15.

[48] M. Henneaux (1988), *Classical Foundations of BRST Invariance* (Naples, Bibliopolis).

[49] M. Henneaux (1990), *Nucl. Phys. Proc. Suppl.* **18A**, 47.

[50] A. Fuster, M. Henneaux, and A. Maas (2005), *Int. J. Geom. Meth. Mod. Phys.* **2**, 939 (arXiv:hep-th/0506098).

[51] R. P. Feynman (1963), *Acta Phys. Polon.* **24**, 697.

[52] B. S. DeWitt (1962), *J. Math. Phys.* **3**, 1073.

[53] S. Mandelstam (1968), *Phys. Rev.* **175**, 1604.

[54] L. D. Fadeev and V. N. Popov (1967), *Phys. Lett.* B **25**, 29.

[55] J. L. Rosner (2003), *Am. J. Phys.* **71**, 302 (arXiv:hep-ph/0206176).

[56] E. Cartan (1913), *Bull. Soc. Math. France* **41**, 53.

[57] H. Weyl (1921), *Raum–Zeit–Materie* (Berlin, Springer-Verlag); English translation *Space–Time–Matter* (New York, Dover, 1950).

[58] O. Klein (1939), "On the theory of charged fields," in *New Theories in Physics* (Paris, International Institute of Intellectual Collaboration), pp. 73–93.

[59] D. Gross (1994), *Oskar Klein and Gauge Theory*, arXiv:hep-th/9411233.

[60] T. D. Lee and C. N. Yang (1955), *Phys. Rev.* **98**, 101.

[61] S. Gerstein and Ya. B. Zel'dovitch (1955), *JETP* **29**, 698.

[62] R. Marshak and E. C. G. Sudarshan (1958), *Phys. Rev.* **109**, 1860.

[63] R. P. Feynman and M. Gell-Mann (1958), *Phys. Rev.* **109**, 193.

[64] C. N. Yang and R. L. Mills (1954), *Phys. Rev.* **96**, 191.

[65] S. Bludman (1958), *Nuovo Cim.* **9**, 433.

[66] J. Schwinger (1957), *Annals Phys.* **2**, 407.

[67] S. Glashow (1961), *Nucl. Phys.* **22**, 579.

[68] J. Schwinger (1962), *Phys. Rev.* **128**, 2425.

[69] P. W. Anderson (1963), *Phys. Rev.* **130**, 439.

[70] P. W. Higgs (1964), *Phys. Lett.* **12**, 132.

[71] P. W. Higgs (1964), *Phys. Rev. Lett.* **13**, 508.

[72] P. W. Higgs (1966), *Phys. Rev.* **145**, 1156.

[73] R. Brout and F. Englert (1964), *Phys. Rev. Lett.* **13**, 321.

[74] G. Guralnik, C. R. Hagen, and T. W. B. Kibble (1967), *Phys. Rev.* **155**, 1554.

[75] S. Weinberg (1967), *Phys. Rev. Lett.* **19**, 1264.

[76] A. Salam (1968), in *Elementary Particle Physics*, ed. N. Svartholm (Stockholm, Almquist and Wiksells).

[77] G. 't Hooft (1971), *Nucl. Phys.* B **35**, 167.

[78] G. 't Hooft and M. J. G. Veltman (1972), *Nucl. Phys.* B **44**, 189.

[79] Y. Nambu (1966), in *Preludes in Theoretical Physics*, ed. A. de Shalit, H. Feshbach, and L. van Hove (Amsterdam, North Holland), p. 133.

[80] O. W. Greenberg (1964), *Phys. Rev. Lett.* **13**, 598.

[81] Han Nambu (1965), *Phys. Rev.* B **139**, 1006.

[82] E. B. Bogomol'nyi (1976), *Sov. J. Nucl. Phys.* **24**, 449.

[83] M. K. Prasad and C. M. Sommerfield (1975), *Phys. Rev. Lett.* **35**, 760.

[84] H. B. Nielsen and P. Olesen (1973), *Nucl. Phys.* B **61**, 45.

[85] A. A. Abrikosov (1957), *Sov. Phys. JETP* **5**, 1174 [*Zh. Éksp. Teor. Fiz.* **32**, 1442 (1957)].

[86] K. G. Wilson (1974), *Phys. Rev.* D **10**, 2445.

[87] G. 't Hooft (1976), in *High Energy Physics: Proceedings of the EPS International Conference, Palermo, June, 1975*, ed. A. Zichichi (Bologna, Editrice Compositori).

[88] S. Mandelstam (1976), in *Extended Systems in Field Theory*, ed. J. L. Gervais and A. Neveu *Phys. Rep.* C **23**, No. 3.

[89] T. Banks, R. Myerson, and J. B. Kogut (1977), *Nucl. Phys.* B **129**, 493.

[90] M. E. Peskin (1978), *Annals Phys.* **113**, 122.

[91] S. Elitzur, R. B. Pearson, and J. Shigemitsu (1979), *Phys. Rev.* D **19**, 3698.

[92] A. Ukawa, P. Windey, and A. H. Guth (1980), *Phys. Rev.* D **21**, 1013.

[93] S. L. Glashow, J. Iliopoulos, and L. Maiani (1970), *Phys. Rev.* D **2**, 1285.

[94] N. Cabibbo (1963), *Phys. Rev. Lett.* **10**, 531.

[95] M. Kobayashi and T. Maskawa (1973), *Prog. Theor. Phys.* **49**, 652.

[96] C. D. Froggatt and H. B. Nielsen (1979), *Nucl. Phys.* B **147**, 277.

[97] T. Banks, Y. Nir, and N. Seiberg, arXiv:hep-ph/9403203.

[98] C. Bernard, C. DeTar, L. Levkova *et al.* (2006), in *Proceedings of the 5th International Workshop on Chiral Dynamics, Theory and Experiment* (CD) (Durham/Chapel Hill, NC) (arXiv:hep-lat/0611024).

[99] C. Vafa and E. Witten (1984), *Nucl. Phys.* B **234**, 173.

[100] J. Steinberger (1949), *Phys. Rev.* **76**, 1180.

[101] J. Schwinger (1951), *Phys. Rev.* **82**, 664.

[102] J. S. Bell and R. Jackiw (1969), *Nuovo Cim.* A **60**, 47.

[103] S. Adler (1969), *Phys. Rev.* **177**, 2426.

[104] W. Bardeen (1969), *Phys. Rev.* **184**, 1848.

[105] S. L. Adler and W. Bardeen (1969), *Phys. Rev.* **182**, 1517.

[106] J. Wess and B. Zumino (1971), *Phys. Lett.* B **37**, 95.

[107] E. Witten (1982), *Phys. Lett.* B **117**, 324.

[108] S. Weinberg and E. Witten (1980), *Phys. Lett.* B **96**, 59.

[109] Y. Frishman, A. Schwimmer, T. Banks, and S. Yankielowicz (1981), *Nucl. Phys.* B **177**, 157.

[110] S. R. Coleman and B. Grossman (1982), *Nucl. Phys.* B **203**, 205.

[111] J. Zinn-Justin (1993), *Quantum Field Theory and Critical Phenomena* (Oxford, Oxford University Press).

[112] S. Weinberg (1970), "Dynamic and algebraic symmetries," in *Lectures on Elementary Particles and Quantum Field Theory*, ed. H. Pendelton and M. Grisaru (Cambridge, MA, MIT Press), p. 283.

[113] I. B. Khriplovich (1969), *Yad. Fiz.* **10**, 409.

[114] D. J. Gross and F. Wilczek (1973), *Phys. Rev. Lett.* **30**, 1343.

[115] D. J. Gross and F. Wilczek (1973), *Phys. Rev.* D **8**, 3633.

[116] D. J. Gross and F. Wilczek (1974), *Phys. Rev.* D **9**, 980.

[117] H. D. Politzer (1974), *Phys. Rep.* **14**, 129.

[118] H. D. Politzer (1973), *Phys. Rev. Lett.* **30**, 1346.

[119] N. Arkani-Hamed, A. G. Cohen, E. Katz, A. E. Nelson, T. Gregoire, and J. G. Wacker (2002), *JHEP* **0208**, 021 (arXiv:hep-ph/0206020).

[120] A. Zee (1973), *Phys. Rev.* D **7**, 3630.

[121] A. Zee (1973), *Phys. Rev.* D **8**, 4038.

[122] S. R. Coleman and D. J. Gross (1973), *Phys. Rev. Lett.* **31**, 851.

[123] E. Farhi and L. Susskind (1981), *Phys. Rep.* **74**, 277.

[124] Yu. L. Dokhitzer, V. A. Khoze, A. H. Mueller, and S. I. Troyan (1991), *Basics of Perturbative QCD* (Paris, Editions Frontiéres).

[125] R. Jackiw (1974), *Phys. Rev.* D **9**, 1686.

[126] J. Kogut and K. Wilson (1974), *Phys. Rep.* **12**, 75–200.

[127] A. D. Linde (1976), *JETP Lett.* **23**, 64 [*Pis'ma Zh. Éksp. Teor. Fiz.* **23**, 73 (1976)].

[128] S. Weinberg (1976), *Phys. Rev. Lett.* **36**, 294.

[129] T. Banks, L. J. Dixon, D. Friedan, and E. J. Martinec (1988), *Nucl. Phys.* B **299**, 613.

[130] M. Carena and H. E. Haber (2003), *Prog. Part. Nucl. Phys.* **50**, 63 (arXiv:hep-ph/0208209).

[131] S. Weinberg (1995), *Quantum Theory of Fields*, Vols. I, II, and III (Cambridge, Cambridge University Press).

[132] H. Georgi, H. R. Quinn, and S. Weinberg (1974), *Phys. Rev. Lett.* **33**, 451.

[133] R. F. Dashen and H. Neuberger (1983), *Phys. Rev. Lett.* **50**, 1897.

[134] H. Neuberger, U. M. Heller, M. Klomfass, and P. M. Vranas, arXiv:hep-lat/9208017.

[135] R. Rajaraman (1982), *Solitons and Instantons: An Introduction to Solitons and Instantons in Quantum Field Theory* (Amsterdam, North-Holland).

[136] M. Shifman (1994), *Instantons in Gauge Theories* (Singapore, World Scientific).

[137] S. Novikov, S. V. Manakov, L. P. Pitaevskii, and V. E. Zakharov (1984), *Theory of Solitons: The Inverse Scattering Method* (New York, Consultants Bureau).

[138] S. Coleman (1995), *Aspects of Symmetry* (Cambridge, Cambridge University Press).

[139] S. R. Coleman, V. Glaser, and A. Martin (1978), *Commun. Math. Phys.* **58**, 211.

[140] E. Witten (1979), *Phys. Lett.* **B 86**, 283.

[141] M. F. Atiyah, N. J. Hitchin, V. G. Drinfeld, and Yu. I. Manin (1978), *Phys. Lett.* **A 65**, 185.

[142] G. 't Hooft (1976), *Phys. Rev.* **D 14**, 3432 [Erratum *Phys. Rev.* **D 18**, 2199 (1978)].

[143] M. F. Atiyah and I. M. Singer (1984), *Proc. Nat. Acad. Sci.* **81**, 2597.

[144] M. F. Atiyah and I. M. Singer (1971), *Annals Math.* **93**, 119.

[145] M. F. Atiyah and I. M. Singer (1968), *Annals Math.* **87**, 546.

[146] M. F. Atiyah and I. M. Singer (1968), *Annals Math.* **87**, 484.

[147] J. L. Gervais, A. Jevicki, and B. Sakita (1975), *Phys. Rev.* **D 12**, 1038.

[148] B. Julia and A. Zee (1975), *Phys. Rev.* **D 11**, 2227.

[149] P. Hasenfratz and G. 't Hooft (1976), *Phys. Rev. Lett.* **36**, 1119.

[150] R. Jackiw and C. Rebbi (1976), *Phys. Rev. Lett.* **36**, 1116.

[151] S. R. Coleman (1988), *Nucl. Phys.* **B 298**, 178.

[152] J. Bagger and J. Wess (1992), *Supersymmetry and Supergravity* (Princeton, MA, Princeton University Press).

[153] S. J. Gates, M. Grisaru, M. Rocek, and W. Siegel (1983), *Superspace: or 1001 Lessons in Supersymmetry* (New York, Benjamin-Cummings).

[154] S. Weinberg (1999), *The Quantum Theory of Fields III* (Cambridge, Cambridge University Press).

[155] S. P. Martin (1996), in *Fields, Strings and Duality* (Boulder, CO, University of Colorado Press) (arXiv:hep-ph/9709356).

[156] J. D. Lykken (1996), *Fields, Strings and Duality* (Boulder, CO, University of Colorado Press) (arXiv:hep-th/9612114).

[157] J. Kapusta (2006), *Finite Temperature Field Theory* (Cambridge, Cambridge University Press).

[158] J. Berges (2005), *AIP Conf. Proc.* **739**, 3 [arXiv:hep-ph/0409233].

[159] N. D. Birrell and P. C. W. Davies (1982), *Quantum Field Theory in Curved Space-time* (Cambridge, Cambridge University Press).

[160] H. Georgi (1964), *Weak Interactions and Modern Particle Theory* (New York, Benjamin-Cummings).

[161] W. Greiner and B. Muller (1996), *Gauge Theories of Weak Interactions* (Berlin, Springer-Verlag).

[162] I. Montvay and G. Muenster (1994), *Quantum Fields on a Lattice* (Cambridge, Cambridge University Press).

[163] C. Itzykson, H. Saleur, and J. B. Zuber (1988), *Conformal Invariance and Applications to Statistical Mechanics* (Singapore, World Scientific).

[164] P. Di Francesco, P. Mathieu, and D. Senechal (1997), *Conformal Field Theory* (Berlin, Springer-Verlag).

[165] T. Banks, W. Fischler, S. H. Shenker, and L. Susskind (1997), *Phys. Rev.* D **55**, 5112 (arXiv:hep-th/9610043).

[166] O. Aharony, S. S. Gubser, J. M. Maldacena, H. Ooguri, and Y. Oz (2000), *Phys. Rep.* **323**, 183 (arXiv:hep-th/9905111).

[167] T. Banks (1999), arXiv:hep-th/991068.

[168] M. Srednicki (2007), *Quantum Field Theory* (Cambridge, Cambridge University Press).

[169] D. Bailin and A. Love (1993), *Introduction to Gauge Field Theory* (Bristol, IOP Press).

[170] P. Ramond (1980), *Field Theory: A Modern Primer*, 2nd edn. (New York, Addison Wesley).

[171] C. Itzykson and J. B. Zuber (1980), *Quantum Field Theory* (New York, McGraw-Hill).

[172] J. M. Drouffe and C. Itzykson (1980), *Statistical Field Theory* (Cambridge, Cambridge University Press).

[173] G. Parisi (1988), *Statistical Field Theory* (New York, Benjamin/Cummings).

[174] S. K. Ma (1976), *Modern Theory of Critical Phenomena* (New York, Benjamin/Cummings).

[175] J. Schwinger (1998), *Particles, Sources and Fields*, Vols I and II (New York, Perseus Books).

[176] J. D. Bjorken and S. Drell (1965), *Relativistic Quantum Field Theory* (New York, McGraw-Hill).

[177] K. Nishijima (1969), *Fields and Particles: Field Theory and Dispersion Relations* (New York, Benjamin).

[178] N. N. Bogoliubov and D. Shirkov (1982), *Quantum Fields* (New York, Benjamin-Cummings).

[179] S. S. Schweber (1961), *Quantum Theory of Fields* (Evanston, Row-Peterson).

[180] R. Balian and J. Zinn-Justin (1976), *Methods in Field Theory: Proceedings*, Les Houches Summer School, Session 28 (Amsterdam, North-Holland).

[181] S. Deser, M. Grisaru, and H. Pendelton (eds.) (1970), *Lectures on Elementary Particles and Quantum Field Theory* (Cambridge, MA, MIT Press).

[182] J. Schwinger (ed.) (1958), *Selected Papers on Quantum Electrodynamics* (New York, Dover).

Author index

Subject index

$D(R)$, 95
R_k gauges, 131
$SU(2) \times SU(2)$, 87
$U_A(1)$ symmetry, 85
$\overline{MS}$, 169

annihilation operators, 9
anomalies, 118, 231
anomaly, 85, 87
anti-commutators, 9
anti-instanton, 208
anti-unitary operators, 49
axion, 200

Baker–Campbell–Hausdorff formula, 121
baryons, 128
Belinfante tensor, 80
block spin (Kadanoff), 146
boost, 9
bosonic zero modes, 209, 212, 232
bosons, 8, 64
bounce solution, 207
bremsstrahlung, 63
BRST symmetry, 97

canonical commutation relations, 14, 18
Cartan subalgebra, 52
center of the gauge group, 110
charge conjugation, 49, 50
Chern–Simons form, 123, 227
chiral symmetry, 47, 85, 87, 91, 172
CKM matrix, 115
Clebsch–Gordan coefficients, 45
cluster property, 32
color, 101
complex, real or pseudo-real representation, 51
Compton scattering, 74
Compton wavelength, 5
confinement, 108
conformal field theory (CFT), 23, 91
conformal transformations, 80, 91
connected component of a group, 13
connected Green functions, 26
connection, 88
continuous spin representations, 39
correlation length, 146

coset generators, 96
coset space, 85, 88
covariant derivative, 88, 94
CP transformation, 53
CPT, 55
creation operators, 9
critical exponents, 146
critical phenomena, 145
critical surface, 152
crossing symmetry, 35
custodial symmetry, 114

diamagnetism, 179
dilute gas approximation, 211, 215
dimensional regularization, 71, 143
dimensional transmutation, 117
Dirac Lagrangian, 48
Dirac mass, 54
Dirac matrices, 48
Dirac picture, 10, 34
Dirac spinor, 45
Dynkin index, 95, 122

ear diagrams, 141
effective field theory, 137
effective potential, 195
electro-weak couplings, 114
electro-weak gauge theory, 113
energy-momentum tensor, 78, 81
extended objects, 214

Faddeev–Popov ghosts, 47, 97
fermion zero modes, 231
fermions, 8, 47
Feynman Green function, 23, 47
Feynman path integral, 18
Feynman slash notation, 48
finite-dimensional representations of the Lorentz
    group, 40
Fock space, 8
Fredholm determinant, 210
functional derivative, 19
functional integrals, 17, 21

gauge coupling unification, 202
gauge equivalence, 88

Printed in the United States
By Bookmasters